ELEMENTARY MATHEMATICS
FROM AN ADVANCED STANDPOINT
ARITHMETIC, ALGEBRA, ANALYSIS

FELIX KLEIN

Translated from the Third German Edition by
E. R. Hedrick
Vice President and Provost
The University of California

and

C. A. Noble
Professor of Mathematics, Emeritus
The University of California

DOVER PUBLICATIONS, INC.
Mineola, New York

Bibliographical Note

This Dover edition, first published in 2004, is an unabridged republication of an earlier Dover reprint (n.d.) of the translation first published by The Macmillan Company, New York, in 1932. The translation follows volume 1 (*Arithmetik, Algebra, Analysis*) of the three-volume third German edition of *Elementarmathematik vom höheren Standpunkte aus*, published by J. Springer, Berlin, in 1924–1928.

Library of Congress Cataloging-in-Publication Data

Klein, Felix, 1849-1925.
 [Elementarmathematik von höheren Standpunkte aus. English]
 Elementary mathematics from an advanced standpoint. Arithmetic, algebra, analysis / Felix Klein ; translated from the third German edition by E.R. Hedrick and C.A. Noble.
  p. cm.
 Originally published: New York : Macmillan, 1932.
 A translation of v. 1 of the author's three-volume work entitled Elementarmathematik vom höheren Standpunkte aus.
 Includes index.
 ISBN 0-486-43480-X (pbk.)
 1. Mathematics. I. Title: Arithmetic, algebra, analysis. II. Title.

QA39.3.K5413 2004
510–dc22

2004045585

Manufactured in the United States of America
Dover Publications, Inc., 31 East 2nd Street, Mineola, N.Y. 11501

Preface to the First Edition.

The new volume which I herewith offer to the mathematical public, and especially to the teachers of mathematics in our secondary schools, is to be looked upon as a first continuation of the lectures *Über den mathematischen Unterricht an den höheren Schulen**, in particular, of those on *Die Organisation des mathematischen Unterrichts*** by *Schimmack* and me, which were published last year by Teubner. At that time our concern was with the different ways in which the problem of instruction can be presented to the mathematician. At present my concern is with developments in the subject matter of instruction. I shall endeavor to put before the teacher, as well as the maturing student, from the view-point of modern science, but in a manner as simple, stimulating, and convincing as possible, both the content and the foundations of the topics of instruction, with due regard for the current methods of teaching. I shall not follow a systematically ordered presentation, as do, for example, Weber and Wellstein, but I shall allow myself free excursions as the changing stimulus of surroundings may lead me to do in the course of the actual lectures.

The program thus indicated, which for the present is to be carried out only for the fields of *Arithmetic, Algebra*, and *Analysis*, was indicated in the preface to Klein-Schimmack (April 1907). I had hoped then that Mr. Schimmack, in spite of many obstacles, would still find the time to put my lectures into form suitable for printing. But I myself, in a way, prevented his doing this by continuously claiming his time for work in another direction upon pedagogical questions that interested us both. It soon became clear that the original plan could not be carried out, particularly if the work was to be finished in a short time, which seemed desirable if it was to have any real influence upon those problems of instruction which are just now in the foreground. As in previous years, then, I had recourse to the more convenient method of *lithographing* my lectures, especially since my present assistant, Dr. Ernst Hellinger, showed himself especially well qualified for this work. One should not underestimate the service which Dr. Hellinger rendered. For it is a far cry from the spoken word of the teacher, influenced as it is by accidental conditions, to the subsequently polished and readable record.

---

* On the teaching of mathematics in the secondary schools.
** The organization of mathematical instruction.

In precision of statement and in uniformity of explanations, the lecturer stops short of what we are accustomed to consider necessary for a printed publication.

I hesitate to commit myself to still further publications on the teaching of mathematics, at least for the field of *geometry*. I prefer to close with the wish that the present lithographed volume may prove useful by inducing many of the teachers of our higher schools to renewed use of independent thought in determining the best way of presenting the material of instruction. This book is designed solely as such a mental spur, not as a detailed handbook. The preparation of the latter I leave to those actively engaged in the schools. It is an error to assume, as some appear to have done, that my activity has ever had any other purpose. In particular, the *Lehrplan der Unterrichtskommission der Gesellschaft Deutscher Naturforscher und Ärzte** (the so-called "Meraner" *Lehrplan*) is not mine, but was prepared, merely with my cooperation, by distinguished representatives of school mathematics.

Finally, with regard to the method of presentation in what follows, it will suffice if I say that I have endeavored here, as always, to combine geometric intuition with the precision of arithmetic formulas, and that it has given me especial pleasure to follow the historical development of the various theories in order to understand the striking differences in methods of presentation which parallel each other in the instruction of today.

Göttingen, June, 1908

Klein.

Preface to the Third Edition.

After the firm of Julius Springer had completed so creditably the publication of my collected scientific works, it offerred, at the suggestion of Professor Courant, to bring out in book form those of my lecture courses which, from 1890 on, had appeared in lithographed form and which were out of print except for a small reserve stock.

These volumes, whose distribution had been taken over by Teubner, during the last decades were, in the main, the manuscript notes of my various assistants. It was clear to me, at the outset, that I could not undertake a new revision of them without again seeking the help of younger men. In fact I long ago expressed the belief that, beyond a certain age, one ought not to publish independently. One is still qualified, perhaps, to direct in general the preparation of an edition, but is not able to put the details into the proper order and to take into proper account recent advances in the literature. Consequently I accepted the

---

* Curriculum prepared by the commission on instruction of the Society of German Natural Scientists and Physicians.

offer of Springer only after I was assured that liberal help in this respect would be provided.

These lithographed volumes of lectures fall into two series. The older ones are of special lectures which I gave from time to time, and were prepared solely in order that the students of the following semester might have at hand the material which I had already treated and upon which I proposed to base further work. These are the volumes on *Non-Euclidean Geometry*, *Higher Geometry*, *Hypergeometric Functions*, *Linear Differential Equations*, *Riemann Surfaces*, and *Number Theory*. In contrast to these, I have published several lithographed volumes of lectures which were intended, from the first, for a larger circle of readers. These are:

a) The volume on *Applications of Differential and Integral Calculus to Geometry*, which was worked up from his manuscript notes by C. H. Müller. This was designed to bridge the gap between the needs of applied mathematics and the more recent investigations of pure mathematicians.

b) and c) Two volumes on *Elementary Mathematics from an Advanced Standpoint*, prepared from his manuscript notes by E. Hellinger. These two were to bring to the attention of secondary school teachers of mathematics and science the significance for their professional work of their academic studies, especially their studies in pure mathematics.

A thoroughgoing revision of the volumes of the second series seemed unnecessary. A smoothing out, in places, together with the addition of supplementary notes, was thought sufficient. With their publication therefore, the initial step is taken. Volumes b), c), a) (in this order) will appear as Parts I, II, III of a single publication bearing the title *Elementary Mathematics from an Advanced Standpoint*. The combining, in this way, of volume a) with volumes b) and c) will meet the approval of all who appreciate the growing significances of applied mathematics for modern school instruction.

Meantime the revision of the volumes of the first series has begun, starting with the volume on *Non-Euclidean Geometry*. But a more drastic recasting of the material will be necessary here if the book is to be a well-rounded presentation, and is to take account of the recent advances of science. So much as to the general plan. Now a few words as to the first part of the *Elementary Mathematics*.

I have reprinted the preface to the 1908 edition of b) because it shows most clearly how the volume came into existence[1]. The second edition (1911), also lithographed, contained no essential changes, and the minor notes which were appended to it are now incorporated into

---

[1] My co-worker, R. Schimmack, who is mentioned there, died in 1912 at the age of thirty-one years, from a heart attack with which he was seized suddenly, as he sat at his desk.

the text without special mention. The present edition retains[1], in the main, the text of the first edition, including such peculiarities as were incident to the time of its origin. Otherwise it would have been necessary to change the entire articulation, with a loss of homogeneity. But during the sixteen years which have elapsed since the first publication, science has advanced, and great changes have taken place in our school system, changes which are still in progress. This fact is provided for in the appendices which have been prepared, in collaboration with me, by Dr. Seyfarth (Studienrat at the local Oberrealschule). Dr. Seyfarth also made the necessary stylistic changes in the text, and has looked after the printing, including the illustrations, so that I feel sincerely grateful to him. My former co-workers, Messrs. Hellinger and Vermeil, as well as Mr. A. Walther of Göttingen, have made many useful suggestions during the proof reading. In particular, I am indebted to Messrs. Vermeil and Billig for preparing the list of names and the index. The publisher, Julius Springer has again given notable evidence of his readiness to print mathematical works in the face of great difficulties.

Göttingen, Easter, 1924

Klein.

Preface to the English Edition.

Professor *Felix Klein* was a distinguished investigator. But he was also an inspiring teacher. With the rareness of genius, he combined familiarity with all the fields of mathematics and the ability to perceive the mutual relations of these fields; and he made it his notable function, as a teacher, to acquaint his students with mathematics, not as isolated disciplines, but as an integrated living organism. He was profoundly interested in the teaching of mathematics in the secondary schools, both as to the material which should be taught, and as to the most fruitful way in which it should be presented. It was his custom, during many years, at the University of Göttingen, to give courses of lectures, prepared in the interest of teachers and prospective teachers of mathematics in German secondary schools. He endeavored to reduce the gap between the school and the university, to rouse the schools from the lethargy of tradition, to guide the school teaching into directions that would stimulate healthy growth; and also to influence university attitude and teaching toward a recognition of the normal function of the secondary school, to the end that mathematical education should be a continuous growth.

These lectures of Professor *Klein* took final form in three printed volumes, entitled *Elementary Mathematics from an Advanced Standpoint*.

---

[1] New comments are placed in brackets.

They constitute an invaluable work, serviceable alike to the university teacher and to the teacher in the secondary school. There is, at present, nothing else comparable with them, either with respect to their skilfully integrated material, or to the fascinating way in which this material is discussed. This English volume is a translation of Part I of the above work. Its preparation is the result of a suggestion made by Professor *Courant*, of the University of Göttingen. It is the expression of a desire to serve the need, in English speaking countries, of actual and prospective teachers of mathematics; and it appears with the earnest hope that, in a rather free translation, something of the spirit of the original has been retained.

The Translators.

Contents

Introduction 1

First Part: Arithmetic

I. Calculating with Natural Numbers 6
  1. Introduction of Numbers in the Schools 6
  2. The Fundamental Laws of Reckoning 8
  3. The Logical Foundations of Operations with Integers 10
  4. Practice in Calculating with Integers 17

II. The First Extension of the Notion of Number 22
  1. Negative Numbers 23
  2. Fractions 28
  3. Irrational Numbers 31

III. Concerning Special Properties of Integers 37

IV. Complex Numbers 55
  1. Ordinary Complex Numbers 55
  2. Higher Complex Numbers, especially Quaternions 58
  3. Quaternion Multiplication — Rotation and Expansion 65
  4. Complex Numbers in School Instruction 75
    Concerning the Modern Development and the General Structure of Mathematics 77

Second Part: Algebra

I. Real Equations with Real Unknowns 87
  1. Equations with one parameter 87
  2. Equations with two parameters 88
  3. Equations with three parameters λ, μ, ν 94

II. Equations in the field of complex quantities 101
  A. The fundamental theorem of algebra 101
  B. Equations with a complex parameter 104
    1. The "pure" equation 110
    2. The dihedral equation 115
    3. The tetrahedral, the octahedral, and the icosahedral equations . 120
    4. Continuation: Setting up the Normal Equation 124
    5. Concerning the Solution of the Normal Equations 130
    6. Uniformization of the Normal Irrationalities by Means of Transcendental Functions 133
    7. Solution in Terms of Radicals 138
    8. Reduction of Genral Equations to Normal Equations 141

Third Part: Analysis

I. **Logarithmic and Exponential Functions** 144
  1. Systematic Account of Algebraic Analysis 144
  2. The Historical Development of the Theory 146
  3. The Theory of Logarithms in the Schools 155
  4. The Standpoint of Function Theory 156

II. **The Goniometric Functions** 162
  1. Theory of the Goniometric Functions 162
  2. Trigonometric Tables 169
    A. Purely Trigonometric Tables 170
    B. Logarithmic—Trigonometric Tables 172
  3. Applications of Goniometric Functions 175
    A. Trigonometry, in particular, spherical trigonometry 175
    B. Theory of small oscillations, especially those of the pendulum . 186
    C. Representation of periodic functions by means of series of goniometric functions (trigonometric series) 190

III. **Concerning Infinitesimal Calculus Proper** 207
  1. General Considerations in Infinitesimal Calculus 207
  2. TAYLORS Theorem 223
  3. Historical and Pedagogical Considerations 234

Supplement

I. **Transcendence of the Numbers e and π** 237

II. **The Theory of Assemblages** 250
  1. The Power of an Assemblage 251
  2. Arrangement of the Elements of an Assemblage 262

Index of Names 269

Index of Contents 271

Introduction

In recent years[1], a far reaching interest has arisen among university teachers of mathematics and natural science directed toward a suitable training of candidates for the higher teaching positions. This is really quite a new phenomenon. For a long time prior to its appearance, university men were concerned exclusively with their sciences, without giving a thought to the needs of the schools, without even caring to establish a connection with school mathematics. What was the result of this practice? The young university student found himself, at the outset, confronted with problems which did not suggest, in any particular, the things with which he had been concerned at school. Naturally he forgot these things quickly and thoroughly. When, after finishing his course of study, he became a teacher, he suddenly found himself expected to teach the traditional elementary mathematics in the old pedantic way; and, since he was scarcely able, unaided, to discern any connection between this task and his university mathematics, he soon fell in with the time honored way of teaching, and his university studies remained only a more or less pleasant memory which had no influence upon his teaching.

There is now a movement to abolish this double discontinuity, helpful neither to the school nor to the university. On the one hand, there is an effort to impregnate the material which the schools teach with new ideas derived from modern developments of science and in accord with modern culture. We shall often have occasion to go into this. On the other hand, the attempt is made to take into account, in university instruction, the needs of the school teacher. And it is precisely in such comprehensive lectures as I am about to deliver to you that I see one of the most important ways of helping. I shall by no means address myself to beginners, but I shall take for granted that you are all acquainted with the main features of the chief fields of mathematics. I shall often talk of problems of algebra, of number theory, of function theory, etc., without being able to go into details. You must, therefore, be moderately familiar with these fields, in order to follow me. My task will always be to show you the *mutual connection between problems in*

[1 Attention is again drawn to the fact that the wording of the text is, almost throughout, that of the lithographed volume of 1908 and that comments which refer to later years have been put into the appendices.]

the various fields, a thing which is not brought out sufficiently in the usual lecture course, and more especially to emphasize the relation of these problems to those of school mathematics. In this way I hope to make it easier for you to acquire that ability which I look upon as the real goal of your academic study: the ability to draw (in ample measure) from the great body of knowledge there put before you a living stimulus for your teaching.

Let me now put before you some documents of recent date which give evidence of widespread interest in the training of teachers and which contain valuable material for us. Above all I think here of the addresses given at the last *Meeting of Naturalists* held September 16, 1907, in Dresden, to which body we submitted the Proposals for the Scientific Training of Prospective Teachers of Mathematics and Science of the Committee on Instruction of the Society of German Naturalists and Physicians. You will find these Proposals as the last section in the Complete Report of this Committee[1] which, since 1904, has been considering the entire complex of questions concerning instruction in mathematics and natural science and has now ended its activity; I urge you to take notice, not only of these Proposals, but also of the other parts of this very interesting report. Shortly after the Dresden meeting there occurred a similar debate at the Meeting of German Philologists and Schoolmen in Basel, September 25, in which, to be sure, the mathematical-scientific reform movement was discussed only as a link in the chain of parallel movements occurring in philological circles. After a report by me concerning our aims in mathematical-natural science reform there were addresses by P. Wendland (Breslau) on questions in *Archeology*, Al. Brandl (Berlin) on *modern languages* and, finally, *Ad. Harnack* (Berlin) on *History and religion*. These four addresses appeared together in one broschure[2] to which I particulary refer you. I hope that this auspicious beginning will develop into further cooperation between our scientists and the philologists, since it will bring about friendly feeling and mutual understanding between two groups whose relations have been unsympathetic even if not hostile. Let us endeavor always to foster such good relations even if we do among ourselves occasionally drop a critical word about the philologists, just as they may about us. Bear in mind that you will later be called upon in the schools to work together with the philologists for the common good and that this requires mutual understanding and appreciation.

Along with this evidence of efforts which reach beyond the borders of our field, I should like to mention a few books which aim in the

---

[1] *Die Tätigkeit der Unterrichtskommission der Gesellschaft deutscher Naturforscher und Ärzte*, edited by A. Gutzmer. Leipzig and Berlin, 1908.

[2] *Universität und Schule*. Addresses delivered by F. Klein, P. Wendland, Al. Brandl, Ad. Harnack. Leipzig 1907.

same direction in the mathematical field and which will therefore be important for these lectures. Three years ago I gave, for the first time, a course of lectures with a similar purpose. My assistant at that time, R. Schimmack, worked the material up and the first part has recently appeared in print[1]. In it are considered the different kinds of schools, including the university, the conduct of mathematical instruction in them, the interests that link them together, and other similar matters. In what follows I shall from time to time refer to things which appear there without repeating them. This makes it possible for me to extend somewhat those considerations. That volume concerns itself with the organization of school instruction. I shall now consider the mathematical content of the material which enters into that instruction. If I frequently advert to the actual conduct of instruction in the schools, my remarks will be based not merely upon indefinite pictures of how the thing might be done or even upon dim recollections of my own school days; for I am constantly in touch with Schimmack, who is now teaching in the Göttingen gymnasium and who keeps me informed as to the present state of instruction, which has, in fact, advanced substantially beyond what it was in earlier years. During this winter semester I shall discuss "the three great A's", that is arithmetic, algebra, and analysis, withholding geometry for a continuation of the course during the coming summer. Let me remind you that, in the language of the secondary schools, these three subjects are classed together as arithmetic, and that we shall often note deviations in the terminology of the schools as compared with that at the universities. You see, from this small illustration, that only living contact can bring about understanding.

As a second reference I shall mention the three volume *Enzyklopädie der Elementarmathematik* by H. Weber and J. Wellstein, the work which, among recent publications, most nearly accords with my own tendencies. For this semester, the first volume, *Enzyklopädie der elementaren Algebra und Analysis*, prepared by H. Weber[2], will be the most important. I shall indicate at once certain striking differences between this work and the plan of my lectures. In Weber-Wellstein, the entire structure of elementary mathematics is built up systematically and logically in the mature language of the advanced student. No account is taken of how these things actually may come up in school instruction. The presentation in the schools, however, should be psychological and not systematic. The teacher so to speak, must be a diplomat. He must take account of the psychic processes in the boy in order to grip his interest;

---

[1] Klein, F., *Vorträge über den mathematischen Unterricht an höheren Schulen*. Prepared by von R. Schimmack. Part 1. *Von der Organisation des mathematischen Unterrichts*. Leipzig 1907. This book is referred to later as "Klein-Schimmack".

[2] Second edition. Leipzig 1906. [Fourth edition, 1922, revised by P. Epstein. — Referred to as "Weber-Wellstein I".

and he will succeed only if he presents things in a form intuitively comprehensible. A more abstract presentation will be possible only in the upper classes. For example: The child cannot possibly understand if numbers are explained axiomatically as abstract things devoid of content, with which one can operate according to formal rules. On the contrary, he associates numbers with concrete images. They are numbers of nuts, apples, and other good things, and in the beginning they can be and should be put before him only in such tangible form. While this goes without saying, one should—mutatis mutandis—take it to heart, that in all instruction, even in the university, mathematics should be associated with everything that is seriously interesting to the pupil at that particular stage of his development and that can in any way be brought into relation with mathematics. It is just this which is back of the recent efforts to give prominence to applied mathematics at the university. This need has never been overlooked in the schools so much as it has at the university. It is just this psychological value which I shall try to emphasize especially in my lectures.

Another difference between Weber-Wellstein and myself has to do with defining the content of school mathematics. Weber and Wellstein are disposed to be conservative, while I am progressive. These things are thoroughly discussed in Klein-Schimmack. We, who are called the reformers, would put the function concept at the very center of instruction, because, of all the concepts of the mathematics of the past two centuries, this one plays the leading role wherever mathematical thought is used. We would introduce it into instruction as early as possible with constant use of the graphical method, the representation of functional relations in the xy system, which is used today as a matter of course in every practical application of mathematics. In order to make this innovation possible, we would abolish much of the traditional material of instruction, material which may in itself be interesting, but which is less essential from the standpoint of its significance in connection with modern culture. Strong development of space perception, above all, will always be a prime consideration. In its upper reaches, however, instruction should press far enough into the elements of infinitesimal calculus for the natural scientist or insurance specialist to get at school the tools which will be indispensable to him. As opposed to these comparatively recent ideas, Weber-Wellstein adheres essentially to the traditional limitations as to material. In these lectures I shall of course be a protagonist of the new conception.

My third reference will be to a very stimulating book: *Didaktik und Methodik des Rechnens und der Mathematik*[1] by Max Simon, who like

---

[1] Second edition, Munich 1908. Separate reprint from Baumeister's *Handbuch der Erziehungs- und Unterrichtslehre für höhere Schulen*, first edition, 1895.

Weber and Wellstein is at Strassburg. Simon is often in agrement with our views, but he sometimes takes the opposite standpoint; and inasmuch as he is a very subjective, temperamental, personality he often clothes these contrasting views in vivid words. To give one example, the proposals of the Committee on Instruction of the Natural Scientists require an hour of geometric propaedeutics in the second year of the gymnasium, whereas at the present time this usually begins in the third year. It has long been a matter of discussion which plan is the better; and the custom in the schools has often changed. But Simon declares the position taken by the Commission, which, mind you, is at worst open to argument, to be "worse than a crime", and that without in the least substantiating his judgment. One could find many passages of this sort. As a precursor of this book I might mention Simon's *Methodik der elementaren Arithmetik in Verbindung mit algebraischer Analysis*[1].

After this brief introduction let us go over to the subject proper, which I shall consider under three headings, as above indicated.

---

[1] Leipzig 1906.

First Part

Arithmetic

I. Calculating with Natural Numbers

We begin with the foundation of all arithmetic, calculation with positive integers. Here, as always in the course of these lectures, we first raise the question as to how these things are handled in the schools; then we shall proceed to the question as to what they imply when viewed from an advanced standpoint.

1. Introduction of Numbers in the Schools

I shall confine myself to brief suggestions. These will enable you to recall how you yourselves learned your numbers. In such an exposition it is, of course, not my purpose to induct you into the practice of teaching, as is done in the Seminars of the secondary schools. I shall merely exhibit the material upon which we shall base our critique.

The problem of teaching children the properties of integers and how to reckon with them, and of leading them on to complete mastery, is very difficult and requires the labor of several years, from the first school year until the child is ten or eleven years old. The manner of instruction as it is carried on in this field in Germany can perhaps best be designated by the words *intuitive* and *genetic*, i. e., the entire structure is gradually erected on the basis of familiar, concrete things, in marked contrast to the customary *logical* and *systematic* method at the university.

One can divide up this material of instruction roughly as follows: The entire first year is occupied with the integers from 1 to 20, the first half being devoted to the range 1 to 10. The integers appear at first as numbered pictures of points or as arrays of all sorts of objects familiar to the children. Addition and multiplication are then presented by intuitional methods, and are fixed in mind.

In the second stage, the integers from 1 to 100 are considered and the Arabic numerals, together with the notion of positional value and the decimal system, are introduced. Let us note, incidentally, that the name "Arabic numerals", like so many others in science, is a misnomer. This form of writing was invented by the Hindus, not by the Arabs. Another principal aim of the second stage is knowledge of the multiplication table. One must know what 5×7 or 3×8 is in one's sleep, so to speak. Consequently the pupil must learn the multiplication table by heart to this degree of thoroughness, to be sure only after it has been made clear to him visually with concrete things. To this end the *abacus* is used to advantage. It consists, as you all know, of 10 wires stretched one above another, upon each of which there are strung ten movable beads. By sliding these beads in the proper way, one can read off the result of multiplication and also its decimal form.

The third stage, finally, teaches calculation with numbers of more than one digit, based on the known simple rules whose general validity is evident, or should be evident, to the pupil. To be sure, this evidence does not always enable the pupil to make the rules completely his own; they are often instilled with the authoritative dictum: "It is thus and so, and if you don't know it yet, so much the worse for you!"

I should like to emphasize another point in this instruction which is usually neglected in university teaching. It is that the application of numbers to practical life is strongly emphasized. From the beginning, the pupil is dealing with numbers taken from real situations, with coins, measures, and weights; and the question, *"What does it cost?"*, which is so important in daily life, forms the pivot of much of the material of instruction. This plan rises soon to the stage of problems, when deliberate thought is necessary in order to determine what calculation is demanded. It leads to the problems in proportion, alligation, etc. To the words *intuitive* and *genetic*, which we used above to designate the character of this instruction, we can add a third word, *applications*.

We might summarize the purpose of the number work by saying: *It aims at reliability in the use of the rules of operation, based on a parallel development of the intellectual abilities involved, and without special concern for logical relations.*

Incidentally, I should like to direct your attention to a contrast which often plays a mischievous role in the schools, viz., the contrast between the university-trained teachers and those who have attended normal schools for the preparation of elementary school teachers. The former displace the latter, as teachers of arithmetic, during or after the sixth school year, with the result that a regrettable discontinuity often manifests itself. The poor youngsters must suddenly make the acquaintance of new expressions, whereas the old ones are forbidden. A simple example is the different multiplication signs, the $\times$ being preferred by the elementary teacher, the point by the one who has attended the university. Such conflicts can be dispelled, if the more highly trained teacher will give more heed to his colleague and will try to meet him on common ground. That will become easier for you, if you will realize what high regard one must have for the performance of the elementary school teachers. Imagine what methodical training is necessary to indoctrinate over and over again a hundred thousand stupid, unprepared children with the principles of arithmetic! Try it with your university training; you will not have great success!

Returning, after this digression, to the material of instruction, we note that after the third year of the gymnasium*, and especially in the fourth year, arithmetic begins to take on the more aristocratic dress of mathematics, for which the transition to operations with letters is characteristic. One designates by a, b, c, or x, y, z any numbers, at first only positive integers, and applies the rules and operations of arithmetic to the numbers thus symbolized by letters, whereby the numbers are devoid of concrete intuitive content. This represents such a long step in abstraction that one may well declare that real mathematics begins with operations with letters. Naturally this transition must not be accomplished rapidly. The pupils must accustom themselves gradually to such marked abstraction.

It seems unquestionably necessary that, for this instruction, the teacher should know thoroughly the logical laws and foundations of reckoning and of the theory of integers.

2. The Fundamental Laws of Reckoning

Addition and multiplication were familiar operations long before any one inquired as to the fundamental laws governing these operations. It was in the twenties and thirties of the last century that particularly English and French mathematicians formulated the fundamental properties of the operations, but I will not enter into historical details here. If you wish to study these, I recommend to you, as I shall often do, the great *Enzyklopädie der Mathematischen Wissenschaften mit Einschluß ihrer Anwendungen*[1], and also the French translation: *Encyclopédie des Sciences mathématiques pures et appliquées*[2] which bears in part the character of a revised and enlarged edition. If a school library has only one mathematical work, it ought to be this encyclopedia, for through it the teacher of mathematics would be placed in position to continue his work in any direction that might interest him. For us, at this place, the article of interest is the first one in the first volume[3] H. Schubert: *"Grundlagen der Arithmetik"*, of which the translation into French is by Jules Tannery and Jules Molk.

---

* The German gymnasium is a nine-year secondary school, following a four-year preparatory school. Hence the third year of the gymnasium is the student's seventh school year.

[1] Leipzig (B. G. Teubner) from 1908 on. Volume I has appeared complete, Volumes II—VI are nearing completion.

[2] Paris (Gauthur-Villars) and Leipzig (Teubner) from 1904 on; unfortunately the undertaking had to be abandoned after the death of its editor J. Molk (1914).

[3] *Arithmetik und Algebra*, edited by W. Fr. Meyer (1896—1904); in the French edition, the editor was J. Molk.

Going back to our theme, I shall enumerate the *five fundamental laws* upon which *addition* depends:

1. $a + b$ *is always again a number*, i. e., *addition is always possible* (in contrast to subtraction, which is not always possible in the domain of positive integers).
2. $a + b$ *is one-valued.*
3. *The associative law holds:*
$$(a + b) + c = a + (b + c),$$
so that one may omit the parentheses entirely.
4. *The commutative law holds:*
$$a + b = b + a.$$
5. *The monotonic law holds:*
$$\text{If } b > c, \text{ then } a + b > a + c.$$

These properties are all obvious immediately if one recalls the process of counting; but they must be formally stated in order to justify logically the later developments.

For multiplication there are *five exactly analogous laws:*
1. $a \cdot b$ *is always a number.*
2. $a \cdot b$ *is one-valued.*
3. *Associative law:* $a \cdot (b \cdot c) = (a \cdot b) \cdot c = a \cdot b \cdot c.$
4. *Commutative law:* $a \cdot b = b \cdot a.$
5. *Monotonic law: If* $b > c$, *then* $a \cdot b > a \cdot c$.

Multiplication together with addition obeys also the following law.
6. *Distributive law:*
$$a \cdot (b + c) = a \cdot b + a \cdot c.$$

It is easy to show that all elementary reckoning can be based upon these eleven laws. It will be sufficient to illustrate this fact by a simple example, say the multiplication of 7 and 12. From the distributive law we have:
$$7 \cdot 12 = 7 \cdot (10 + 2) = 70 + 14,$$
and if we separate 14 into $10 + 4$ (carrying the tens), we have, by the associative law of addition,
$$70 + (10 + 4) = (70 + 10) + 4 = 80 + 4 = 84.$$

You will recognize in this procedure the steps of the usual decimal reckoning. It would be well for you to construct for yourselves more complicated examples. We might summarize by saying *that ordinary reckoning with integers consists in repeated use of the eleven fundamental laws together with the memorized results of the addition and multiplication tables.*

But where does one use the monotonic laws? In ordinary formal reckoning, to be sure, they are superfluous, but not in certain other problems. Let me remind you of the process called *abridged multiplication and division* with decimal numbers[1]. That is a thing of great practical importance which unfortunately is too little known in the schools, as well as among university students, although it is sometimes mentioned in the second year of the gymnasium. As an example, suppose that one wished to compute $567 \cdot 134$, and that the units digit in each number was of questionable accuracy, say as a result of physical measurement. It would be unnecessary work, then, to determine the product *exactly*, since one could not guarantee an exact result. It is, however, important to know the *order* of *magnitude* of the product, i. e., to know between which tens or between which hundreds the exact value lies. The monotonic law supplies this estimate at once; for it follows by that law that the desired value lies between $560 \cdot 134$ and $570 \cdot 134$ or between $560 \cdot 130$ and $570 \cdot 140$. I leave to you the carrying out of the details; at least you see that *the monotonic law is continually used in abridged reckoning.*

A systematic exposition of these fundamental laws is, of course, not to be thought of in the secondary schools. After the pupils have gained a concrete understanding and a secure mastery of reckoning with numbers, and are ready for the transition to operations with letters, the teacher should take the opportunity to state, at least, the associative, commutative, and distributive laws and to illustrate them by means of numerous obvious numerical examples.

3. The Logical Foundations of Operations with Integers

While instruction in the schools will naturally not rise to still more difficult questions, present mathematical investigation really begins with the question: *How does one justify the above-mentioned fundamental laws, how does one account for the notion of number at all?* I shall try to explain this matter in accordance with the announced purpose of these lectures to endeavor to get new light upon school topics by looking at them from another point of view. I am all the more willing to do this because these modern thoughts crowd in upon you from all sides during your academic years, but not always accompanied by any indication of their psychological significance.

First of all, so far as the notion of number is concerned, it is very difficult to discover its origin. Perhaps one is happiest if one decides to ignore these most difficult things. For more complete information as to these questions, which are so earnestly discussed by the philosophers, I must refer you to the article, already mentioned, in the French encyclopedia, and I shall confine myself to a few remarks. A widely accepted belief is that the notion of number is closely connected with the notion of time, with temporal succession. The philosopher Kant

---

[1] The monotonic laws will be used later, also, in the theory of irrational numbers.

The Logical Foundations of Operations with Integers. 11

and the mathematician Hamilton represent this view. Others think that number has more to do with space perception. They base the notion of number upon the *simultaneous perception of different objects which are near each other*. Still others see, in number concepts, the expression of a *peculiar faculty of the mind* which exists independently of, and coordinate with, or even above, perception of space and time. I think that this conception would be well characterized by quoting from Faust the lines which Minkowski, in the preface of his book on *Diophantine Approximation*, applies to numbers:

"Göttinnen thronen hehr in Einsamkeit,
Um sie kein Ort, noch weniger eine Zeit."

While this problem involves primarily questions of psychology and epistemology, the justification of our eleven laws, at least the recent researches regarding their compatibility, implies questions of logic. We shall distinguish the following four points of view.

1. According to the first of these, best represented perhaps by Kant, the rules of reckoning are immediate necessary results of perception, whereby this word is to be understood, in its broadest sense, as "inner perception" or intuition. It is not to be understood by this that mathematics rests throughout upon experimentally controllable facts of external experience. To mention a simple example, the commutative law is established by examining the accompanying picture, which consists of two rows of three points each, that is, $2 \cdot 3 = 3 \cdot 2$. If the objection is raised that in the case of only moderately large numbers, this immediate perception would not suffice, the reply is that we call to our assistance the *theorem of mathematical induction*. *If a theorem holds for small numbers, and if an assumption of its validity for a number n always insures its validity for $n + 1$, then it holds generally for every number*. This theorem, which I consider to be really an intuitive truth, carries us over the boundary where sense perception fails. This standpoint is more or less that of Poincaré in his well known philosophical writings.

If we would realize the significance of this question as to the source of the validity of our eleven fundamental rules of reckoning, we should remember that, along with arithmetic, mathematics as a whole rests ultimately upon them. Thus it is not asserting too much to say, that, according to the conception of the rules of reckoning which we have just outlined, *the security of the entire structure of mathematics rests upon intuition, where this word is to be understood in its most general sense*.

2. The second point of view is a modification of the first. According to it, one tries to separate the eleven fundamental laws into a larger number of shorter steps of which one need take only the simplest directly from intuition, while the remainder are deduced from these by rules of logic without any further use of intuition. Whereas, before,

the possibility of logical operation began *after* the eleven fundamental laws had been set up, it can start earlier here, after the simpler ones have been selected. *The boundary between intuition and logic is displaced in favor of the latter.* Hermann Grassmann did pioneer work in this direction in his *Lehrbuch der Arithmetik*[1] in 1861. As an example from it, I mention merely that the commutative law can be derived from the associative law by the aid of the principle of mathematical induction. Because of the precision of his presentation, one might place by the side of this book of Grassmann one by the Italian Peano, *Arithmetices principia nova methodo exposita*[2]. Do not assume, however, because of this title, that the book was written in Latin! It is written in a peculiar symbolic language designed by the author to display each logical step of the proof and emphasize it as such. Peano wishes to have a guarantee in this way, that he is making use only of the principle which he specifically mentions, with nothing whatever coming from intuition. He wishes to avoid the danger that countless uncontrollable associations of ideas and reminders of perception might creep in if he used our ordinary language. Note, too, that Peano is the leader of an extensive Italian school which is trying in a similar way to separate into small groups the premises of each individual branch of mathematics, and, with the aid of such a symbolic language, to investigate their exact logical connections.

3. We come now to a *modern extension of these ideas*, which has, moreover, been influenced by Peano. I refer to that treatment of the foundations of arithmetic which puts the theory of point sets into the foreground. You will be able to form a notion of the wide range of the idea of a point set if I tell you that the totality of all integers, as well as that of all points on a line segment, are special examples of point sets. Georg Cantor, as is generally known, was the first to make this general idea the object of orderly mathematical speculation. The *theory of point sets*, which he created, is now claiming the profound attention of the younger generation of mathematicians. Later I shall endeavor to give you a cursory view of this subject. For the present, it is sufficient to characterize as follows the tendency of the new foundation of arithmetic which have been based upon it: *The properties of integers and of operations with them are to be deduced from the general properties and abstract relations of point sets*, in order that the foundation may be as sound and general as possible.

---

[1] With the addition to the title *"für höhere Lehranstalten"* (Berlin 1861). The corresponding chapters are reprinted in H. Grassmann's *Gesammelten mathematischen und physikalischen Werken* (edited by F. Engel), Vol. II, 1, pp. 295—349. Leipzig 1904.

[2] Augustae Taurinorum. Torino 1889. [There is a more comprehensive presentation in Peano's *Formulaire de Mathématiques* (1892—1899).]

The Logical Foundations of Operations with Integers. 13

One of the pioneers along this path was Richard Dedekind, who, in his small but important book *Was sind und was sollen die Zahlen?*[1], attempted such a foundation for integers. H. *Weber* inclines to this point of view in the first part of Weber-Wellstein, volume I (See p. 3). To be sure, the deduction is quite abstract and offers, still, certain grave difficulties, so that Weber, in an Appendix to Volume III[2], gave a more elementary presentation, using only finite point sets. In later editions, this appendix is incorporated into Volume I. Those of you who are interested in such questions are especially referred to this presentation.

4. Finally, I shall mention the *purely formal theory of numbers*, which, indeed, goes back to Leibniz and which has recently been brought into the foreground again by Hilbert. His address *Über die Grundlagen der Logik und Arithmetik** at the Heidelberg Congress in 1904 is important for arithemtic[3]. His fundamental conception is as follows: Once one has the eleven fundamental rules of reckoning, one can operate with the letters $a, b, c, \ldots$, which actually represent arbitrary integers, without bearing in mind that they have a real meaning as numbers. In other words: let $a, b, c, \ldots$, be things devoid of meaning, or things of whose meaning we know nothing; let us agree only that one may combine them according to those eleven rules, but that these combinations need not have any real known meaning. Obviously one can than operate with $a, b, c, \ldots$, precisely as one ordinarily does with actual numbers. Only the question arises here *whether these operations could lead one to contradictions*. Now ordinarily one says that intuition shows us the existence of numbers for which these eleven laws hold, and that it is consequently impossible for contradictions to lurk in these laws. But in the present case, where we are not thinking of the symbols as having definite meaning, such an appeal to perception is not permissible. *In fact, there arises the entirely new problem, to prove logically that no operations with our symbols which are based on the eleven fundamental laws can ever lead to a contradiction, i. e., that these eleven laws are consistent, or compatible.* While we were discussing the first point of view, we took the position that the certainty of mathematics rests upon the existence of intuitional things which fit its theorems. The adherents of this formal standpoint, on the other hand, must hold that *the certainty of mathematics rests upon the possibility of showing that the fundamental laws considered formally and without reference to their intuitional content, constitute a logically consistent system.*

---

[1] Braunschweig 1888; third edition 1911.
[2] *Angewandte Elementarmathematik.* Revised by H. Weber, J. Wellstein, R. H. Weber. Leipzig 1907.
* *On the foundations of logic and arithmetic.*
[3] *Verhandlungen des 3. internationalen Mathematikerkongresses in Heidelberg* August 8—13, 1904, p. 174 et seq., Leipzig 1905.

I shall close this discussion with the following remarks:

a) Hilbert indicated all of these points of view in his Heidelberg address, but he followed none of them through completely. Afterwards he pushed them somewhat farther in a course of lectures, but then abandoned them. We can thus say that *here is a field for investigation*[1].

b) The tendency to crowd intuition completely off the field and to attain to really *pure* logical investigations seems to me not completely feasible. It seems to me that *one must retain something, albeit a minimum, of intuition.* One must always use a certain intuition in the most abstract formulation with the symbols one uses in operations, in order to recognize the symbols again, even if one thinks only about the shape of the letters.

c) Let us even assume that the proposed problem has been solved in a way free from objection, that the compatibility of the eleven fundamental laws has been proved logically. Precisely at this point an opening is offered for a remark which I should like to make with the utmost emphasis. One must *see clearly that the real arithmetic, the theory of actual integers, is neither established, nor can ever be established, by considerations of this nature.* It is impossible to show in a purely logical way that the laws whose consistency is established in that manner are actually valid for the numbers with which we are intuitively familiar; that the undefined things of which we speak, and the operations which we apply to them, can be identified with actual numbers and with the processes of addition and multiplication in their intuitively clear significance. What is accomplished is, rather, that the tremendous *problem of building the foundations of arithmetic, unassailable in its complexity, is split into two parts*, and that the first, the purely logical problem, *the setting up of independent fundamental laws or axioms and the investigation of them as to independence and consistency* has been made available to study. The second, the more epistemological part of the problem, which has to do with the justification for the *application of these laws to actual conditions*, is not even touched, although it must of course be solved also if one will really build the foundations of arithmetic. This second part presents, in itself, an extremely profound problem, whose difficulties lie in the general field of epistemology. I can characterize its standing most clearly perhaps, by the somewhat paradoxical remark that anyone who tolerates only pure logic in investigations in pure mathematics must, to be consistent, look upon the second part of the problem of the foundation of arithmetic, and hence upon arithmetic itself, as belonging to *applied* mathematics.

[1 Concerning more recent developments in these investigations, see the preceding footnote.]

I have felt obliged to go into detail here very carefully, in as much as misunderstandings occur so often at this point, because people simply overlook the existence of the second problem. This is by no means the case with Hilbert himself, and neither my disagreement nor my agreement with him is a warranted conclusion if it be based on such an assumption.

Thomae of Jena, coined the neat expression "thoughtless thinkers" for those persons who confine themselves exclusively to these abstract investigations concerning things that are devoid of meaning, and to theorems that tell nothing, and who forget not only that second problem but often also all the rest of mathematics. This facetious term cannot apply, of course, to people who carry on those investigations alongside of many others of a different sort.

In connection with this brief survey of the foundation of arithmetic, I shall bring to your notice a few general matters. Many have thought that one could, or that one indeed must, teach all mathematics deductively *throughout*, by starting with a definite number of axioms and deducing everything from these by means of logic. This method, which some seek to maintain upon the authority of Euclid, certainly does not correspond to the historical development of mathematics. In fact, mathematics has grown like a tree, which does not start at its tiniest rootlets and grow merely upward, but rather sends its roots deeper and deeper at the same time and rate that its branches and leaves are spreading upward. Just so — if we may drop the figure of speech —, mathematics began its development from a certain standpoint corresponding to normal human understanding, and has progressed, from that point, according to the demands of science itself and of the then prevailing interests, now in the one direction toward new knowledge, now in the other through the study of fundamental principles. For example, our standpoint today with regard to foundations is different from that of the investigators of a few decades ago; and what we today would state as ultimate principles, will certainly be outstripped after a time, in that the latest truths will be still more meticulously analyzed and referred back to something still more general. *We see, then, that as regards the fundamental investigations in mathematics, there is no final ending, and therefore, on the other hand, no first beginning, which could offer an absolute basis for instruction.*

Still another remark concerning the relation between the logical and the intuitional handling of mathematics, between pure and applied mathematics. I have already emphasized the fact that, in the schools, applications accompany arithmetic from the beginning, that the pupil learns not only to understand the rules, but to do something with them. And it should always be so in the teaching of mathematics! Of course, the logical connections, one might say *the rigid skeleton in the mathematical organism*, must remain, in order to give it its peculiar trustworthiness. But the living thing in mathematics, its most important stimulus, its effectiveness in all directions, depends entirely upon the applications, i. e., upon the mutual relations between those purely logical things and all other domains. To banish applications from mathematics would be comparable to seeking the essence of the living animal in the skeleton alone, without considering muscles, nerves and tissues, instincts, in short, the very life of the animal.

In scientific *investigation* there is often, to be sure, a *division of labor* between pure and applied science, but when this happens, provision must be made otherwise for maintaining their connection if conditions are to remain sound. In any case, and this should be especially emphasized here, *for the school such a division of labor, such a fareaching specialization of the individual teacher, is not possible.* To put the matter crassly, imagine that at a certain school a teacher is appointed who treats numbers only as meaningless symbols, a second teacher who knows how to bridge the gap from these empty symbols to actual numbers, a third, a fourth, a fifth, finally, who understands the application of these numbers to geometry, to mechanics, and to physics; and that these different teachers are all turned lose upon the pupils. You see that such an organization of teaching is impossible. In this way, the things could not be brought to the comprehension of the pupils, neither would the individual teachers be able even to understand each other. The needs of school instruction itself require precisely a certain many sidedness of the individual teacher, a comprehensive orientation in the field of pure and applied mathematics, in the broadest sense, and include thus a desirable remedy against a too extensive splitting up of science.

In order to give a practical turn to the last remarks I refer again to our above mentioned *Dresden Proposals*. There we recommend outright that applied mathematics, which since 1898 has been a special subject in the examination for prospective teachers, be made a required part in all normal mathematical training, so that competence to teach pure and applied mathematics should always be combined. In addition to this, it should be noted that, in the Meran Curriculum[1] of the Commission of Instruction, the following three tasks are announced as the *purpose of mathematical instruction in the last school year:*

1. *A scientific survey of the systematic structure of mathematics.*

2. *A certain degree of skill in the complete handling, numerical and graphical, of problems.*

---

[1] *Reformvorschläge für den mathematischen und naturwissenschaftlichen Unterricht, überreicht der Versammlung der Naturforscher und Ärzte zu Meran.* Leipzig, 1905. — See also a reprint in the *Gesamtbericht der Kommission*, p. 93, as well as in Klein-Schimmack, p. 208.

3. *An appreciation of the significance of mathematical thought for a knowledge of nature and for modern culture.*
All these formulations I approve with deep conviction.

4. Practice in Calculating with Integers

Turning from discussions which have been chiefly abstract, let us give our attention to more concrete things by considering the *carrying out of numerical calculation*. As suitable literature for collateral reading, I should mention first of all, the article on *Numerisches Rechnen* by R. Mehnicke[1] in the *Enzyclopädie*. I can best give you a general view of the things that belong here by giving a brief account of this article. It is divided into two parts: A. *Die Lehre vom genauen Rechnen**, and B. *Die Lehre vom genäherten Rechnen***. Under A occur all methods for simplifying exact calculation with large integers. *Convenient devices for calculating, tables of products and squares*, and in particular, *calculating machines*, which we shall discuss soon. Under B, on the other hand, one finds a discussion of the methods and devices for all calculating in which only the *order of magnitude of the result* is important, especially *logarithmic tables and allied devices*, the *slide rule*, which is only an expecially well-arranged graphical logarithmic table; finally, also, the numerous important *graphical methods*. In addition to this reference I can recommend the little book by J. Lüroth, *Vorlesungen über numerisches Rechnen*[2]***, which, written in agreeable form by a master of the subject, gives a rapid survey of this field.

From the many topics that have to do with calculating with integers, I shall select for discussion only the calculating machine, which you will find in use, in a great variety of ingenious forms, by the larger banks and business houses, and which is really of the greatest practical significance. We have in our mathematical collection one of the most widely used types, the "Brunsviga", manufactured by the firm Brunsviga-Maschinenwerke Grimme, Natalis & Co. A.-G. in Braunschweig. The design originated with the Swedish engineer Odhner, but it has been much changed and improved. Is hall describe the machine here in some detail, as a typical example. You will find other kinds described in the books mentioned above[3]. My description of course can give you a real understanding of the

---

[1] Enzyklopädie der mathematischen Wissenschaften, Band I, Teil II. See also *v. Sanden, H.*, Practical Mathematical Analysis (Translation by Levy), Dutton & Co. — *Horsburgh, E. M.*, Modern Instruments and Methods of Calculation. Bell & Sons.
* The *Theory of Exact Calculation*.
** The *Theory of Approximate Calculation*.
[2] Leipzig 1900.
*** *Lectures on Numerical Calculation*.
[[3] Concerning other types of calculating machines, see also A. Galle, *Mathematische Instrumente*, Leipzig 1912.]

machine only if you examine it afterwards personally and if you see, by actual use, how it is operated. The machine will be at your disposal, for that purpose, after the lecture.

So far as the *external appearance* of the Brunsviga is concerned, it presents schematically a picture somewhat as follows (see Fig. 1, p. 18). There is a fixed frame, the *"drum"*, below which and sliding on it, is a smaller longish case, the *"slide"*. A handle which projects from the drum on the right, is operated by hand. On the drum there is a series of parallel slits, each of which carries the digits 0, 1, 2, ..., 9, read downwards; a peg s projects from each slit and can be set at pleasure at any one of the ten digits. Corresponding to each of these slits there is an opening on the slide under which a digit can appear. Figure 3, p. 19 gives a view of a newer model of the machine.

I think that the arrangement of the machine will be clearer if I describe to you the process of carrying out a definite calculation, and the way in which the machine brings it about. For this I select *multiplication*.

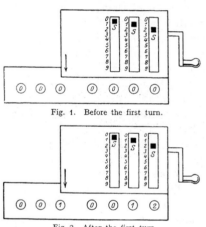

Fig. 1. Before the first turn.

Fig. 2. After the first turn.

The procedure is as follows: One *first sets the drum pegs on the multiplicand*, i. e., beginning at the right, one puts the first lever at the one's digit, the second at the ten's digit of the multiplicand, etc. If, for example, the multiplicand is 12, one sets the first lever at 2, the second lever at 1; all the other levers remain at zero (see Fig. 1). *Now turn the handle once around, clockwise.* The multiplicand appears under the openings of the slide, in our case a 2 in the first opening from the right, a 1 in the second, while zeros remain in all the others. Simultaneously, however, in the first of a series of openings in the slide, at the left, the digit 1 appears to indicate that we have turned the handle once (Fig. 2). *If now one has to do with a multiplier of one digit, one turns the handle as many times as this digit indicates; the multiplier will then be exhibited on the slide to the left, while the product will appear on the slide to the right.* How does the apparatus bring this result about? In the first place there is attached to the under side of the slide, at the left, a cogwheel which carries, equally spaced on its rim, the digits 0, 1, 2, ..., 9. By means of a driver, this cogwheel is rotated through one tenth of its perimeter with every turn of the handle, so that a digit becomes visible through the opening in the slide, which actually indicates

Practice in Calculating with Integers.

the number of revolutions, in other words the multiplier. Now as to the *obtaining of the product*, it is brought about by similar cogwheels, one under each opening at the right of the slide. But how is it that by one and the same turning of the handle, one of these wheels, in the above case, moves by one unit, the other by two? This is where the peculiarity in construction of the Brunsviga appears. Under each slit of the drum there is a flat wheel-shaped disc (driver) attached to the axle of the handle, upon which there are nine teeth which are movable in a radial direction (see Fig. 4). By means of the projecting peg S, mentioned above, one can turn a ring R which rests upon the periphery of the disc, so that, according to the mark upon which one sets S in the slit, 0, 1, 2, ..., 9 of the movable teeth spring outward (in Fig. 4, two teeth). These teeth engage the cogs under the corresponding openings of the slide, so that *with one turn of the handle each driver thrusts forward the corresponding cogwheel by as many units as there are teeth pushed out*, i. e., *by as many teeth as one has set with the corresponding peg S*. Accordingly, in the above illustration, when we start at the zero position, and turn the handle once, the units wheel must jump to 2, the ten's wheel to 1, so that 12 appears. A second turn of the handle moves the units wheel another 2 and the tens wheel another 1, so that 24 appears, and similarly, we get, after 3 or 4 times, $3 \cdot 12 = 36$ or $4 \cdot 12 = 48$, respectively.

Fig. 3.

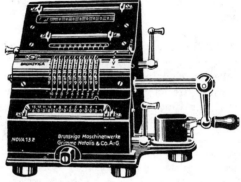

Fig. 3a.

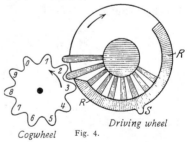

Cogwheel   Fig. 4.   Driving wheel

20   Arithmetic: Calculating with Natural Numbers.

But now turn the handle a fifth time: Again, according to the account above, the units wheel should jump again by two units, in other words back to 0, the tens wheel by one, or to 5, and we should have the false result $5 \cdot 12 = 50$. In the actual turning, however, the slide shows 50, to be sure, until just before the completion of the turn; but at the last instant the 5 changes into 6, so that the correct result appears. Something has come into action now that we have not yet described, and which is really the most remarkable point of such machines: the so called *carrying the tens*. Its principle is as follows: *when one of the number bearing cogwheels under the slide* (e. g., *the units wheel*) *goes through zero, it presses an otherwise inoperative tooth of the neighboring driver* (*for the tens*) *into position, so that it engages the corresponding cogwheel* (*the tens wheel*) *and pushes this forward one place farther than it would have gone otherwise.* You can understand the details of this construction only by examining the apparatus itself. There is the less need for my going into particulars here because it is just the method of carrying the tens that is worked out in the greatest variety of ways in the different makes of machines, but I recommend a careful examination of our machine as an example of a most ingenious model. Our collection contains separately the most important parts of the Brunsviga —which are for the most part invisible in the assembled machine—so that you can, by examining them, get a complete picture of its arrangement.

We can best characterize the operation of the machine, so far as we have made its acquaintance, by the words *adding machine, because, with every turn of the handle, it adds, once, to the number on the slide at the right, the number which has been set on the drum.*

Finally, I shall describe in general that arrangement of the machine which permits convenient operation with *multipliers of more than one digit*. If we wish to calculate, say, $15 \cdot 12$ we should have to turn the handle fifteen times, according to the plan already outlined; moreover, if one wished to have the multiplier indicated by the counter at the left of the slide, it would be necessary to have, there also, a device for carrying the tens. Both of these difficulties are avoided by the following arrangement[1]. We first perform the multiplication by five, so that 5 appears on the slide at the left and 60 at the right (see Fig. 5). *Now we push the slide one place to the*

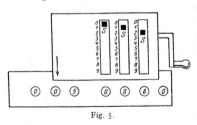

Fig. 5.

---

[1] In the newer models the cogwheel device for "carrying over" is likewise very complete.

right, so that, as shown in Fig. 5, its units cogwheel is cut out, its tens cogwheel is moved under the units slit of the drum, its hundreds cogwheel under the tens slit, etc., while, at the left, this shift brings it about that the tens cogwheel, instead of the units, is connected with the driver which the handle carries. If we now turn the handle once, 1 appears at the left, in ten's place, so that we read 15; at the right, however, we do not get the addition $\begin{cases} 60 \\ +12 \end{cases}$ but $\begin{cases} \cdot 60 \\ +12 \cdot \end{cases}$ or, in other words, 60 + 120, since the 2 is "carried over" to the tens wheel, the 1 to the hundreds wheel. Thus we get correctly $15 \cdot 12 = 180$. It is, as you see, *the exact mechanical translation of the customary process of written multiplication*, in which one writes down under one another, the products of the multiplicand by the successive digits of the multiplier, each product moved to the left one place farther than the preceding, and then adds. *In just the same way one proceeds quite generally when the multiplier has three or more digits, that is, after the usual multiplication by the ones, one moves the slide 1, 2, ... places to the right and turns the handle in each place as many times as the digit in the tens, hundreds, ... place of the multiplier indicates.*

Direct examination of the machine will disclose how one can perform other calculations with it; the remark here will suffice that *subtraction and division are effected by turning the handle in the direction opposite to that employed in addition.*

Permit me to summarize by remarking that *the theoretical principle of the machine is quite elementary and represents merely a technical realization of the rules which one always uses in numerical calculation.* That the machine really functions reliably, that all the parts engage one another with unfailing certainty, so that there is no jamming, that the wheels do not turn farther than is necessary, is, of course, the remarkable accomplishment of the man who made the design, and the mechanician who carried it out.

Let us consider for a moment the *general significance of the fact that there really are such calculating machines,* which relieve the mathematician of the purely mechanical work of numerical calculation, and which do this work faster, and, to a higher degree free from error, than he himself could do it, since the errors of human carelessness do not creep into the machine. In the existence of such a machine we see an outright confirmation *that the rules of operation alone, and not the meaning of the numbers themselves, are of importance in calculating;* for it is only these that the machine can follow; it is constructed to do just that; it could not possibly have an intuitive appreciation of the *meaning* of the numbers. We shall not, then, wish to consider it as accidental that such a man as Leibniz, who as both an abstract thinker of first rank and a man of the highest practical gifts, was, at the same time, both the father of purely formal mathematics and the inventor of a calculating machine. His machine is, to this day, one of the most prized possessions of the Kästner Museum in Hannover. Although it is not historically authenticated, still I like to assume that when Leibniz invented the calculating machine, he not only followed a useful purpose, but that he also wished to exhibit, clearly, the purely formal character of mathematical calculation.

With the construction of the calculating machine Leibniz certainly did not wish to minimize the *value of mathematical thinking*, and yet it is just such conclusions which are now sometimes drawn from the existence of the calculating machine. If the activity of a science can be supplied by a machine, that science cannot amount to much, so it is said; and hence it deserves a subordinate place. The answer to such arguments, however, is that the mathematician, even when he is himself operating with numbers and formulas, is by no means an inferior counterpart of the errorless machine, "thoughtless thinker" of Thomae; but rather, he sets for himself his problems with definite, interesting, and valuable ends in view, and carries them to solution in appropriate and original manner. He turns over to the machine only certain operations which recur frequently in the same way, and it is precisely the mathematician—one must not forget this—who invented the machine for his own relief, and who, for his own intelligent ends, designates the tasks which it shall perform.

Let me close this chapter with the wish that the calculating machine, in view of its great importance, may become known in wider circles than is now the case. Above all, every teacher of mathematics should become familiar with it, and it ought to be possible to have it demonstrated in secondary instruction.

II. The First Extension of the Notion of Number

With the last section we leave operations with integers, and shall treat, in a new chapter, the *extension of the number concept.* In the schools it is customary, in this field, to take in order the following steps:

1. *Introduction of fractions and operations with fractions.*

2. *Treatment of negative numbers*, in connection with the beginnings of operations with letters.

3. *More or less complete presentation of the notion of irrational numbers by examples that arise upon different occasions*, which leads, then, gradually, to the notion of the *continuum of real numbers.*

It is a matter of indifference in which order we take up the first two points. Let us discuss negative numbers before fractions.

1. Negative Numbers

Let us first note, as to terminology, that in the schools, one speaks of positive and negative numbers, inclusively, as *relative numbers* in distinction from the *absolute* (positive) numbers, whereas, in universities this language is not common. Moreover, in the schools one speaks of "algebraic numbers"[1] along with relative numbers, an expression which we in universities employ, as you know, in quite another sense.

Now, as to the origin and introduction of negative numbers, I can be brief in my reference to source material; these things are already familiar to you, or you can at least easily make them so with the help the references I shall give. You will find a complete treatment, for example, in Weber-Wellstein; also, in very readable form, in H. Burkhardt's Algebraischer Analysis[2]. This book, moreover, you might well purchase, as it is of moderate size.

The creation of negative numbers is motivated, as you know, by the demand *that the operation of subtraction shall be possible in all cases*. If $a < b$ then $a - b$ is meaningless in the domain of natural integers; a number $c = b - a$ does exist, however, and we write

$$a - b = -c$$

which we call a *negative number*. This definition at once justifies the representation of all integers by means of the scale of equidistant points

on a straight line the "axis of abscissas" which extends in both directions from an origin. One may consider this picture as a common possession of all educated persons today, and one can, perhaps, assume that it owes its general dissemination, chiefly, to the thermometer scale. The commercial balance, with its reckoning in debits and credits, affords likewise a graphic and familiar picture of negative numbers.

Let us, however, realize at once and emphatically how extraordinarily difficult in principle is the step, which is taken in school when negative numbers are introduced. Where the pupil before was accustomed to represent visually by concrete numbers of things the numbers, and, later, the letters, with which he operated, as well as the results which he obtained by his operations, he finds it now quite different. He has to do with something new, the "negative numbers", which have, immediately, nothing in common with his picture of numbers of things, but he must operate with them as though they had, although the operations

---

[1] See, e. g. Mehler, *Hauptsätze der Elementarmathematik*, Nineteenth edition, p. 77, Berlin, 1895.

[2] Leipzig 1903. [Third edition, revised by G. Faber, 1920.] — See also Fine, H., *The Number-System of Algebra treated Theoretically and Historically*, Heath.

have graphically a meaning much less clear than the old ones. Here, for the first time, we meet the transition from concrete to formal mathematics. The complete mastery of this transition requires a high order of ability in abstraction.

We shall now inquire in detail what happens to the operations of calculation when negative numbers are introduced. The first thing to notice is that addition and subtraction coalesce, substantially: The addition of a positive number is the subtraction of the equal and opposite negative number. In this connection, Max Simon makes the amusing remark that, whereas negative numbers were created to make the operation of subtraction possible without any exception, subtraction as an independent operation ceased to exist by virtue of that creation. For this new operation of addition (including subtraction) in the domain of positive and negative numbers the five formal laws stated before hold without change. These are, in brief (see p. 9 et seq.):

1. Always possible.
2. Unique.
3. Associative law.
4. Commutative law.
5. Monotonic law.

Notice, in connection with 5, that $a < b$ means, now, that a lies to the left of b in the geometric representation, so that we have, for example $-2 < -1$, $-3 < -2$.

The chief point in the *multiplication* of positive and negative numbers is the *rule of signs*, that $a \cdot (-c) = (-c) \cdot a = -(a \cdot c)$, and $(-c)(-c') = +(c \cdot c')$. Especially the latter rule: "Minus times minus gives plus" is often a dangerous stumbling block. We shall return presently to the inner significance of these rules; just now we shall combine them into a statement defining multiplication of a series of positive and negative numbers: *The absolute value of a product is equal to the product of the absolute values of the factors; its sign is positive or negative according as an even or an odd number of factors is negative.* With this convention, multiplication in the domain of positive and negative numbers has again the following properties:

1. Always possible.
2. Unique.
3. Associative.
4. Commutative.
5. Distributive with respect to addition.

There is a change only in the monotonic law; in its place one has the following law:

6. If $a > b$ then $a \cdot c \gtreqless b \cdot c$ according as $c \gtreqless 0$.

Let us inquire, now, whether these laws, considered again purely formally, are consistent. We must admit at once, however, that a purely logical proof of consistency is as yet much less possible here than it is in the case of integers. Only a reduction is possible, in the sense that the present laws are consistent if the laws for integers are consistent. But until this has been completed by a logical consistency proof for integers, one will have to hold that the *consistency of our laws is based solely on the fact that there are intuitive things, with intuitive relations, which obey these laws.* We noted above, as such, the series of integral points on the axis of abscissas and we need only indicate what the rules of operation signify there: The addition $x' = x + a$, where a is fixed, assigns to each point x a second point x', so that the infinite straight line is simply displaced along itself by an amount a, to the right or to the left, according as a is positive or negative. In an analogous manner, the multiplication $x' = a \cdot x$ represents a similarity transformation of the line into itself, a pure stretching for $a > 0$, a stretching together with a reflexion in the origin for $a < 0$.

Permit me now to explain how, historically, all these things arose. One must not think that the negative numbers are the invention of some clever man who menufactured them, together with their consistency perhaps, out of the geometric representation. Rather, during a long period of development, the use of negative numbers forced itself, so to speak, upon mathematicians. Only in the nineteenth century, after men had been operating with them for centuries, was the consideration of their consistency taken up.

Let me preface the history of negative numbers with the remark that the ancient Greeks certainly had no negative numbers, so that one cannot yield them the first place, in this case, as so many people are otherwise prone to do. One must attribute this invention to the Hindus, who also created our system of digits and in particular our zero. In Europe, negative numbers came gradually into use at the time of the Renaissance, just as the transition to operating with letters had been completed. I must not omit to mention here that this completion of operations with letters is said to have been accomplished by Vieta in his book *In Artem Analyticam Isagoge*[1].

From the present point of view, we have the so called parenthesis rules for operations with positive numbers, which are, of course, contained in our fundamental formulas, provided one includes the correponding laws for subtraction. But I should like to take them up somewhat in detail, by means of two examples, in order, above all, to show the possibility of extremely simple intuitive proofs for them, proofs which need consist only of the representation and of the word "Look"!, as was the custom with the ancient Hindus.

1. Given $a > b$ and $c > a$, where a, b, c are positive. Then $a - b$ is a positive number and is smaller than c, that is, $c - (a - b)$ must

---
[1] Tours 1591.

exist as a positive number. Let us represent the numbers on the axis of abscissas and note that the segment between the points b and a has the length $a - b$. A glance at the representation shows that, if we take away from c the segment $a - b$, the result is the same as though we first took away the entire segment a and then restored the part b, i. e.,

(1) $$c - (a - b) = c - a + b.$$

2. *Given $a > b$ and $c > d$;* then $a - b$ and $c - d$ are positive integers. We wish to examine the product $(a - b) \cdot (c - d)$; for that purpose

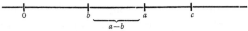

draw the diagonally hatched rectangle (Fig. 6) with sides $a - b$ and $c - d$ whose area is the number sought, $(a - b) \cdot (c - d)$, and which is part of the rectangle with sides a and c. In order to obtain the former rectangle from the latter, we take away first the horizontally hatched rectangle $a \cdot d$, then the vertically hatched one $b \cdot c$; in doing this we have removed twice the double-hatched rectangle $b \cdot d$, and we must put it back. But these operations express precisely the known formula

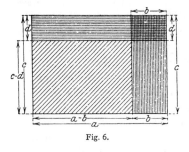

Fig. 6.

(2) $$(a - b)(c - d) = ac - ad - bc + bd.$$

As the most important psychological moment to which the introduction of negative numbers, upon this basis of operations with letters, gave rise, that general peculiarity of human nature shows itself, by virtue of which we *are involuntarily inclined to employ rules under circumstances more general than are warranted by the special cases under which the rules were derived and have validity*. This was first claimed as a guiding principle in arithmetic by Hermann Hankel, in his *Theorie der komplexen Zahlsysteme*[*][1], under the name "Prinzip von der Permanenz der formalen Gesetze"[**]. I can recommend to your notice this most interesting book. For the particular case before us, of transition to negative numbers, the above principle would declare that one desired to forget, in formulas like (1) and (2) the expressed assumptions as to the relative magnitude of a and b and to employ them in other cases. If one applies (2), for example, to $a = c = 0$, for which the formulas were not proved at all, one obtains $(-b) \cdot (-c) = + bd$, i. e., the *sign rule for multiplication of negative numbers*. In this manner we may derive, in fact almost unconsciously, all the rules, which we must now

---

[*] *Theory of Complex Number Systems.*
[1] Leipzig 1867.
[**] Principle of the permance of formal laws.

designate, following the same line of thought, as *almost necessary assumptions, necessary insofar as one would have validity of the old rules for the new concepts*. To be sure, the old mathematicians were not happy with this abstraction, and their uneasy consciences found expression in names like *invented numbers, false numbers*, etc., which they gave to the negative numbers on occasion. But in spite of all scruples, the negative numbers found more and more general recognition in the sixteenth and seventeenth centuries, because they justified themselves by their usefulness. To this end, the development of analytic geometry without doubt contributed materially. Nevertheless the doubts persisted, and were bound to persist, so long as one continued to seek for a representation in the concept of a number of things, and had not recognized the leading role of formal laws when new concepts are set up. In connection with this stood the continually recurring attempts to prove the rule of signs. The simple explanation, which was brought out in the nineteenth century, is that it is *idle to talk of the logical necessity of the theorem*, in other words, *the rule of signs is not susceptible of proof*; *one can only be concerned with recognizing the logical permissibility of the rule*, and, at the same time, that it is arbitrary, and regulated by considerations of expedience, such as the principle of permanence.

In this connection one cannot repress that oft recurring thought that things sometimes seem to be more sensible than human beings. Think of it: one of the greatest advances in mathematics, the introduction of negative numbers and of operations with them, was not created by the conscious logical reflection of an individual. On the contrary, its slow organic growth developed as a result of intensive occupation with things, so that it almost seems as though men had learned from the letters. The rational reflection that one devised here something correct, compatible with strict logic, came at a much later time. And, after all, the function of pure logic, when it comes to setting up new concepts, is only to *regulate* and *never to act as the sole guiding principle*; for there will always be, of course, many other conceptual systems which satisfy the single demand of logic, namely, freedom from contradiction.

If you desire still other *literature* concerning questions about the history of negative numbers, let me recommend Tropfkes *Geschichte der Elementarmathematik*[1]*, as an excellent collection of material containing, in lucid presentation, a great many details about the development of elementary notions, views, and names.

---

[1] Two volumes, Leipzig 1902/03. [Second edition revised and much enlarged, to appear in seven volumes, of which six had appeared by 1924.]—See also Cajori, F., *History of Mathematics*, Macmillan.

* *History of Elementary Mathematics.*

If we now look critically at the way in which negative numbers are presented in the schools, we find frequently the error of trying to prove the logical necessity of the rule of signs, corresponding to the above noted efforts of the older mathematicians. One is to derive $(-b)(-d) = +bd$ heuristically, from the formula $(a - b)(c - d)$ and to think that one has a proof, completely ignoring the fact that the validity of this formula depends on the inequalities $a > b, c > d$[1]. Thus the proof is fraudulent, and the psychological consideration which would lead us to the rule by way of the principle of permanence is lost in favor of quasi-logical considerations. Of course the pupil, to whom it is thus presented for the first time, cannot possibly comprehend it, but in the end he must nevertheless believe it; and if, as it often happens, the repetition in a higher class does not supply the corrective, the conviction may become lodged with some students that the whole thing is mysterious, incomprehensible.

In opposition to this practice, I should like to urge you, in general, never to attempt to make impossible proofs appear valid. One should convince the pupil by simple examples, or, if possible, let him find out for himself that, in view of the actual situation, *precisely these conventions, suggested by the principle of permanence, are appropriate in that they yield a uniformly convenient algorithm, whereas every other convention would always compel the consideration of numerous special cases.* To be sure, one must not be precipitate, but must allow the pupil time for the revolution in his thinking which this knowledge will provoke. And while it is easy to understand that other conventions are not advantageous, one must emphasize to the pupil how really wonderful the fact is that a general useful convention really *exists*; it should become clear to him that this is by no means self-evident.

With this I close my discussion of the theory of negative numbers and invite you now to give similar consideration to the second extension of the notion of number.

2. Fractions.

Let us begin with the treatment of fractions in the schools. There the fraction a/b has a thoroughly concrete meaning from the start. In contrast to the graphic picture of the integer, there has been only a change of base: We have passed from the *number* of things to their *measure*, from the consideration of *countable things* to *measurable things*. The system of *coins*, or of *weights*, affords, with some restriction, and the system of *lengths* affords completely, an example of *measurable manifolds*. These are the examples with which the idea of the fraction is

---

[1] See, for example, E. Heis, *Sammlung von Beispielen und Aufgaben aus der Arithmetik und Algebra*. Edition 1904, p. 46, 106—108.

given to every pupil. No one has great difficulty in grasping the meaning of $1/3$ meter oder $1/2$ pound. The *relations* $=$, $>$, $<$, between fractions can be immediately developed by means of the same concrete intuition, and likewise the operations of *addition* and *subtraction*, as well as the multiplication of a fraction *by an integer*. After this, *general multiplication* can easily be made comprehensible: *To multiply a number by a/b means to multiply it by a and then to divide by b; in other words: the product is derived from the multiplicand just as a/b is derived from* 1. *Division* by a fraction is then presented as the *operation inverse* to multiplication: *a divided by 2/3 is the number which multiplied by 2/3 gives a*. These notions of operations with fractions combine with that of negative numbers so that one finally has the totality of all rational numbers. I cannot enter into the details of this building-up process, which, in the school, takes, of course, a long time. Let us rather compare it at once with the perfected presentation of modern mathematics, using for this purpose the above mentioned books of Weber-Wellstein and Burkhardt[1].

Weber-Wellstein emphasizes primarily the formal point of view which, from the multiplicity of possible interpretations, selects what is of necessity common to all. According to this view, the fraction a/b is a symbol, a "number-pair" with which one can operate according to certain rules. These rules, which in our discussion above arose naturally from the meaning of fraction, have here the character of arbitrary conventions. For example, that which, to the pupil, is an obvious theorem concerning the multiplication or division of both terms of a fraction by the same number, appears here as a definition of equality: *two fractions a/b, c/d are called equal when* $ad = bc$. Similarly, *greater than* and *smaller than* are defined, and one *agrees* that *the fraction* $\left(\frac{ad+bc}{bd}\right)$ *shall be called the sum of the two fractions a/b, c/d, etc.* It is thus *proved* that the operations, so defined in the new domain of numbers, possess formally exactly the properties of addition and multiplication for integers, i. e., they satisfy the eleven fundamental laws which have been repeatedly enumerated.

Burkhardt does not proceed quite so formally as does Weber-Wellstein, whose presentation we have sketched in its essentials. He looks upon the *fraction a/b as a sequence of two operations in the domain of integers: a multiplication by a and a division by b*, in which the object upon which these operations are performed is an arbitrarily chosen integer. If one undertakes two such *"pairs of operations" a/b, c/d*, this is said to correspond to *multiplication of the fractions*, and one sees easily that the operation so resulting is none other than multiplication by $a \cdot c$ and division by $b \cdot d$, so that the *rule for the multiplication of fractions*,

---

[1] In what follows, the first editions of these books have been used.

$\left(\frac{a}{b}\right) \cdot \left(\frac{c}{d}\right) = \left(\frac{a \cdot c}{b \cdot d}\right)$, *is obtained out of the clear meaning of the fractions,* but not determined merely as an arbitrary convention. One can, of course, treat division in the same way. Addition and subtraction, on the other hand, do not admit of such a simple explanation with this representation; thus the formula $\frac{a}{b} + \frac{c}{d} = \frac{(ad + bc)}{bd}$ remains, with Burkhardt also, only a convention for which he adduces only reasons of plausibility.

Let us now compare the older presentation in the schools, with the modern conception just sketched. According to the latter, in the one book as well as in the other, we are left really completely in the field of integers, in spite of the extension of the notion of number. It is merely assumed that the totality of whole numbers is intuitively grasped, or that the rules of operation with them are known; the things newly defined as number-pairs, or as operations with whole numbers, fit completely into this frame. The school treatment, on the other hand, is based entirely on the newly acquired conception of measurable quantities, which supplies an immediate intuitive picture of fractions. We can best grasp this difference if we imagine a being who has the notion of whole numbers, but no conception of measurable quantities. For him the school presentation would be wholly unintelligible, whereas he could well comprehend the discussions of either Weber-Wellstein or Burkhardt.

Which of the two methods is the better? What does each accomplish? The answer to this will be like the one we gave recently when we put the analogous question concerning the different conceptions of integers. The modern presentation is surely purer, but it is also less rich. For, of that which the traditional curriculum supplies as a unit, it gives really only one part: the abstract and logically complete introduction of certain arithmetic concepts, called *"fractions"*, and of operations with them. But it leaves unexplained an entirely independent and no less important question: Can one really apply the theoretical doctrine so derived, to the concrete measurable quantities about us? Again one could call this a problem of "applied mathematics", which admits an entirely independent treatment. To be sure, it is questionable whether such a separation would be desirable pedagogically. In Weber-Wellstein, moreover, this splitting of the problem into two parts finds characteristic expression. After the abstract introduction of operations with fractions, of which alone we have thus far taken account, they devote a special (the fifth) division—called "ratios"—to the question of applying *rational numbers to the external world*. The presentation is, to be sure, rather abstract than intuitive.

I shall now close this discussion of fractions with a general remark concerning the totality of rational numbers, where, for the sake of clearness, I shall make use of the representation upon a straight line.

Think of all points with rational abscissas marked upon this line; we designate them briefly as rational points. We say, then, that the totality of these rational points on the axis of abscissas is "dense", meaning that in every interval, however small, there are still infinitely many

rational points. If we wish to avoid putting anything new into the notion of rational numbers, we might say, more abstractly, that between any two rational points there is always another rational point. It follows that one can separate from the totality of rational points, finite parts which contain neither a smallest nor a largest element. The totality of all rational points between 0 and 1, these points excluded, is an example. For, given any number between 0 and 1, there would still be a number between it and 0, i. e., a smaller, and a number between it and 1, i. e., a larger. In their systematic development, these concepts belong to the *theory of point sets* of Cantor. In fact, we shall make use later of the totality of rational numbers, together with the property just mentioned, as an important example of a point set.

I shall pass now to the third extension of the number system: the irrational numbers.

3. Irrational Numbers.

Let us not spend any time in discussing how this field is usually treated in the schools, for there one does not get much beyond a few examples. Let us rather proceed at once to the historical development. Historically, the origin of the concept of irrational numbers lies certainly in geometric intuition and in the requirements of geometry. If we consider, as we did just now, that the set of rational points is *dense* on the axis of abscissas, then there are still other points on it. Pythagoras is said to have shown this in a manner somewhat as follows. Given a right

Fig. 7.

triangle with each leg of length 1, then the hypotenuse is of length $\sqrt{2}$, and this is certainly not a rational number; for if one puts $\sqrt{2} = \frac{a}{b}$ where a and b are integers, prime to each other, one is led easily by the laws of divisibility of integers to a contradiction. *If we now lay off geometrically on the axis of abscissas, beginning at zero, the segment thus constructed, we obtain a non-rational point which is not one of the original set that is dense on the axis.* Furthermore, the Pythagoreans certainly were aware that, in most cases, the hypotenuse, $\sqrt{m^2 + n^2}$, of a right triangle with legs m and n, is irrational. The discovery of this extraordinarily essential fact was indeed worth the sacrifice of one hundred oxen with which Pythagoras is said to have celebrated it. We know also that the Pythagorean School was fond of searching out those special pairs of values for m and n for which the right triangle has *three commensurable sides*, whose lengths, in an appropriately selected unit of measure, can be expressed in integers (so called Pythagorean numbers). The simplest example of one of these number-triples is 3, 4, 5.

Later Greek mathematicians studied, in addition to these simplest irrationalities, others that were more complicated; thus one finds in Euclid types such as $\sqrt{\sqrt{a}+\sqrt{b}}$, and the like. We may say, however, in general, that they confined themselves essentially to such irrationalities as one obtains by repeated extraction of square root, and which can therefore be constructed geometrically with ruler and compasses. The general idea of irrational number was not yet known to them.

I must, modify this remark somewhat, however, in order to avoid misunderstanding. The more precise statement is that the Greeks possessed no method for producing or defining, arithmetically, the general irrational number in terms of rational numbers. This is a result of modern development and will soon engage our attention. Nevertheless, from another point of view they were familiar with the notion of the general real number which was not necessarily rational; but the concept had an entirely different appearance to them because they did not use letters for general numbers. In fact they studied, and Euclid developed very systematically, *ratios of two arbitrary segments*. They operated with such ratios precisely as we do today with arbitrary real numbers. Indeed we find in Euclid definitions which suggest strongly the modern theory of irrational numbers. Moreover the name used is different from that of the natural number; the latter is called ἀριϑμος, whereas the line ratio, the arbitrary real number, is called λόγος.

I should like to add a remark concerning the word *"irrational"*. It is without doubt the translation into Latin of the Greek "ἄλογος". The Greek word, however, meant presumably "inexpressible" and implied that the new numbers, or line ratios, could not, like the rational numbers, be expressed by the ratio of two whole numbers[1]. The misunderstanding put upon the Latin "ratio", that it could convey only the meaning "reason", gave to "irrational" the meaning "unreasonable", which seems still to cling to the term *irrational number*.

The general idea of the irrational number appeared first at the end of the sixteenth century as a consequence of the introduction of decimal fractions, the use of which became established at that time in connection with the appearance of logarithmic tables. If we transform a rational number into a decimal, we may obtain *infinite decimals*[2], as well as finite

---

[1] See Tropfke, second edition, Vol. 2, p. 71.
[2] For complete treatment of this subject see, p. 40 et seq.

decimals, but they will always be *periodic*. The simplest example is $\frac{1}{3} = 0.333\ldots$, i.e., a decimal whose period of one digit begins immediately after the decimal point. Now there is nothing to prevent our thinking of an aperiodic decimal whose digits proceed according to any definite law whatever, and anyone would instinctively consider it as a definite, and hence a non-rational, number. By this means the general notion of irrational number is established. It arose to a certain extent automatically, by the consideration of decimal fractions. Thus, historically, the same thing happened with irrational numbers that, as we have seen, happened with negative numbers. Calculation forced the introduction of the new concepts, and without being concerned much as to their nature or their motivation, one operated with them, the more particularly since they often proved to be extremely useful.

It was not until the sixth decade of the nineteenth century that the need was felt for a more precise arithmetic formulation of the foundations of irrational numbers. This occurred in the lectures which Weierstrass delivered at about that date. In 1872, a general foundation was laid simultaneously by G. Cantor of Halle, the founder of the theory of point sets, and independently by R. Dedekind of Braunschweig. I will explain Dedekind's point of view in a few words. Let us assume a knowledge of the totality of rational numbers, but let us exclude all space perception, which would force upon us forthwith the notion of the continuity of the number series. With this understanding, in order to attain to a purely arithmetic definition of the irrational number, Dedekind sets up the notion of a *"cut"* in the domain of rational numbers. If r is any rational number, it separates the totality of rational numbers into two parts A and B *such that every number in A is smaller than any number in B and every rational number belongs to one of these two classes*. A is the totality of all rational numbers which are smaller than r, B those that are larger, whereby r itself may be thought of indifferently as belonging to the one or to the other. Besides these *"proper cuts"* there are also *"improper cuts"*, these being separations of all rational numbers into two classes having the same properties except that they are not brought about by a rational number, i. e., separations such that there is neither a smallest rational number in B nor a largest in A. An example of such an improper cut is supplied by, say, $\sqrt{2} = 1.414\ldots$ In fact, every infinite decimal fraction defines a cut, provided one assigns to B every rational number which is larger than every approximation to the infinite decimal, and to A every other rational number; each number in A would thus be equalled or exceeded by at least one approximation (and hence by infinitely many). One can easily show that this cut is proper if the decimal is periodic, improper if it is not periodic.

With these considerations as his basis, Dedekind sets up his definition, which, from a purely logical standpoint, must be looked upon as an arbitrary convention: *A cut in the domain of rational numbers is called a rational number or an irrational number according as the cut is proper or improper.* A definition of equality follows from this at once: *Two numbers are said to be equal if they yield the same cut in the domain of rational numbers.* From this definition we can immediately *prove* for example, that, $1/_3$ is equal to the infinite decimal 0.3333 .... If we accept this standpoint, we must demand a proof, i. e., a process of reasoning depending upon the definition given, although this would appear quite unnecessary to one approaching the subject naively. Moreover, such a proof is immediate, if one reflects that every rational number smaller than $1/_3$ will be exceeded ultimately by the decimal approximations, whereas these are smaller than every rational number which exceeds $\frac{1}{3}$. The corresponding definition in the lectures of Weierstrass appears in the following form: *Two numbers are called equal if they differ by less than any preassigned constant, however small.* The connection with the preceding explanation is clear. The last definition becomes striking if one reflects why 0.999 ... is equal to 1; the difference is certainly smaller than 0.1, smaller than 0.01, etc., that is, it is exactly zero according to the definition.

If we enquire how it happens that we can admit the irrational numbers into the system of ordinary numbers and operate with them in just the same way, the answer is to be found in the validity of the monotonic law for the four fundamental operations. The principle is as follows: *If we wish to perform upon irrational numbers the operation of addition, multiplication, etc., we can enclose them between ever narrowing rational limits and perform upon these limits the desired operations; then, because of the validity of the monotonic law, the result will also be enclosed between ever narrowing limits.*

It is hardly necessary for me to explain these things in greater detail, since very readable presentations of them are easily available in many books, especially in Weber-Wellstein and in Burkhardt. I hope that you will read more fully than I could tell you here in these books, about the definition of irrational numbers.

I should prefer, rather, to talk about something which you will hardly find in the books, namely, how, after establishing this arithmetic theory, we can pass to the applications in other fields. This applies in particular, to analytic geometry, which to the naive perception appears to be (and psychologically really is) the source of irrational numbers. If we think of the axis of abscissa, with the origin and also the rational points marked on it, as above, then these applications depend upon the following fundamental principle: *Corresponding to every rational or irrational number there is a point which has this number as abscissa and, conversely, corresponding to every point on the line there is a rational or an irrational number, viz., its abscissa.* Such a fundamental principle, which stands at the head of a branch of knowledge, and from which all that follows is logically deduced, while it itself cannot be logically proved, may properly be called an *axiom*. Such an axiom will appear intuitively obvious or will be accepted as a more or less arbitrary convention, by each person according to his gifts. This axiom concerning the one-to-one correspondence between real numbers on one hand, and the points of a straight line on the other, is usually called the *Cantor axiom* because G. Cantor was the first to formulate it specifically (in the Mathematische Annalen, vol. 5, 1872).

This is the proper place to say a word about the nature of space perception. It is variously ascribed to two different sources of knowledge. One the sensibly immediate, the empirical intuition of space, which we can control by means of measurement. The other is quite different, and consists in a subjective idealizing intuition, one might say, perhaps, our *inherent idea of space*, which goes beyond the inexactness of sense observation. I pointed out to you an analogous difference when we were discussing the notion of number. We may characterize it best as follows: It is immediately clear to us what a small number means, like 2 or 5, or even 7, whereas we do not have such immediate intuition of a larger number, say 2503. Immediate intuition is replaced here by the subjective intuition of an ordered number series, which we derive from the first numbers by mathematical induction. There is a similar situation regarding space perception. Thus, if we think of the distance between two points, we can estimate or measure it only to a *limited degree of exactness*, because our eyes cannot recognize as different two line-segments whose difference in length lies below a certain limit. This is the concept of the *threshold* of perception which plays such an important role in psychology. This phenomenon still persists, in its essentials, when we aid the eye with instruments of the highest precision; for there are physical properties which prohibit our exceeding a certain degree of exactness. For instance, optics teaches that the wave-length of light, which varies with the color, is of the order of smallness of $1/1000$ mm. ($= 1$ micron); it shows also that objects whose dimensions are of this order of smallness cannot be seen distinctly with the best microscopes because diffraction enters then and hence no optical image can give exact reproductions of the details. *The result of this is the impossibility, by direct optical means, of getting measures of length that are finer than to within one micron, so that, when measured lengths are given in millimeters, only the first three decimals can have an assured meaning.* In the same way, in all physical observations and measurements, one meets such threshold values which cannot be passed, which determine the extreme limits of possible exactness of lengths which have been measured and expressed in millimeters. Statements beyond this limit have no meaning, and are an evidence of ignorance or of attempted deception. One often finds such excessively exact numbers in the advertisements of medicinal springs, where the percentage of salt, which really varies with the time, is given to a number of decimal places which could not possibly be determind by weighing.

In contrast with this property of empirical space perception which is restricted by limitations on exactness, *abstract, or ideal space perception demands unlimited exactness, by virtue of which, in view of Cantor's axiom, it corresponds exactly to the arithmetic definition of the number concept.*

In harmony with this division of our perception, it is natural to divide mathematics also into two parts, which have been called *mathematics of approximation* and the *mathematics of precision*. If we desire to explain this difference by an interpretation of the equation $f(x) = 0$, we may note that, in the mathematics of approximation, just as in our empirical space perception, one is not concerned that $f(x)$ should be *exactly* zero, but merely that its absolute value $|f(x)|$ should *remain below the attainable threshold of exactness* ε. The symbol $f(x) = 0$ is merely an abbreviation for the inequality $|f(x)| < \varepsilon$, with which one is really concerned. It is only in the mathematics of precision that one insists that the equation $f(x) = 0$ be exactly satisfied. Since mathematics of approximation alone plays a rôle in applications, one might say, somewhat crassly, that one needs only this branch of mathematics, whereas the mathematics of precision exists only for the intellectual pleasure of those who busy themselves with it, and to give valuable and indeed indispensable support for the development of mathematics of approximation.

In order to return to our real subject, I add here the remark that the *concept of irrational number belongs certainly only to mathematics of precision*. For, the assertion that two points are separated by an irrational number of millimeters cannot possibly have a meaning, since, as we saw, when our rigid scales are measured in meters, all decimal places beyond the sixth are devoid of meaning. *Thus in practice we can, without concern, replace irrational numbers by rational ones.* This may seem, to be sure, to be contradicted by the fact that, in crystallography, one talks of the law of rational indices, or by the fact that in astronomy, one distinguishes different cases according as the periods of revolution of two planets have a rational or an irrational ratio. In reality, however, this form of expression only exhibits the many-sidedness of language; for one is using here rational and irrational in a sense entirely different from that hitherto used, namely, in the sense of *mathematics of approximation*. In this sense, one says that two magnitudes have a rational ratio when they are to each other as two *small* integers, say 3/7; whereas one would call the ratio 2021/7053 irrational. We cannot say how large numerator and denominator in this second case must be, in general, since that depends upon the problem in hand. I discussed all these interesting relations in a course of lectures in the Summer Semester of 1901, which was lithographed in 1902 and which will constitute the third volume of the present work (see the preface to the third edition, p. V): *Applications of Differential and Integral Calculus to Geometry, a Revision of Principles* [Elaborated by C. H. Müller].

In conclusion let me say, in a few words, how I would have these matters handled in the schools. An exact theory of irrational numbers would hardly be adapted either to the interest or to the power of comprehension of most of the pupils. The pupil will usually be content with results of limited exactness. He will look with astonished approval upon correctness to within $1/_{1000}$ mm and will not demand unlimited exactness. For the average pupil it will be sufficient if one makes the irrational number intelligible in general by means of examples, and this is what is usually done. To be sure, especially gifted individual pupils will demand a more complete explanation than this, and it will be a laudable exercise of pedagogical skill on the part of the teacher to give such students the desired supplementary explanation without sacrificing the interests of the majority.

III. Concerning Special Properties of Integers

We shall now begin a new chapter which will be devoted to the *actual theory of integers*, to the *theory of numbers*, or *arithmetic in its narrower sense*. I shall first recall in tabular form the individual questions from this science which appear in the school curriculum.

1. The first problem of the theory of numbers is that of *divisibility*: Is one number divisible by another or not?

2. Simple rules can be given which enable us easily to decide as to the *divisibility of any given number by smaller numbers*, such as 2, 3, 4, 5, 9, 11, etc.

3. There are *infinitely many prime numbers*, that is, numbers which have no *integral divisors* except one and themselves): 2, 3, 4, 5, 9, 11, etc.

4. We are in control of all of the properties of given integers if we know their decomposition *into prime factors*.

5. In the *transformation of rational fractions into decimal fractions* the theory of numbers plays an important role; it shows why the decimal fraction must be *periodic* and how large the period is.

Although such questions may be considered in secondary schools, when the pupils are between the ages of eleven and thirteen, the theory of numbers comes up only in isolated places during the later years, and, at most, the following points are considered.

6. *Continued fractions* are taught occasionally, although not in all schools.

7. Sometimes instruction is given also in Diophantine equations, that is, equations with several unknowns which can take only integral values.

The *Pythagorean numbers* of which we spoke (see p. 32), furnish an example; here one has to do with triplets of integers which satisfy the equation
$$a^2 + b^2 = c^2.$$

8. The *problem of dividing the circle into equal parts* is closely related to the theory of numbers, although the connection is hardly ever worked out in the schools. If we wish to divide the circle into n equal parts, using, of course, *only ruler and compasses*, it is easy to do it for $n = 2, 3, 4, 5, 6$. It cannot be done, however, if $n = 7$, hence we stop respectfully when we come to this problem in the school. To be sure, it is not always stated definitely that this construction is really impossible when $n = 7$, — a fact whose explanation lies somewhat deep in number-theoretic considerations. In order to forestall misunderstandings, which unfortunately often arise, let me say, with emphasis, that one is concerned here again with a *problem of mathematics of precision*, which is devoid of meaning for the applications. In practice, even in cases where an "exact" construction is possible, it would not be used ordinarily; for, in the field of mathematics of approximation, the circle can be divided into any desired number of equal parts more suitably by simple skillful experiment; and any prescribed, practically possible, degree of exactness can be attained. Every mechanician who makes instruments that carry divided circles proceeds in this way.

9. The higher theory of numbers is touched by the school curriculum in one other place, namely, when π *is calculated*, during the study of the *quadrature* of the circle. We usually determine the first decimal places for π, by some method or other, and we mention incidentally, perhaps, the *modern proof of the transcendence of π which sets at rest the old problem of the quadrature of the circle with ruler and compasses*. At the end of this course I shall consider this proof in detail. For the present I shall give merely a prescise formulation of the fact, namely, *that the number π does not satisfy any algebraic equation with integral coefficients*:
$$a\pi^n + b\pi^{n-1} + \cdots + k\pi + 1 = 0.$$

It is especially important that the coefficients be integers, and it is for this reason that the problem belongs to the theory of numbers. Of course here, again, one is concerned solely with a *problem of the mathematics of precision*, because it is only in this sense that the number-theoretic character of π has any significance. The mathematics of approximation is satisfied with the determination of the first few decimals, which permit us to effect the quadrature of the circle with any desired degree of exactness.

I have sketched for you the place of the theory of numbers in the schools. Let us consider now its proper place in *university instruction* and in *scientific investigation*. In this connection I should like to divide research mathematicians, according to their attitude toward theory of numbers, into two classes, which I might call the *enthusiastic* class and the *indifferent* class. For the former there is no other science so beautiful and so important, none which contains such clear and precise proofs, theorems of such impeccable rigor, as the theory of numbers. Gauss said "If mathematics is the queen of sciences, then the theory of numbers is the queen of mathematics". On the other hand, theory of numbers lies remote from those who are indifferent; they show little interest in its development, indeed they positively avoid it. The majority of students might, as regards their attitude, be put into the second class.

I think that the *reason for this remarkable division* can be summarized as follows: On the one hand the theory of numbers is *fundamental for all more thoroughgoing mathematical research*; proceeding from entirely different fields, one comes at last, with extraordinary frequency, upon relatively simple arithmetic facts. On the other hand, however, the *pure theory of numbers is an extremely abstract thing*, and one does not often find the gift of ability to understand with pleasure anything so abstract. The fact that most textbooks are at pains to present the subject in the most abstract way tends to accentuate this unattractiveness of the subject. I believe that *the theory of numbers would be made more accessible, and would awaken more general interest, if it were presented in connection with graphical elements and appropriate figures*. Although its theorems are logically independent of such aids, still one's comprehension would be helped by them. I attempted to do this in my lectures in 1895/96[1] and a similar plan is followed by H. Minkowski in his book on *Diophantische Approximationen*[2]. My lectures were of a more elementary introductory character, whereas Minkowski considers at an early point special problems in a detailed manner.

As to *textbooks in the theory of numbers*, you will often find all you need in the textbooks in algebra. Among the large number of books on the theory of real numbers, I would mention especially Bachman's *Grundlagen der neueren Zahlentheorie*[3].

In the *more special number-theoretic discussions* which I shall give here, I shall keep touch with the points mentioned above and I shall endeavor especially to present the matter as graphically as possible. While I shall restrict myself to material that is *valuable for the teacher*, I shall by no means put it into a form suitable for immediate presentation to the pupils. The necessity for this arises from my *experiences in*

---

[1] *Ausgewähltes Kapitel der Zahlentheorie* (mimeographed lectures written up by A. Sommerfeld and Ph. Furtwängler). Second printing (already exhausted). Leipzig 1907.

[2] With an appendix: *Eine Einführung in die Zahlentheorie*. Leipzig 1907.

[3] Sammlung Schubert No. 53. Leipzig 1907. [Second edition published by R. Hauszner 1921.] — See also Carmichael, R. D., *Theory of Numbers*. Wiley.

examinations, which show me that the number-theoretic information of candidates is often confined to catchwords which have no thorough knowledge back of them. Every candidate can tell me that π is "transcendental"; but many of them do not know what that means; I was told, once, that a transcendental number was neither rational nor irrational. Likewise I often find candidates who tell me that the number of primes is infinite, but who have no notion as to the proof, although it is so simple.

I shall start my number-theoretic discussion with this proof, assuming that you are acquainted with the first two points metioned in our list. As a matter of history I remind you that this proof was handed on to us by *Euclid*, whose "elements" (Greek στοιχεῖα) contained not only his system of geometry, but also *algebraic and arithmetic information in geometric language*. Euclid's transmitted *proof of the existence of infinitely many prime numbers* is as follows: Assuming that the sequence of prime numbers is finite, let it be $1, 2, 3, 5, \ldots, p$; then the number $N = (1 \cdot 2 \cdot 3 \cdot 5 \ldots p) - 1$ is not divisible by any of the numbers $2, 3, 5, \ldots p$ since there is always the remainder 1; hence N must either itself be a prime number or there are prime numbers larger than p. Either of these alternatives contradicts the hypothesis, and the proof is complete.

In connection with the *fourth point*, the *separation into prime factors*, I should like to call to your attention one of the older *factor tables*: *Chernac, Cribum Arithmeticum*[1], a large, meritorious work which deserves, historically, all the more attention because it is so reliable. The name of the table suggests the sieve of *Eratosthenes*. The idea on which it was based is that we should discard gradually from the series of all integers those which are divisible by $2, 3, 5, \ldots$, so that only the prime numbers would remain. Chernac gives the decomposition into prime factors of all integers up to 1 020 000 which are not divisible by 2, 3, or 5; all the prime numbers are marked with a bar. It was in the Chernac work that all the prime numbers lying within the limits stated above were first given. During the nineteenth century the determination was extended to all prime numbers as far as nine million.

I turn now to the *fifth point*, the transformation *of ordinary fractions into decimal fractions*. For the complete theory I shall refer you to Weber-Wellstein, and I shall explain here only the principle of the method by means of a typical example. Let us consider the fraction $1/p$, where p is a prime number different from 2 and 5. We shall show *that $1/p$ is equal to an infinite periodic decimal, and that the number δ of places in the period is the smallest exponent for which 10^δ, when divided by p, leaves 1*

---

[1] Deventer 1811.

as a remainder, or that, in the language of number theory, δ *is the smallest exponent which satisfies the "congruence"*:

$$10^\delta \equiv 1 \pmod{p}.$$

The proof requires, in the first place, the knowledge that this congruence always has a solution. This is supplied by the *theorem of Fermat*, which states that for every prime number p except 2 and 5:

$$10^{p-1} \equiv 1 \pmod{p}.$$

We shall omit here the proof of this fundamental theorem, which is one of the permanent tools of every mathematician. Secondly, we must borrow from the theory of numbers the theorem that the smallest exponent in question, δ, is *either $p-1$ itself or a divisor of $p-1$*. We can apply this to the given value p and find that $\dfrac{10^\delta - 1}{p}$ is an integer N so that one has:

$$\frac{10^\delta}{p} = \frac{1}{p} + N.$$

If we now think of $10^\delta/p$, as well as $1/p$, converted into a decimal, the digits in the two decimals must be identical, since the difference is an integer. But since $10^\delta/p$ is got from $1/p$ by moving the decimal point δ places to the right, it follows that the digits in the decimal expression of $1/p$ are unaltered by this operation, in other words *that the decimal fraction $1/p$ consists of continued repetition of the same "period" of δ digits*.

In order now to see that there *cannot be a smaller period of $\delta' < \delta$ digits* one needs only to prove that the digit number δ' of *every* period must satisfy the congruence $10^{\delta'} \equiv 1$; for we know that δ was the *smallest* solution of this congruence. This proof will result if we pursue the preceding argument in the reverse direction. It follows from our assumption that $1/p$ and $10^{\delta'}/p$ coincide in their decimal places, hence that $\dfrac{10^{\delta'}}{p} - \dfrac{1}{p}$ is an integer N', and therefore that $10^{\delta'} - 1$ is divisible by p, or, in other words, that $10^{\delta'} \equiv 1 \pmod{p}$. This completes the proof.

I will give you a few of the simplest instructive *examples*, which will show that δ can take widely different values, both smaller than and equal to $p-1$. Notice first that for:

$$\tfrac{1}{3} = 0.333\ldots$$

the number of digits in the period is 1, and that in fact, $10^1 \equiv 1 \pmod{3}$. Similarly we find

$$\tfrac{1}{11} = 0.0909\ldots,$$

whence $\delta = 2$, and correspondingly $10^1 \equiv 10, 10^2 \equiv 1 \pmod{11}$. The maximum value $= p-1$ appears in the example:

$$\frac{1}{7} = 0.142857142857\ldots.$$

Here $\delta = 6$ and we have, in fact, $10^1 \equiv 3$, $10^2 \equiv 2$, $10^3 \equiv 6$, $10^4 \equiv 4$, $10^5 \equiv 5$, and $10^6 \equiv 1 \pmod 7$.

Now let us take up, in a similar way, the *sixth point* of my list, *continued fractions*. I shall not present this, however, in the usual abstract arithmetic manner, since you will find it given elsewhere, e. g., in Weber-Wellstein. I shall take this opportunity to show you how number-theoretic things take on a clear and easily intelligible form through geometric and graphical presentation. In this use of geometric aids in number theory we are really only retracing the steps followed by *Gauss* and *Dirichlet*. It was the later mathematicians, say from 1860 on, who banished geometric methods from the theory of numbers. Of course, I can give here only the most important trains of thought and theorems, without proof, and I shall assume that you are not entire strangers to the elementary theory of continued fractions. My lithographed lectures on number theory[1] contain a thoroughgoing account.

You know how the *development of a given positive number ω into a continued fraction* arises. We separate out the largest positive integer n_0 contained in ω and write:

$$\omega = n_0 + r_0, \qquad \text{where } 0 \leq r_0 < 1,$$

then, if $r_0 \neq 0$, we treat $1/r_0$ as we did ω:

$$1/r_0 = n_1 + r_1, \qquad \text{where } 0 \leq r_1 < 1,$$

and continue in the same way:

$$1/r_1 = n_2 + r_2, \qquad \text{where } 0 \leq r_2 < 1,$$
$$1/r_2 = n_3 + r_3, \qquad \text{where } 0 \leq r_3 < 1,$$
$$\dots\dots\dots\dots\dots\dots\dots\dots\dots\dots\dots\dots\dots$$

The process *terminates after a finite number of steps if ω is rational*, because a vanishing remainder r_ν must appear in that case; otherwise the process goes on indefinitely. In any case, we write, as the *development of ω into a continued fraction*:

$$\omega = n_0 + \cfrac{1}{n_1 + \cfrac{1}{n_2 + \cfrac{1}{n_3 + \cdots}}}$$

As an example, the continued fraction for π is

$$\pi = 3{\cdot}14159265 \cdots = 3 + \cfrac{1}{7 + \cfrac{1}{15 + \cfrac{1}{1 + \cfrac{1}{292 + \cdots}}}}$$

---

[1] See also Klein, F., *Gesammelte Mathematische Abhandlungen*, Vol. II, pp. 209 to 211.

If we stop the development after the first, second, third, ... partial denominator, we obtain rational fractions, called *convergents*:

$$n_0 = \frac{p_0}{q_0}, \qquad n_0 + \frac{1}{n_1} = \frac{p_1}{q_1}, \qquad n_0 + \cfrac{1}{n_1 + \cfrac{1}{n_2}} = \frac{p_2}{q_2}, \ldots;$$

these give remarkably good approximations to the number ω, or, to speak more exactly, each one of them gives an *approximation which is closer than that given by any other rational fraction which does not have a larger denominator*. Because of this property, continued fractions are of practical importance where one seeks the best possible approximation to an irrational number, or to a fraction with a large denominator (e. g. a many-place decimal) by means of a fraction having the smallest possible denominator. The following convergents of the continued fraction for π, converted into decimals, enable one to see how close the approximations are to the value $\pi = 3{,}14159265 \ldots$.

$$\frac{p_0}{q_0} = 3, \qquad \frac{p_1}{q_1} = \frac{22}{7} = 3{,}14285\ldots,$$

$$\frac{p_2}{q_2} = \frac{333}{106} = 3{,}141509\ldots, \qquad \frac{p_3}{q_3} = \frac{355}{113} = 3{,}14159292\ldots.$$

You will observe, moreover, in this example, that the convergents are alternately less than and greater than π. This is true in general, as is well known, that is *the successive convergents of the continued fraction for ω are alternately less than and greater than ω, and enclose it between ever narrowing limits*.

Let us now enliven these considerations with geometric pictures. Confining our attention to positive numbers, let us *mark all those points in the positive quadrant of the xy plane* (see Fig. 8) *which have integral coordinates*, forming thus a so called *point lattice*. Let us examine this lattice, I am tempted to say this "firmament" of points, with our point of view at the origin. The radius vector from 0 to the point $(x = a, y = b)$ has for its equation

$$\frac{x}{y} = \frac{a}{b},$$

and conversely, there are upon every such ray, $x/y = \lambda$, where $\lambda = a/b$ is rational, infinitely many integral points (ma, mb), where m is an arbitrary whole number. Looking from 0, then, one sees points of the lattice *in all rational directions and only in such directions*. The field of view is everywhere "densely" but not completely and continuously filled with "stars". One might be inclined to compare this view with that of the milky way. With the exception of 0 itself there is *not a single integral point lying upon an irrational ray $x/y = \omega$, where ω is irrational*, which is very remarkable. If we recall Dedekind's definition of irrational number, it becomes obvious that such a ray makes a *cut in the field*

44 Arithmetic: Concerning Special Properties of Integers.

of integral points by separating the points into two point sets, one lying to the right of the ray and one to the left. If we inquire how these point sets converge toward our ray $x/y = \omega$, we shall find a very simple relation to the continued fraction for ω. By marking each point ($x = p_\nu$, $y = q_\nu$), corresponding to the convergent p_ν/q_ν, we see that the rays to these points approximate to the ray $x/y = \omega$ better and better, alternately from the left and from the right, just as the numbers p_ν/q_ν approximate to the number ω. Moreover, if one makes use of the known number-theoretic properties of p_ν, q_ν, one finds the following theorem: *Imagine pegs or needles affixed at all the integral points, and wrap a tightly drawn string about the sets of pegs to the right and to the left of the ω-ray, then the vertices of the two convex string-polygons which bound our two point sets will be precisely the points (p_ν, q_ν) whose coordinates are the numerators and denominators of the successive convergents to ω, the left polygon having the even convergents, the right one the odd.* This gives a new,

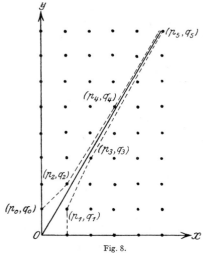

Fig. 8.

and, one may well say, an extremely graphic definition of a continued fraction. The representation in Fig. 8 corresponds to the example

$$\omega = \frac{\sqrt{5}-1}{2} = \cfrac{1}{1+\cfrac{1}{1+\cfrac{1}{1+\cdots}}}$$

which is the irrationality associated with the regular decagon. In this example, the first few vertices of the two polygons are

left: $p_0 = 0$, $q_0 = 1$; $p_2 = 1$, $q_2 = 2$; $p_4 = 3$, $q_4 = 5$; ...

right: $p_1 = 1$, $q_1 = 1$; $p_3 = 2$, $q_3 = 3$; $p_5 = 5$, $q_5 = 8$; ...

The values p_ν, q_ν for π grow much more rapidly, so that one could hardly draw the corresponding representation. The proof of our theorem, which I cannot give here, can be found in detail on page 43 of in my lithographed lectures.

I shall now pass on to the treatment of the *seventh point*, the *Pythagorean numbers*, where we shall use space perception in a somewhat different form. Instead of the equation:

(1) $$a^2 + b^2 = c^2,$$

whose integral solutions are sought, let us set:

(2) $\quad\quad\quad\quad a/c = \xi, \quad\quad b/c = \eta$

and consider the equation:

(3) $\quad\quad\quad\quad \xi^2 + \eta^2 = 1,$

with the problem of finding all the *rational number-pairs* ξ, η which satisfy it. Accordingly, we start from the representation of *all rational points* ξ, η (i.e. all points with rational coordinates ξ, η), which will fill the $\xi\eta$-plane "densely". $\xi^2 + \eta^2 = 1$ is the equation of the *unit circle* about the origin in this plane. It is our task to see how this *circle threads its way through the dense set of rational points, in particular, to see which of these points it contains.* We know a few such points of old, such as the intercepts with the axes, one of which, $S\,(\xi = -1, \eta = 0)$, we shall consider (see Fig. 9). All rays through S are given by the equation

(4) $\quad\quad\quad\quad \eta = \lambda(\xi + 1);$

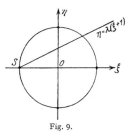

Fig. 9.

we call such a ray *rational* or *irrational* according as the parameter λ is rational or not. We have now the double theorem *that every rational point of the circle is projected from S by a rational ray and that every rational ray* (4) *meets the circle in a rational point.* The first half of the theorem is obvious. We prove the second half by substituting from (4) in (3). This gives for the abscissas of the points of intersection the equation

$$\xi^2 + \lambda^2(\xi + 1)^2 = 1$$

or

$$(1 + \lambda^2)\,\xi^2 + 2\lambda^2\xi + \lambda^2 - 1 = 0.$$

We know one solution of this equation, $\xi = -1$, which corresponds to the intersection S; for the other, one gets by easy calculation

(5a) $\quad\quad\quad\quad \xi = \dfrac{1-\lambda^2}{1+\lambda^2},$

and from (4) the corresponding ordinate

(5b) $\quad\quad\quad\quad \eta = \dfrac{2\lambda}{1+\lambda^2}.$

From (5a) and (5b) it follows that the second intersection is a rational point if λ is rational.

Our double theorem, now fully proved, can be stated also as follows. *All the rational points of the circle are represented by formulas* (5) *if λ is an arbitrary rational number.* This solves our problem and we need only to transform to whole numbers. For this purpose we put

$$\lambda = n/m,$$

where n, m are integers and obtain from (5):

$$\xi = \frac{m^2 - n^2}{m^2 + n^2}, \qquad \eta = \frac{2mn}{m^2 + n^2},$$

as the totality of rational solutions of (3). All integral solutions of the original equation (1), i.e., *all Pythagorean numbers are therefore given by the equations*

$$a = m^2 - n^2, \qquad b = 2mn, \qquad c = m^2 + n^2;$$

and one obtains the totality of solutions which have no common divisor if m and n take all pairs of relatively prime integral values. We have thus a graphic deduction of a result which usually appears very abstract.

In this connection I should like to discuss the *great Fermat theorem*. It is quite after the manner of the geometers of antiquity that one should generalize the question regarding Pythagorean numbers, from the plane to space of three and more dimensions in the following manner. Is it possible that the sum of the cubes of two integers should be a cube? Or that the sum of two fourth powers should be a fourth power, etc.? In general, *has the equation*

$$x^n + y^n = z^n,$$

where n is an arbitrary integer, solutions which are whole numbers? To this question Fermat gave the answer *no*, in the theorem named after him: *The equation $x^n + y^n = z^n$ has no integral solutions for integral values of n except when $n = 1$ and $n = 2$.* Let me begin with a few historical notes. Fermat lived from 1601 to 1665 and was a parliamentary councillor, i.e., a jurist, in Toulouse. He devoted himself, however, extensively and most fruitfully to mathematics so that he may counted as one of the greatest of mathematicians. Fermat's name deserves a prominent place among those of the founders of analytic geometry, of infinitesimal calculus, and of the theory of probability. Of special significance however, are his *attainments in the theory of numbers*. All of his results in this field appear as marginal notes on his *copy of Diophantus*, the famous ancient master of number-theory who lived in Alexandria probably about 300 A. D., i. e., about 600 years after Euclid. In this form they were published by his son five years after Fermat's death. Fermat himself had published nothing, but he had, by means of voluminous correspondance with the most significant of his contemporaries, made his discoveries known, although only in part. It was in that edition of Diphantus that the famous theorem with which we are now concerned was found. Fermat wrote concerning it that "he had found a really wonderful proof, but the margin was too narrow to accommodate it"[1]. To this day, no one has succeeded in finding a proof of this theorem!

---

[1] See the edition issued by the Paris Academy: *Œuvres de Fermat*, vol. I, p. 291. Paris 1891, and vol. III, p. 241. Paris 1896.

In order to orient ourselves somewhat as to its purport, let us inquire, as in the case of $n = 2$, in the first place about the *rational* solutions of the equation:

$$\xi^n + \eta^n = 1,$$

i. e., about the relation of the curve which represents this equation to the totality of the rational points in the ξ η-plane. For $n = 3$ and $n = 4$ the curves have approximately the appearance indicated in Fig. 10, 11 They contain, at least, the points $\xi = 0$, $\eta = 1$ and $\xi = 1$, $\eta = 0$ when

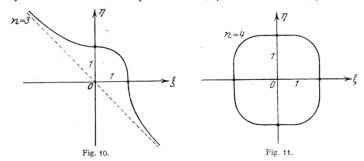

Fig. 10. Fig. 11.

$n = 3$, and the points $\xi = 0$, $\eta = \pm 1$ and $\xi = \pm 1$, $\eta = 0$ when $n = 4$. The assertion of Fermat means, now, that these curves, unlike the circle considered above, thread through the dense set of the rational points without passing through a single one, except those just noted.

The interest in this theorem rests on the fact that *all efforts to find a complete proof of it have been, thus far, in vain*. Among those who have attempted proof, one should, above all, mention *Kummer*, who advanced the problem materially by *bringing it into relation with the theory of algebraic numbers*, in particular with the theory of the *n-th roots of unity* (cyclotomic numbers). By using the *n-th* root of 1, $\varepsilon = e^{\frac{2i\pi}{n}}$, we can, indeed, separate $z^n - y^n$ into n linear factors, and we may write the Fermat equation in the form

$$x^n = (z - y)(z - \varepsilon y)(z - \varepsilon^2 y) \ldots (z - \varepsilon^{n-1} y).$$

The problem is therefore reduced to the separation of the *n-th* power of the integer x into n linear factors which shall be built up from two integers z and y and the number ε, in the manner indicated. Kummer developed, for such numbers, theories quite similar to those which have long been known for the case of ordinary integers, theories, that is, which depend on the notions of divisibility and factorization. One speaks, accordingly, of *integral algebraic numbers*, and here, in particular, of cyclotomic numbers, because of the relation of the number ε to the division of the circle. *Fermat's theorem is, then, for Kummer, a theorem on factorization in the domain of algebraic cyclotomic numbers*. From this theory he tried to deduce a proof of the theorem. He succeeded, in fact, for a very large number of values of n, for example for all values of n below 100. Among the larger numbers, however, there appeared *exceptional values* for which no proof has been found, either by him or by the later mathematicians who continued his investigations.

I must content myself with these remarks. You will find particulars concerning the state of the problem, and concerning Kummer's publications in the Encyclopedia, Vol. I_2, p. 714, at the end of the report by *Hilbert, Theorie der Algebraischen Zahlkörper*. Hilbert himself is among those who have continued and extended the investigations of Kummer[1].

It can indeed hardly be assumed that Fermat's "wonderful proof" lay in this direction. For it is not very likely that he could have operated with algebraic numbers at a time when one was not even certain about the meaning of the imaginary. At that time, also, the theory of numbers was quite undeveloped. It received at the hands of Fermat himself far-reaching stimulation. On the other hand, one cannot assume that a mathematician of Fermat's rank made an error in his proof, although such errors have occurred with the greatest mathematicians. Thus we must indeed believe that he succeeded in his proof by virtue of an especially fortunate simple idea. But as we have not the slightest indication as to the direction in which one could search for that idea, *we shall probably expect a complete proof of Fermat's theorem only through systematic extension of Kummer's work.*

These questions assumed new signifance when our Göttingen Science Association offered a *prize of 100000 marks for the proof of Fermat's theorem.* This was a foundation of the mathematician *Wolfskehl*, who died in 1906. He had probably been interested all his life in Fermat's theorem, and he bequeathed from his large fortune this sum for the fortunate person who should either establish the truth of the theorem of Fermat, or by means of a single example, exhibit its untruth[2]. Such a refutation would, be no simple matter, of course, because the theorem is already proved for exponents below 100 and one would have to start one's calculations with very large numbers.

It will be clear, from my foregoing remarks, how difficult the winning of this prize must seem to the mathematician, who understands the situation and who knows what efforts have been made by Kummer and his successors to prove the theorem. But the *great public* thinks

---

[1] A summarized account of the elementary investigations about Fermat's theorem is given in P. Bachmann, *Das Fermatsche Problem*. Berlin 1919.]

[2] The detailed conditions governing competition for this prize (long since become valueless) were published in the Nachrichten d. Ges. d. Wissenschaften zu Göttingen, business announcements 1908, p. 103 et seq., and copied into many other mathematical journals (Sec. e. g. Math. Ann. vol. 66, p. 143; Journal für Mathematik, vol. 134, p. 313).

otherwise. Since the summer of 1907, when the news of the prize was published in the papers (without authorization, by the way) we have received a prodigious heap of alleged "proofs". People of all walks of life, engineers, schoolteachers, clergymen, one banker, many women, have shared in these contributions. The common thing about them all is that they have *no idea of the serious mathematical nature, of the problem*. Moreover, they have made no attempt to inform themselves regarding it, but have trusted to finding the solution by a sudden flash of thought, with the inevitable result that their work is nonsense. One can see what absurdities are brought forth if one reads the numerous critical discussions of such proofs by A. Fleck (who is a practising physician by profession), Ph. Maennchen, and O. Perron, in *Archiv für Mathematik und Physik*[1]. It is amusing to read these wholesale slaughterings, sad as it is that they are necessary. I should like to mention one example, which is related to our treatment of the case $x^2 + y^2 = z^2$. The author seeks a rational parameter representation for the function $x^n + y^n = z^n$ ($n > 2$), and finds the result, long known from the theory of algebraic functions, that this, unlike the case $n = 2$, is not possible. Now this person overlooks the fact that a non-rational function can very well take on rational values for single rational values of the argument, and he therefore believes that he has proved the Fermat theorem.

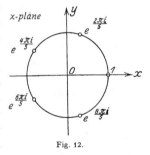

Fig. 12.

With this I close my remarks about Fermat's theorem and come to the *eighth point* of my list, the *problem of the division of the circle*. I shall make use here of operations with complex numbers, $x + iy$, assuming that they are familiar to you, although we shall consider them systematically later on. The *problem is to divide the circle into n equal parts*, or *to construct a regular polygon of n sides*. We identify the circle with the unit circle about the origin of the complex xy-plane and take $x + iy = 1$ as the first of the n points of division (see Fig. 12), in which n is chosen equal to five); then the n complex numbers belonging to the n vertices:

$$z = x + iy = \cos\frac{2k\pi}{n} + i\sin\frac{2k\pi}{n} = e^{\frac{2k\pi i}{n}} \quad (k = 0, 1, \ldots, n-1)$$

satisfy, according to De Moivre's theorem, the equation:

$$z^n = 1,$$

and with this *the problem of the division of the circle is resolved into the solving of this simple algebraic equation*. Since it has the rational root

---

[1] Vols. XIV, XV, XVI, XVII, XVIII (1901—1911).

$z = 1$, $z^n - 1$ is divisible by $z - 1$, and there remains for the $n - 1$ other roots the so called *cyclotomic equation*

$$z^{n-1} + z^{n-2} + \cdots + z^2 + z + 1 = 0,$$

an equation of degree $n - 1$, all of whose coefficients are $+1$.

Since ancient times, interest has centered in the question as to *what regular polygons can be constructed with ruler and compasses*. It was known to the ancients that this construction was possible for the numbers $n = 2^h$, 3, 5 (h an arbitrary integer), and likewise for the composite values $n = 2^h \cdot 3 \cdot 5$. Here the problem rested until the end of the eighteenth century when the young Gauss undertook its solution. He found the *desired construction was possible with ruler and compasses for all prime numbers of the form* $p = 2^{(2^\mu)} + 1$, *but for no others*. For the first values $\mu = 0, 1, 2, 3, 4$ this formula yields, in fact, prime numbers, namely

$$3, 5, 17, 257, 65537,$$

of which the first two cases were already known, while the others were new. Of these the *regular polygon of seventeen sides* is especially famous. The fact that it can be constructed with ruler and compasses was first established by Gauss. Moreover, it is not known for what values of μ the above formula yields prime numbers. It has been known, for example, since Euler's time, that for $\mu = 5$ the number is composite. I shall not go farther into details, but rather outline the general conditions, and the significance of this discovery. You will find in Weber-Wellstein details concerning the regular polygon of seventeen sides.

I should like to call to your attention especially the reprint of *Gauss' diary* in the fifty-seventh volume of the *Mathematische Annalen* (1903) and in Volume X, 1 (1917) of Gauss' Works. It is a small, insignificant looking book, which Gauss kept from 1796 on, beginning shortly before his nineteenth birthday. It was precisely the first entry which had to do with the possibility of constructing the polygon of seventeen sides (March 30, 1796); and it was this early important discovery which led Gauss to decide to devote himself to mathematics. The perusal of this diary is of the highest interest for every mathematician, since it permits one, farther on, to follow closely the genesis of Gauss' fundamental discoveries in the field of number theory, of elliptic functions, etc.

The publication of that first great discovery of Gauss appeared as a short communication in the "Jenaer Literaturzeitung" of June 1, 1796, instigated by Gauss' teacher and patron, Hofrat Zimmermann, of Braunschweig, and accompanied by a short personal note by the latter[1]. Gauss published the proof later in his fundamental number-theoretic work,

---

[1] Also reprinted in Mathematische Annalen, vol. 57, p. 6 (1903); and in Gauss' Works, vol. 10, p. 1 (1917).

Disquisitiones Arithmeticae[1] in 1801; here one finds for the first time the negative part of the theorem, which was lacking in his communication, *that the construction with ruler and compasses is not possible for prime numbers other than those of the form* $2^{2^\mu} + 1$, e.g., *for* $p = 7$. I shall put before you here an example of this important *proof of impossibility*—the more willingly because there is such a lack of *understanding for proofs of this sort* by the great public. By means of such proofs of impossibility modern mathematics has settled an entire series of famous problems, concerning the solution of which many mathematicians had striven in vain since ancient times. I shall mention, besides the construction of the polygon of seven sides, only the *trisection of an angle* and the *quadrature of the circle* with ruler and compasses. Nevertheless there are surprisingly many persons who devote themselves to these problems without having a glimmering of higher mathematics and without even knowing or understanding the nature of the proof of impossibility. According to their knowledge, which is mostly limited to elementary geometry, they make trials, by drawing, as a rule, auxiliary lines and circles, and multiply these finally in such number that no human being, without undue expenditure of time, can find his way out of the maze and show the author the error in his construction. A reference to the arithmetic proof of impossibility avails little with such persons, since they are amenable, at best, only to a direct consideration of their own "proof" and a direct demonstration of its falsity. Every year brings to every even moderately known mathematician a heap of such consignments, and you also, when you are at your posts, will get such proofs. It is well for you to be prepared in advance for such experiences and to know how to hold your ground. Perhaps it will be well for you, then, if you are master of a definite proof of impossibility in its simplest form.

Accordingly, I should like to give you, in detail, the proof that it is *impossible to construct the heptagon with ruler and compasses in the sense of geometry of precision*. It is well known that every construction with ruler and compasses finds its arithmetic equivalent in a succession of square roots, placed one above another, and, conversely, that one can represent geometrically every such square root by the intersection of lines and circles. This you can easily verify for yourselves. We can formulate our assertion analytically, then, by saying that *the equation of degree six*

$$z^6 + z^5 + z^4 + z^3 + z^2 + z + 1 = 0,$$

which characterizes the regular heptagon, cannot be solved by a succession of square roots in finite number. Now this is a so-called *reciprocal equation*,

---

[1] Reprinted Works, Vol. 1.

i. e., it has, for every root z, also $1/z$ as a root. This becomes obvious if we write it in the form:

(1) $$z^3 + z^2 + z + 1 + \frac{1}{z} + \frac{1}{z^2} + \frac{1}{z^3} = 0.$$

We can reduce by half the degree of such an equation, if we take

$$z + \frac{1}{z} = x$$

as a new unknown. By easy calculation, we obtain for x the cubic equation

(2) $$x^3 + x^2 - 2x - 1 = 0,$$

and one sees at once that the equations (1) and (2) are, or are not, both solvable by square roots. Moreover, we can represent x geometrically in connection with the construction of the heptagon. For, if we consider the unit circle in the complex plane, we see easily that the following relations are obvious. If one designates by $\varphi = \frac{2\pi}{7}$ the central angle of the regular heptagon, and remembers that $z = \cos\varphi + i\sin\varphi$ and $\frac{1}{z} = \cos\varphi - i\sin\varphi$ are the two vertices of the heptagon nearest to $x = 1$, then $x = z + \frac{1}{z} = 2\cos\varphi$ (Fig. 13). Thus, if one knows x, one can at once construct the heptagon.

We must now show *that the cubic equation* (2) *cannot be solved by square roots*. The proof falls into an arithmetic and an algebraic part.

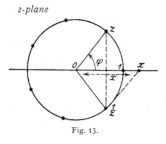

Fig. 13.

We shall start by showing that the equation (2) is *irreducible*, i.e. that its left side cannot be separated into two factors whose coefficients are rational numbers. Let us assume that the equation is reducible. Then its left side must have a linear factor with rational coefficients, and hence it must vanish for a rational number p/q, where p and q are integers without a common divisor. But that means that $p^3 + p^2 q - 2pq^2 - q^3 = 0$, or that p^3, and therefore p itself, is divisible by q. In the same way it follows that q^3, and hence q, must be divisible by p. Consequently $p = \pm q$ and the equation (2) must have the root $x = \pm 1$. But inspection shows that this is not the case.

The *second part of the proof* consists, in showing *that an irreducible cubic equation with rational coefficients is not solvable by square roots*. It is essentially *algebraic* in nature, but because of the connection I shall give it here. Let us make the assertion in positive form. *If a cubic equation with rational coefficients A, B, C:*

(8) $$f(x) = x^3 + Ax^2 + Bx + C = 0$$

can be solved by square roots, it must have a rational root, i. e., it is reducible. For the existence of a rational root α is equivalent to the existence of a *rational* factor $x - \alpha$ of $f(x)$ and thus to reducibility. It is most important that this proof be preceded by a *classification of all expressions that can be built up with square roots,* or, more precisely, of *all expressions that can be built up with square roots and rational numbers, in finite number, by means of rational operations.* A concrete example of such a number is

$$x = \frac{\sqrt{a + \sqrt{b}} + \sqrt{c}}{\sqrt{d + \sqrt{e + \sqrt{f}}}},$$

where $a, b, \ldots, f$ are rational numbers. Of course we are talking only about square roots which cannot be extracted rationally. All others must be simplified. Every such expression is a rational function of a certain number of square roots. In our example there are three. We shall first consider *a single such square root*, whose radicand, however, may have a form as complicated as one pleases. *By its "order" we shall understand the largest number of root signs which appear in it, one above another.* In the preceding example, α, the roots of the numerator have the orders 2 and 1, respectively, while that of the denominator has the order 3.

In the case of a *general square root expression* we examine the *orders of the different "simple square root expressions" of the sort just discussed,* out of which the general expression is rationally constructed, and we *designate the largest among them as the order μ of the expression in question.* In our example, $\mu = 3$. Now several "simple square root expressions" of order μ might appear in our expression and we consider *their number, n, the "number of terms"* of order μ, as a second characteristic. This number is thought of as so determined that *no one of the n simple expressions of order μ can be rationally expressed in terms of the others of order μ, or of lower order.* For example, the expression of order 1

$$\sqrt{2} + \sqrt{3} + \sqrt{6}$$

has 2, not 3, as the "number of terms" since $\sqrt{6} = \sqrt{2} \cdot \sqrt{3}$. The example α given above has $n = 1$.

We have thus assigned to every square root expression two finite numbers μ, n which we combine in the *symbol (μ, n) as the "characteristic" or "rank" of the root expression. When two root expressions have different orders we assign a lower rank to the one of lower order; when the orders are the same, the lower number of terms determines the lower rank.*

Now let us suppose that a *root x_1* of the cubic equation (8) is expressible by means of square roots; and, to be explicit, by means of an *expression of rank (μ, n)*. Selecting one of the n terms $\sqrt{R}$ of rank μ, let x be written in the form

$$x_1 = \frac{\alpha + \beta \sqrt{R}}{\gamma + \delta \sqrt{R}},$$

where α, β, γ, δ contain at most $n-1$ terms of order μ and where R is of order $\mu - 1$. Here $\gamma - \delta \sqrt{R}$ is certainly different from zero; for $\gamma - \delta \sqrt{R} = 0$ would imply either $\delta = \gamma = 0$, which is obviously impossible, or $\sqrt{R} = \gamma : \delta$, i.e., $\sqrt{R}$ would be rationally expressible by means of the other $(n-1)$ terms of order μ, which appear in x, and hence it would be superfluous. Multiplying numerator and denominator by $\gamma - \delta \sqrt{R}$, we find

$$x_1 = \frac{(\alpha + \beta \sqrt{R})(\gamma - \delta \sqrt{R})}{\gamma^2 - \delta^2 \cdot R} = P + Q\sqrt{R},$$

where P, Q are *rational functions of* $\alpha, \beta, \gamma, \delta$, that is, they contain at most $(n-1)$ terms of order μ, and, besides, only those of order $\mu - 1$, so that they have *at most the rank* $(\mu, n-1)$. Substituting this value of x in (8), we get

$$f(x_1) = (P + Q\sqrt{R})^3 + A(P + Q\sqrt{R})^2 + B(P + Q\sqrt{R}) + C = 0,$$

and when we remove parentheses we obtain a relation of the form

$$f(x_1) = M + N\sqrt{R} = 0,$$

where M, N are polynomials in P, Q, R, that is, rational functions of α, β, γ, δ, R. If $N \neq 0$, we should have $\sqrt{R} = -M/N$, i.e., $\sqrt{R}$ would be expressible rationally in terms of $\alpha, \beta, \gamma, \delta, R$, that is, by means of the other $(n-1)$ terms of order μ and others of lower order. But that is impossible, as remarked above, according to the hypothesis. Thus it follows necessarily that $N = 0$ and hence also $M = 0$. From this we may conclude, that

$$x_2 = P - Q\sqrt{R}$$

is also a root of the cubic equation (8). For a comparison with the last equations yields at once

$$f(x_2) = M - N\sqrt{R} = 0.$$

The proof may now be finished very simply and surprisingly. If x_3 is *the third root* of our cubic equation, we have

$$x_1 + x_2 + x_3 = -A,$$

and hence $\quad x_3 = -A - (x_1 + x_2) = -A - 2P$

is of the same rank as P and *therefore certainly of lower rank than* x_1.

If x_3 is itself rational, our theorem is proved. If not, we can make it the starting point of the same series of deductions. It appears that, in the case of the other roots, the higher rank must have been an illusion, so *that, in particular, one of them has, actually, lower rank than* x_3. If we keep this up, back and forth among the roots, we see, each time, that the rank is really lower than we had thought. We must, then, of necessity, come finally to a root with the order $\mu = 0$. *This demon-*

strates *the existence of a rational root of the cubic equation*. We cannot continue our procedure beyond this point. The two other roots must then be, either themselves rational, or else of the form $P = Q\sqrt{R}$, where P, Q, R are rational numbers. Hence we have shown *that $f(x)$ separates into a quadratic and a linear rational factor and is therefore reducible. Every irreducible cubic equation, and in particular, our equation for the regular heptagon, is insoluble by means of square roots*. The proof is therefore complete *that the regular heptagon cannot be constructed with ruler and compasses*.

You observe how simply and obviously this proof proceeds, and how little knowledge it really presupposes. For all that, some of the steps, especially the explanation of the classification of square root expressions, demand a certain measure of mathematical abstraction. Whether the proof is simple enough to convince one of those mathematical laymen, mentioned above, of the futility of his attemps at an elementary geometric proof, I do not presume to decide. Nevertheless one should try to explain the proof slowly and clearly to such a person.

In conclusion, I shall mention some of the *literature on the question of regular polygons* together with some, on the broader question of geometric constructibility in general which we have touched upon on this occasion. First of all, there is again *Weber-Wellstein* I (Sections 17 and 18 in the fourth edition). Next let me mention the souvenir booklet *Vorträge über ausgewählte Fragen der Elementargeometrie*[1]* which I prepared in 1895, on the occasion of a gathering of teachers in Göttingen. I might mention, as a more detailed and comprehensive substitute for this little book (which is out of print) the German translation, *Fragen der Elementargeometrie*[2]**, of a compilation by *F. Enriques* in Bologna, where you will find information on all allied questions.

I leave now the discussion of number theory, reserving the last point, the transcendence of π, for the conclusion of this course of lectures, and turn, in the next chapter, to our final extension of the number system.

IV. Complex Numbers.

1. Ordinary Complex Numbers

Let me give, as a preliminary, *some historical facts*. Imaginary numbers are said to have been used first, incidentally, to be sure, by *Cardan* in 1545, in his solution of the cubic equation. As for the further

---

[1] Worked up by F. Tägert. Leipzig 1895.

[2] Teil II: *Die geometrischen Aufgaben, ihre Lösung und Lösbarkeit*. Deutsch von H. Fleischer. Leipzig 1907. [2. Aufl. 1923.] — See also *Young, J. W. A., Monographs on Topics in Modern Mathematics*.

* Translation by Beman and Smith: *Famous Problem of Geometry*. Ginn, reprinted by Stechert, New York.

** *Problems of Elementary Geometry*.

development, we can make the same statement as in the case of negative numbers, *that imaginary numbers made their own way into arithmetic calculation without the approval, and even against the desires of individual mathematicians, and obtained wider circulation only gradually and to the extent to which they showed themselves useful.* Meanwhile the mathematicians were not altogether happy about it. Imaginary numbers long retained a somewhat *mystic* coloring, just as they have today for every pupil who hears for the first time about that remarkable $i = \sqrt{-1}$. As evidence, I mention a very significant utterance by *Leibniz* in the year 1702, "Imaginary numbers are a fine and wonderful refuge of the divine spirit, almost an amphibian between being and non-being". In the eighteenth century, the notion involved was indeed by no means cleared up, although *Euler*, above all, *recognized their fundamental significance for the theory of functions*. In 1748 Euler set up that remarkable relation:

$$e^{ix} = \cos x + i \sin x$$

by means of which one recognizes the fundamental relationship among the kinds of functions which appear in elementary analysis. The *nineteenth century finally brought the clear understanding of the nature of complex numbers*. In the first place, we must emphasize here the *geometric interpretation* to which various investigators were led about the end of the century. It will suffice if I mention the man who certainly went deepest into the essence of the thing and who exercised the most lasting influence upon the public, namely Gauss. As his diary, mentioned above, proves incontrovertibly, he was, in 1797, already in full possession of that interpretation, although, to be sure, it was published very much later. The second achievement of the nineteenth century is the creation of a *purely formal foundation* for complex numbers, which reduces them to dependence upon real numbers. This originated with English mathematicians of the thirties, the details of which I shall omit here, but which you will find in Hankel's book, mentioned above.

Let me now explain these *two prevailing foundation methods*. We shall take first the *purely formal standpoint*, from which the consistency of the rules of operation among themselves, rather than the meaning of the objects, guarantees the correctness of the concepts. According to this view, complex numbers are introduced in the following manner, which precludes every trace of the mysterious.

1. The complex number $x + iy$ is the *combination of two real numbers* x, y, that is, a *number-pair*, concerning which one adopts the conventions which follow.

2. Two complex numbers $x + iy$, $x' + iy'$ are called *equal* when $x = x'$, $y = y'$.

3. Addition and subtraction are defined by the relation
$$(x + iy) \pm (x' + iy') = (x \pm x') + i(y \pm y').$$
All the *rules of addition* follow from this, as is easily verified. The *monotonic law* alone loses its validity in its original form, since complex numbers, by their nature, do not have the same simple order in which natural or real numbers appear by virtue of their magnitude. For the sake of brevity I shall not discuss the modified form which this gives to the monotonic law.

4. We stipulate that in *multiplication* one operates as with ordinary letters, except that one always puts $i^2 = -1$; in particular, that
$$(x + iy)(x' + iy') = (xx' - yy') + i(xy' + x'y).$$
It is easy to see that, with this, *all the laws of multiplication hold, with the exception of the monotonic law, which does not enter into consideration.*

5. *Division* is defined as the *inverse of multiplication*; in particular, we may easily verify that
$$\frac{1}{x + iy} = \frac{x}{x^2 + y^2} - i\frac{y}{x^2 + y^2}.$$
This number always exists except for $x = y = 0$, i.e., *division by zero has the same exceptional place here as in the domain of real numbers.*

It follows from this that operations with complex numbers cannot lead to contradictions, since they depend exclusively upon real numbers and known operations with them. We shall assume here that these are devoid of contradiction.

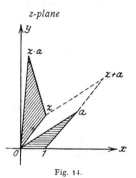

Fig. 14.

Besides this purely formal treatment, we should of course like to have a geometric, or otherwise visual, interpretation of complex numbers and of operations with them, in which we might see a *graphical foundation of consistency*. This is supplied by common geometric interpretation, which, as you all know and as we have already mentioned, *looks upon the totality of points (x, y) of the plane in an xy-coordinate system as representing the totality of complex numbers $z = x + iy$*. The sum of two numbers z, a follows by means of the familiar *parallelogram construction* with the two corresponding points and the origin 0, while the product $z \cdot a$ is obtained by constructing on the segment $0z$ a triangle similar to $a01$, where 1 is the point $(x = 1, y = 0)$ (Fig. 14). In brief, *addition $z' = z + a$ is represented by a translation of the plane into itself, multiplication $z' = z' a$ by a similarity transformation*, i.e., *by a turning and a stretching, the origin remaining fixed*. From the order of the points in the plane, considered as representatives of complex numbers, one sees at once what takes the place here of the monotonic laws for real numbers. These suggestions will suffice, I hope, to recall the subject clearly to your memory.

I must call to your attention the place in Gauss in which this foundation of complex numbers, by means of their geometric interpretation, is set out with full emphasis, since it was this which first exhibited the general importance of complex numbers. In the year 1831 Gauss' researches carried him into the theory especially of *integral complex numbers* $a + ib$, where a, b are real integers, in which he developed for the new numbers the theorems of ordinary *number theory* concerning prime factors, quadratic and biquadratic residues, etc. We mentioned such generalizations of number theory, in connection with our discussion of Fermat's theorem. In his own abstract[1] of this paper Gauss expresses himself concerning what he calls the "true metaphysics of imaginary numbers". For him, the right to operate with complex numbers is justified by the geometric interpretation which one gives to them and to the operations with them. Thus he takes *by no means the formal standpoint*. Moreover, these long, beautifully written expositions of Gauss are extremely well worth reading. I mention here, also, that Gauss proposes the clearer word "complex", instead of "imaginary", a name that has, in fact, been adopted.

2. Higher Complex Numbers, especially Quaternions

It has occurred to everyone who has worked seriously with complex numbers to ask if we cannot set up other, higher, complex numbers, with more ne wunits than the one i and if we cannot operate with them logically. Positive results in this direction were obtained about 1840 by *H. Grassmann,* in Stettin, and *W. R. Hamilton,* in Dublin, independently of each other. We shall examine the invention of Hamilton, the *calculus of quaternions,* somewhat carefully later on. For the present let us look at the general problem.

We can look upon the ordinary complex number $x + iy$ as a *linear combination*

$$x \cdot 1 + y \cdot i$$

formed from two different *"units"* 1 and i, by means of the *real parameters* x and y. Similarly, let us now imagine an arbitrary number, n, of units $e_1, e_2, \ldots, e_n$ all different from one another, and let us call the totatily of combinations of the form $x = x_1 e_1 + x_2 e_2 + \ldots, + x_n e_n$ a *higher complex number system* formed from them with n arbitrary real numbers $x_1, x_2, \ldots, x_n$. If there are given two such numbers, say x, defined above, and

$$y = y_1 e_1 + y_2 e_2 + \ldots, + y_n e_n$$

---

[1] See Werke, vol. II.

it is nearly obvious that we should call them *equal when, and only when, the coefficients of the individual units, the so called "components" of the number, are equal in pairs*

$$x_1 = y_1, \ x_2 = y_2, \ \ldots \ x_n = y_n.$$

The definition of addition and subtraction, which reduces these operations simply to the *addition and subtraction of the components*,

$$x \pm y = (x_1 \pm y_1)e_1 + (x_2 \pm y_2)e_2 + \ldots, + (x_n \pm y_n)e_n,$$

is equally obvious.

The matter is more difficult and more interesting in the case of *multiplication*. To start with, we shall proceed according to the general rule for multiplying letters, i.e., multiply each i-th term of x by every k-th term of y $(i, k = 1, 2, \ldots, n)$. This gives:

$$x \cdot y = \sum_{(i, \, k = 1, \, \ldots, \, n)} x_i y_k e_i e_k.$$

In order that this expression should be a number in our system, one must have a rule which represents the *products $e_i \cdot e_k$ as complex numbers of the system*, i.e., as linear combinations of the units. Thus one must have n^2 equations of the form:

$$e_i e_k = \sum_{(l = 1, \, \ldots, \, n)} c_{ikl} \cdot e_l. \qquad (i, \, k = 1, \ldots, n)$$

Then we may say that the number

$$x \cdot y = \sum_{(l = 1, \, \ldots, \, n)} \left\{ \sum_{(i, \, k = 1, \, \ldots, \, n)} x_i y_k c_{ikl} \right\} e_l$$

will always belong to our complex number system. *Each particular complex number system is characterized by the method of determining this rule for multiplication, i.e., by the table of the coefficients C_{ikl}*.

If one now defines *division as the operation inverse to multiplication*, it turns out that, under this general arrangement, division *is not always uniquely possible*, even when the divisor does not vanish. For, the determination of y from $x \cdot y = z$ requires the solution of the n linear equations $\sum_{i, k} x_i y_k C_{ikl} = z_l$ for the n unknowns $y_1, \ldots, y_n$, and these would have either no solution, or infinitely many solutions, if their determinant happened to vanish. Moreover, all the z_l may be zero even when not all the x_i or not all the y_k vanish, i.e., *the product of two numbers can vanish without either factor being zero*. It is only by a skillful special choice of the numbers C_{ikl} that one can bring about accord here with the behavior of ordinary numbers. To be sure, a closer investigation shows, when $n > 2$, that, to attain this, we must sacrifice one of the other *rules of operation*. We choose as the rule that fails to be satisfied, one which appears less important under the circumstances.

Let us now follow up these general explanations by a more detailed discussion of *quaternions* as the example which, by reason of its applications in physics and mathematics, constitutes the *most important higher complex number system*. As the name indicates, these are *four-term numbers* ($n = 4$); as a sub-class, they include the *three-term vectors*, which are generally known today, and which are sometimes discussed in the schools.

As the first of the four units with which we shall construct quaternions, we shall select the *real unit* 1, (as in the case of ordinary complex numbers). We ordinarily denote the other three units, as did Hamilton, by i, j, k, so that the general from of the quaternion is

$$p = d + ia + jb + kc,$$

where a, b, c, d are real parameters, *the coefficients of the quaternion*. We call the first component, the one which is multiplied by 1, and which corresponds to the real part of the common complex number, the *"scalar part" of the quaternion*, the aggregate $ai + bj + ck$ of the other three terms its *"vector part"*.

The addition of quaternions follows from the preceding general remarks. I shall give an obvious *geometric interpretation*, which goes back to that interpretation of vectors which is familiar to you. We imagine the *segment*, corresponding to the vector part of p, and having the projections a, b, c on the coordinate axes, as loaded with a *weight* equal to the scalar part. Then addition of p and $p' = d' + ia' + jb' + kc'$ is accomplished by constructing the resultant of the two segments, according to the well known parallelogram law of vector addition (see Fig. 15), and then loading it with the sum of the weights, for this would then in fact represent the quaternion:

(1) $\quad p + p' = (d + d') + i(a + a') + j(b + b') + k(c + c')$.

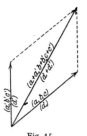

Fig. 15.

We come first to specific properties of quaternions when we turn to *multiplication*. As we saw in the general case, these properties must be implicit in the *conventions adopted as to the products of the units*. To begin with, I shall indicate the quaternions to which *Hamilton* equated the sixteen products of two units each. As its symbol indicates, we shall operate with the first unit 1 as with the real number 1, so that:

(2a) $\quad 1^2 = 1, \; i \cdot 1 = 1 \cdot i = i, \; j \cdot 1 = 1 \cdot j = j, \; k \cdot 1 = 1 \cdot k = k.$

As something essentially new, however, we agree that, for the squares of the other units:

(2b) $\quad\quad\quad\quad\quad\quad i^2 = j^2 = k^2 = -1,$

and for their binary products:

(2c) $$jk = +i, \quad ki = j, \quad ij = +k$$

whereas for the inverted position of the factors:

(2d) $$kj = -i, \quad ik = -j, \quad ji = -k.$$

One is struck here by the fact that the *commutative law for multiplication is not obeyed*. This is the inconvenience in quaternions which one must accept in order to rescue the uniqueness of division, as well as the theorem that a product should vanish only when one of the factors vanishes. *We shall show at once that not only this theorem but also all the other laws of addition and multiplication remain valid, with this one exception, in other words, that these simple agreements are very expedient.*

We construct, first, the *product of two general quaternions*

$$p = d + ia + jb + kc \quad \text{and} \quad q = w + ix + jy + kz.$$

Let us start from the equation

$$q' = p \cdot q = (d + ia + jb + kc) \cdot (w + ix + jy + kz);$$

and let us multiply out term by term. In carrying out this multiplication, we must note the order in the case of the units i, j, k. We must follow the commutative law for products composed of the components a, b, c, d, and for products of components and one unit, we must replace the products of units in accordance with our multiplication table, and we must then collect the terms having the same unit. We must then collect the terms having the same unit. We then have

(3)
$$\begin{aligned} q' = pq = w' + ix' + jy' + kz' &= (dw - ax - by - cz) \\ &+ i(aw + dx + \underline{bz - cy}) \\ &+ j(bw + dy + \underline{cx - az}) \\ &+ k(cw + dz + \underline{ay - bx}). \end{aligned}$$

The components of the product quaternion are thus definite simple *bilinear combinations* of the components of the two factors. If we invert the order of the factors, the six underscored terms change their signs, so that $q \cdot p$, *in general, is different from* $p \cdot q$, and the difference is more than a change of sign as was the case with the individual units.

Although the commutative law fails for multiplication, the *distributive and associative laws hold without change*. For, if we construct on the one hand $p(q + q_1)$, on the other $pq + pq_1$ by multiplying out formally without replacing the products of the units, we must, of necessity, get identical results, and no change can be brought about by then using the multiplication table. Further, the associative law must hold in general, if it holds for the multiplication of the units.

But this follows at once from the multiplication table, as the following example shows:
$$(ij)k = i(jk).$$
In fact, we have:
$$(ij)k = k \cdot k = -1,$$
and
$$i(jk) = i \cdot i = -1.$$

We shall now take up *division*. It will suffice to show *that for every quaternion $p = d + ia + jb + kc$ there is a definite second one, q, such that*:
$$p \cdot q = 1.$$

We shall denote q appropriately by $1/p$. Division in general can be reduced easily to this special case, as we shall show later. In order to determine q, let us put, in equation (3),
$$q' = 1 = 1 + 0 \cdot i + 0 \cdot j + 0 \cdot k,$$
and obtain, by equating components, the following four equations for four unknown components x, y, z, w of q:

$$dw - ax - by - cz = 1$$
$$aw + dx - cy + bz = 0$$
$$bw + cx + dy - az = 0$$
$$cw - bx + ay + dz = 0.$$

The solvability of such a system of equations depends, as is well known, upon its determinant, which, in the case before us, is a skew symmetric determinant, in which all the elements of the principal diagonal are the same, and all the pairs of elements which are symmetrically placed with respect to that diagonal are equal and opposite in sign. According to the theory of determinants, such determinants are easily calculated; and we find

$$\begin{vmatrix} d & -a & -b & -c \\ a & d & -c & b \\ b & c & d & -a \\ c & -b & a & d \end{vmatrix} = (a^2 + b^2 + c^2 + d^2)^2.$$

By direct calculation this result can be easily verified. The real *elegance of Hamilton's conventions* depends upon this result, that the determinant is a power of the sum of squares of the four components of p; for it follows *that the determinant is always different from zero except when $a = b = c = d = 0$*. With this one self evident exception ($p = 0$), the equations are *uniquely solvable* and the *reciprocal quaternion q is uniquely determined*.

The quantity
$$T = \sqrt{a^2 + b^2 + c^2 + d^2}$$
plays an important role in the theory, and is called the *tensor of p*. It is easy to show that these unique solutions are
$$x = -\frac{a}{T^2}, \qquad y = -\frac{b}{T^2}, \qquad z = -\frac{c}{T^2}, \qquad w = \frac{d}{T^2}$$
so that we have as the final result
$$\frac{1}{p} = \frac{1}{d + ia + jb + kc} = \frac{d - ia - jb - kc}{a^2 + b^2 + c^2 + d^2}.$$
If we introduce the *conjugate value* of p, as in ordinary complex numbers:
$$\bar{p} = d - ia - jb - kc,$$
we can write the last formula in the form
$$\frac{1}{p} = \frac{\bar{p}}{T^2}$$
or
$$p \cdot \bar{p} = T^2 = a^2 + b^2 + c^2 + d^2.$$
These formulas which are immediate generalizations of certain properties of ordinary complex numbers. Since p is also the number conjugate to $\bar{p}$, it follows also that:
$$\bar{p} \cdot p = T^2,$$
so that the commutative law holds in this special case.

The general problem of division can now be solved. For, from the equation
$$p \cdot q = q',$$
it follows, by multiplication by $1/p$, that
$$q = \frac{1}{p} \cdot q' = \frac{\bar{p}}{T^2} \cdot q',$$
whereas the equation
$$q \cdot p = q',$$
which one gets by changing the order of the factors, has the solution
$$q = q' \cdot \frac{1}{p} = q' \cdot \frac{\bar{p}}{T^2}.$$
This solution is different, in general, from the other.

Now we must inquire whether there is a *geometric interpretation of quaternions* in which these operations, together with their laws, appear in a natural form. In order to arrive at it, we start with the special case in which *both factors reduce to simple vectors*, i.e., in which the

scalar parts w, d, are zero. The formula (3) for multiplication then becomes
$$q' = p \cdot q = (ia + jb + kc)(ix + jy + kz)$$
$$= -(ax + by + cz) + i(bz - cy) + j(cx - az) + k(ay - bx),$$

i. e., *when each of two quaternions reduces to a vector, their product consists of a scalar and a vector part.* We can easily bring these two parts into relation with the different kinds of *vector multiplication* which are in use. The notions of vector calculus, which is far more wide spread than quaternion calculus, go back to *Grassmann*, although the word *vector* is of English origin. The two kinds of vector product with which one usually operates are designated now, mostly, by *inner (scalar) product* $ax + by + cz$ (i. e., the scalar part of the above quaternion product, except for the sign), and *outer (vector) product* $i(bz - cy) + j(cx - az) + k(ay - bx)$, (i. e., the vector part of the quaternion product. We shall give a geometric interpretation of each part separately.

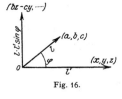

Fig. 16.

Let us lay off both vectors (a, b, c) and (x, y, z), as segments, from the origin O (Fig. 16). They terminate in the points (a, b, c) and (x, y, z) respectively, and have the lengths $l = \sqrt{a^2 + b^2 + c^2}$ and $l' = \sqrt{x^2 + y^2 + z^2}$. If φ is the angle between these two segments, then, according to well known formulas of analytic geometry, which I do not need to develop here, the *inner product* is:
$$ax + by + cz = l \cdot l' \cdot \cos\varphi;$$

and the *outer product*, on the other hand, is itself a *vector*, which, as is easily seen, is *perpendicular to the plane of l and l'* and has the *length* $l \cdot l' \cdot \sin\varphi$.

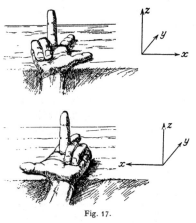

Fig. 17.

It is essential now to decide as to the *sense of the product vector*, i. e., toward which side of the plane determined by l and l' one is to lay off this vector. This sense is *different according to the coordinate system* which one chooses. As you know, one can choose *two rectangular coordinate systems which are not congruent*, i. e., *which cannot be made to coincide with one another*, by holding, say, the y- and the z-axis fixed and reversing the sense of the x-axis. These systems are then *symmetric to each other, like the right and the left hand* (Fig. 17). The distinction between them can be borne in mind by the following rule: *In the one* system, *the x, y, and z axis lie like the outstretched thumb, fore finger and middle finger, respectively, of the right hand*; in the other, *like the same fingers of the left hand*. These two systems are used confusedly in the literature; different habits obtain in different countries, in different fields, and, finally, with different writers, or even with the same writer. Let us now examine the simplest case, where $p = i, q = j$, these being the unit lengths laid off on the x and y axis. Then, since $i \cdot j = k$, the outer vector product is the unit length laid off on the z-axis. (See Fig. 18.) Now one can transform i and j continuously into two arbitrary vectors p and q so that k transforms continuously into the vector component of $p \cdot q$ without going through zero. Consequently the *first factor, the second factor, and the vector product must always lie, with respect to each other, like the x, y, and z-axis of the system of coordinates*, i.e., right-handed (as in Fig. 18) or left-handed (as in Fig. 16), *according to the choice of coordinate system*. (In Germany, now, the choice indicated in Fig. 18 is customary.)

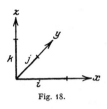

Fig. 18.

I should like to add a few words concerning the much disputed *question of notation in vector analysis*. There are, namely, a great many different symbols used for each of the vector operations, and it has been impossible, thus far, to bring about a generally accepted notation. At the meeting of natural scientists at Kassel (1903) a commission was set up for this purpose. Its members, however, were not able even to come to a complete understanding among themselves. Since their intentions were good, however, each member was willing to meet the others part way, so that the only result was that about three new notations came into existence! My experience in such things inclines me to the belief that real agreement could be brought about only if important material interests stood behind it. It was only after such pressure that, in 1881, the uniform system of measures according to volts, amperes, and ohms was generally adopted in electrotechnics and afterward settled by public legislation, due to the fact that industry was in urgent need of such uniformity as a basis for all of its calculations. But there are no such strong material interests behind vector calculus, as yet, and hence one must agree, for better or worse, to let every mathematician cling to the notation which he finds the most convenient, or—if he is dogmatically inclined—the only correct one.

3. Quaternion Multiplication — Rotation and Expansion

Before we proceed to the consideration of the geometric meaning of multiplication of general quaternions, let us consider the following question. Let us consider the product $q' = p \cdot q$ of two quaternions p and q, and let us replace p and q by their conjugates $\bar{p}$ and $\bar{q}$, that is, let us change the signs of a, b, c, x, y, z. Then the scalar part of the product, as given in (3), p. 61, remains unchanged, and only those factors of i, j, k which are not underscored will change sign. On the other hand, if we also reverse the order of the factors $\bar{p}$ and $\bar{q}$, the factors of i, j, k which are underscored will change sign. Hence the product $\bar{q}' = \bar{q} \cdot \bar{p}$ is precisely the conjugate of the original product q'; and we have

$$q' = p \cdot q, \quad \bar{q}' = \bar{q} \cdot \bar{p},$$

where $\bar{q}'$ is the conjugate of q'. If we multiply these two equations together, we obtain

$$q' \cdot \bar{q}' = p \cdot q \cdot \bar{q} \cdot \bar{p}.$$

In this equation the order of the factors is essential, since the commutative law does not hold. We may apply the associative law, however, and we may write

$$q' \cdot \bar{q}' = p \cdot (q \cdot \bar{q}) \cdot \bar{p}.$$

Since we have, by p. 63,

$$q \cdot \bar{q} = x^2 + y^2 + z^2 + w^2,$$

we may write

$$w'^2 + x'^2 + y'^2 + z'^2 = p(w^2 + x^2 + y^2 + z^2)\bar{p}.$$

The middle factor on the right is a scalar, and the commutative law does hold for multiplication of a scalar by a quaternion, since $M \cdot p = Md + i(Ma) + j(Mb) + k(Mc) = pM$. Hence we have

$$w'^2 + x'^2 + y'^2 + z'^2 = p\bar{p}(w^2 + x^2 + y^2 + z^2),$$

and, since $p \cdot \bar{p}$ is the square of the tensor of p, we find[1]

(I) $\quad w'^2 + x'^2 + y'^2 + z'^2 = (d^2 + a^2 + b^2 + c^2)(w^2 + x^2 + y^2 + z^2),$

that is, *the tensor of the product of two quaternions is equal to the product of the tensors of the factors.* This formula can be obtained also by direct calculation, by taking the values of w', x', y', z' from the formula for a product given on p. 61.

We shall now represent a quaternion as the segment joining the origin of a four-dimensional space to the point (x, y, z, w), in a manner exactly analogous to the representation of a vector in three-dimensional space. It is no longer necessary to apologize for making use of four-dimensional space, as was the custom when I was a student. All of you are fully aware that no metaphysical meaning is intended, and that higher dimensional space is nothing more than a convenient mathematical expression which permits us to use terminology analogous to that of

---

[1] This formula, in all that is essential, occurs in Lagrange's works.

actual space representation. If we regard p as a constant, that is, if we regard a, b, c, d as constants, the quaternion equation

$$q' = p \cdot q$$

represents a certain *linear tranformation* of the points (x, y, z, w) of the four-dimensional space into the points (x', y', z', w'), since the equation assigns to every four-dimensional vector q another vector q' linearly. The explicit equations for this transformation, i.e., the expressions for x', y', z', w' as linear functions of x, y, z, w, may be obtained by comparison of the coefficients of the product formula (3), p. 61. The tensor equation (I) shows that the distance of any point from the origin, $\sqrt{x^2 + y^2 + z^2 + w^2}$, is multiplied by the same constant factor $T = \sqrt{a^2 + b^2 + c^2 + d^2}$, for all points of the space. Finally, by p. 62, the determinant of the linear transformation is surely positive.

It is shown in analytic geometry of three-dimensional space that if a linear transformation of the coordinates x, y, z is *orthogonal* (that is, if it carries the expression $x^2 + y^2 + z^2$ into itself), and if the determinant of the transformation is positive, the transformation represents a *rotation about the origin*. Conversely, *any* rotation can be obtained in this manner. If the linear transformation carries $x^2 + y^2 + z^2$ into the similar expression in x', y', z' multiplied by a constant factor T^2, however, and if the determinant is positive, the transformation represents a *rotation about the origin combined with an expansion in the ratio T about the origin*, or, briefly, *a rotation and expansion*.

The facts just mentioned for three-dimensional space may be extended to four-dimensional space. We shall say that our transformation of four-dimensional space represents in precisely the same sense a *rotation and expansion about the origin*. It is easy to see, however, that in this case we do *not* obtain the most general rotation and expansion about the origin. For our transformation contains only four arbitrary constants, namely, the components a, b, c, d of p, whereas, as we shall show immediately, the most general rotation and expansion about the origin in the four-dimensional space R_4 contains seven arbitrary constants. Indeed, in order that the general linear transformation should be a rotation and expansion, we must have

$$x'^2 + y'^2 + z'^2 + w'^2 = T^2(x^2 + y^2 + z^2 + w^2).$$

If we replace x', y', z', w' by linear integral functions of x, y, z, w, we obtain a quadratic form in four variables, which contains $(4 \cdot 5)/2 = 10$ terms. Equating coefficients, we obtain ten equations. Since T is still arbitrary, these reduce to nine equations among the sixteen coefficients of the transformation. Hence there remain seven arbitrary constants.

It is remarkable that in spite of this *the most general rotation and expansion can be obtained by quaternion multiplication*. Let $\pi = \delta + i\alpha$

$+ j\beta + k\gamma$ be another constant quaternion. Then we may show, just as before, that the transformation $q' = q \cdot \pi$, which differs from the preceding one only in that the order is reversed, represents a rotation and expansion of R_4. Hence the combined transformation

(II) $\quad q' = p \cdot q \cdot \pi = (d + ia + jb + kc) \cdot q \cdot (\delta + i\alpha + j\beta + k\gamma)$

also represents such a rotation and expansion. This transformation contains only seven (not eight) arbitrary constants, for the transformation remains unchanged if we multiply a, b, c, d by any real number and divide $\alpha, \beta, \gamma, \delta$ by the same number. It is therefore plausible that this combined transformation represents the general rotation and expansion of four-dimensional space. This beautiful result is actually true, as was shown by Cayley. I shall restrict myself to the mention of the historical fact, in order not to be drawn into too great detail. The formula is given in Cayley's paper *on the homographic transformation of a surface of the second order into itself*[1], in 1854, and also in certain other papers of his[2].

This formula of Cayley's has the great advantage that it enables us to grasp at once the combination of two rotations and expansions. Thus, if a second rotation and expansion be given by the equation

$$q'' = w'' + ix'' + jy'' + kz'' = p' \cdot q' \cdot \pi',$$

where p' and π' are new given quaternions, we find, by (II),

$$q'' = p' \cdot (p \cdot q \cdot \pi) \cdot \pi',$$

whence, by the associative law,

$$q'' = (p' \cdot p) \cdot q \cdot (\pi \cdot \pi')$$

or

$$q'' = r \cdot q \cdot \varrho$$

where $r = p' \cdot p$ and $\varrho = \pi \cdot \pi'$ are definite new quaternions. We have therefore obtained an expression for the rotation and expansion that carries q into q'' in precisely the old form, and we see that the multipliers which precede and follow q in the quaternion product are, respectively, the products of the corresponding multipliers of q in the separate transformations which were combined, the order of the factors being necessarily as shown in the formula.

This four-dimensional representation may seem unsatisfactory, and there may be a desire for something more tangible which can be represented in ordinary three-dimensional space. We shall therefore show that we can obtain similar formulas for the similar three-dimensional

---

[1] Journal für Mathematik, 1855. Reprinted in Cayley's Collected Papers, vol. 2, p. 133. Cambridge 1889.

[2] See, for example, *Recherches ultérieures sur les déterminants gauches*, loc. cit., p. 214.

operations by a simple specialization of the formulas just given. Indeed the importance of quaternion multiplication for ordinary physics and mechanics is based upon these very formulas. I have said "ordinary", because I do not desire at this point to explain those generalizations of these science for which the preceding formulas apply without any modification. These generalizations are more immediate, however, than you may suppose. The new developments of electrodynamics which are associated with the *principle of relativity*, are essentially nothing else than the logical use of rotations and expansions in a four-dimensional space. These ideas have been presented and enlarged upon recently by Minkowski[1].

Let us remain, however, in three-dimensional space. In such a space, a rotation and expansion carries a point (x, y, z) into a point (x', y', z') in such a way that

$$x'^2 + y'^2 + z'^2 = M^2(x^2 + y^2 + z^2),$$

where M denotes the ratio of expansion of every length. Since the general linear transformation of (x, y, z) into (x', y', z') contains nine coefficients, and since the left-hand side of the preceding equation, after the insertion of the values of x', y', z', becomes a quadratic form in x, y, z with six terms, the comparison of coefficients in the preceding equation leads to six equations, which reduce to five if the value of M is supposed arbitrary. Therefore the nine original coefficients of the linear transformation, which are subject to these five conditions, are reduced to four arbitrary constants. (Compare p. 67.) If such a linear transformation has a positive determinant, it represents, as was stated on p. 67, a rotation of space about the origin, together with an expansion in the ratio $1/M$. If the determinant is negative, however, the transformation represents a rotation and expansion, combined with a *reflection*, such as, for example, the reflection defined by the equations $x = -x', y = -y', z = -z'$. Moreover, it can be shown that the determinant of the transformation must have one of the two values $\pm M^3$.

In order to represent these relationships by means of quaternions, let us first reduce the variable quaternions q and q' to their vectorial parts:

$$q' = ix' + jy' + kz', \qquad q = ix + jy + kz,$$

which we shall think of as the three-dimensional vectors joining the origin to the positions of the point before and after the transformation, respectively. We shall show that *the general rotation and expansion*

---

[1] Since this was written, an extensive literature on the special theory of relativity mentioned above has appeared. Let me mention here my address *Über die geometrischen Grundlagen der Lorentzgruppe*, Jahresbericht der deutschen Mathematiker-Vereinigung, vol. 19 (1910), p. 299, reprinted in Klein's *Gesammelte mathematische Abhandlungen*, vol. 1, p. 533.

of the three-dimensional space is given by the formula (II) if p and π have conjugate values, that is, if we write $q' = p \cdot q \cdot \bar{p}$; or, in expanded form,

(1) $\begin{cases} ix' + jy' + kz' \\ = (d + ia + jb + kc)(ix + jy + kz)(d - ia - jb - kc). \end{cases}$

In order to prove this, we must show first that the scalar part of the product on the right vanishes; that is, that q' is indeed a *vector*. To do this, we first mutiply p by q according to the rule for quaternion multiplication, and we find

$$q' = [-ax - by - cz + i(dz + bz - cy) \\ + j(dy + cx - az) + k(dz + ay - bx)] \cdot [d - ia - jb - kc].$$

After another quaternion multiplication, we actually find the scalar part of q' to be zero, whereas we find for the components of the vector part the expressions

(2) $\begin{cases} x' = (d^2 + a^2 - b^2 - c^2)x + \quad 2(ab - cd)y + \quad 2(ac + bd)z \\ y' = \quad 2(ba + cd)x + (d^2 + b^2 - c^2 - a^2)y + \quad 2(bc - ad)z \\ z' = \quad 2(ca - bd)x + \quad 2(cb + ad)y + (d^2 + c^2 - a^2 - b^2)z \end{cases}$

That these formulas actually represent a rotation and expansion becomes evident if we write the tensor equation for (1), which, by (I), is

$$x'^2 + y'^2 + z'^2 = (d^2 + a^2 + b^2 + c^2)(x^2 + y^2 + z^2)(d^2 + a^2 + b^2 + c^2),$$

or

$$x'^2 + y'^2 + z'^2 = T^4 \cdot (x^2 + y^2 + z^2),$$

where $T = \sqrt{d^2 + a^2 + b^2 + c^2}$ denotes the tensor of p. Hence, our transformation is precisely a rotation and expansion (see p. 69), provided the determinant is positive; otherwise it is such a transformation combined with a reflection. In any case, the ratio of expansion is $M = T^2$. As remarked above, the determinant must have one of the two values $\pm M^3 = \pm T^6$. If we consider the transformation for all possible values of the parameters a, b, c, d which correspond to the same tensor value T, which must obviously be different from zero, we see that the determinant must *always* have the value $+T^6$ if it has that value for any *single* system of values of a, b, c, d; for the determinant is a continuous function of a, b, c, d, and therefore it cannot suddenly change in value from $+T^6$ to $-T^6$ without taking on intermediate values. One set of values for which the determinant is positive is $a = b = c = 0$, $d = T$, since, by (2), the value of the determinant for these values of a, b, c, d, is

$$\begin{vmatrix} d^2 & 0 & 0 \\ 0 & d^2 & 0 \\ 0 & 0 & d^2 \end{vmatrix} = d^6 = +T^6.$$

Quaternion Multiplication — Rotation and Expansion.

It follows that the sign is always positive, and hence (1) always represents actually a rotation and expansion. It is easy to write down a transformation which combines a reflection with a rotation and an expansion, for we need only combine the preceding transformation with the reflection $x' = -x, y' = -y, z' = -z$, which is equivalent to writing the quaternion equation $\bar{q}' = p \cdot q \cdot \bar{p}$.

We shall now show that, conversely, every rotation and expansion may be written in the form (1), or in the equivalent form (2). In the first place, this formula contains the four arbitrary constants which, as we saw on p. 69, are necessary for the general case. That we can actually obtain any desired value of the expansion-ratio $M = T^2$, any desired position of the axis of rotation, and any desired angle of rotation, by a suitable choice of the four arbitrary constants, can be seen by means of the following formulas. Let ξ, η, ζ denote the direction cosines of the axis of rotation, and let ω denote the angle of rotation. We have, of course, the well known relation

(3) $$\xi^2 + \eta^2 + \zeta^2 = 1.$$

I shall now prove that a, b, c, d are given by the equations

(4) $$\begin{cases} d = T \cdot \cos\frac{\omega}{2}; \\ a = T \cdot \xi \cdot \sin\frac{\omega}{2}, \quad b = T \cdot \eta \cdot \sin\frac{\omega}{2}, \quad c = T \cdot \zeta \cdot \sin\frac{\omega}{2}, \end{cases}$$

which, by (3), obviously satisfy the condition

$$d^2 + a^2 + b^2 + c^2 = T^2.$$

When these relations have been proved, we can evidently obtain the correct values of a, b, c, d for any given values of $T, \xi, \eta, \zeta, \omega$.

To prove the relations (4), let us remark first that if a, b, c, d are given, the quantities ω, ξ, η, ζ are determined, and in such a way that (3) is satisfied. For, squaring and adding the equations (4), since T is the tensor of the quaternion $p = d + ia + jb + kc$, we have

$$1 = \cos^2\frac{\omega}{2} + \sin^2\frac{\omega}{2}(\xi^2 + \eta^2 + \zeta^2),$$

whence we see that (3) holds. It follows that ξ, η, ζ are fully determined by the relations

(4') $$a : b : c = \xi : \eta : \zeta,$$

which appear directly from (4). These equations express the fact that the point (a, b, c) lies on the axis of revolution of the transformation. This fact is easy to verify, for if we put $x = a, y = b, z = c$ in (2), we find

$$x' = (d^2 + a^2 + b^2 + c^2) a = T^2 \cdot a,$$
$$y' = (d^2 + a^2 + b^2 + c^2) b = T^2 \cdot b,$$
$$z' = (d^2 + a^2 + b^2 + c^2) c = T^2 \cdot c$$

that is, the point (a, b, c) remains on the same ray through the origin, which identifies it as a point on the axis of revolution. It remains only to prove that the angle ω defined by (4) is actually the angle of rotation. This demonstration requires extended discussion which I can avoid now by remarking that the transformation (2) for $T = 1$ reduces precisely to the transformation given by Euler for the revolution of the axes through the angle ω about an axis of revolution whose direction cosines are ξ, η, ζ. This is to be found, for example, in Klein-Sommerfeld, *Theorie des Kreisels*, volume 1[1], where explicit mention of the theory of quaternions is given, or in Baltzer, *Theorie und Anwendung der Determinanten*[2].

Finally, if we substitute the values given by (4) in the equation (1), we obtain the very brief and convenient equation in quaternion form for the revolution through an angle ω about an axis whose direction cosines are ξ, η, ζ, combined with an expansion of ratio T^2:

$$(5) \quad \begin{cases} ix' + jy' + kz' = T^2 \left\{ \cos\frac{\omega}{2} + \sin\frac{\omega}{2}(i\xi + j\eta + k\zeta) \right\} \cdot \{ix + jy + kz\} \\ \qquad \cdot \left\{ \cos\frac{\omega}{2} - \sin\frac{\omega}{2}(i\xi + j\eta + k\zeta) \right\}. \end{cases}$$

This formula expresses in a form that is easy to remember Euler's formulas for rotation: the multipliers which precede and follow the vector $ix + jy + kz$, are, respectively, the two conjugate quaternions whose tensor is unity (so-called *versor*, that is, "rotator", in contradistinction to *tensor*, "stretcher"), and then the whole result is to be multiplied by a scalar factor which is the expansion-ratio.

We shall proceed now to show that when we specialize these formulas still further to two-dimensions, they become the well known formulas for the representation of a rotation and expansion of the xy plane by means of the multiplication of two complex numbers. (See p. 57.) For this purpose, let us choose the axis of rotation as the z axis ($\xi = \eta = 0, \zeta = 1$). Then the formula (5), for $z = z' = 0$, may be written in the form

$$ix' + jy' = T^2 \left(\cos\frac{\omega}{2} + k\sin\frac{\omega}{2} \right)(ix + jy)\left(\cos\frac{\omega}{2} - k\sin\frac{\omega}{2} \right),$$

or, upon multiplication with due regard to the rules for products of the units,

$$\begin{aligned} ix' + jy' &= T^2 \left\{ \cos\frac{\omega}{2}(ix + jy) + \sin\frac{\omega}{2}(jx - iy) \right\}\left\{ \cos\frac{\omega}{2} - k\sin\frac{\omega}{2} \right\} \\ &= T^2 \left\{ \cos^2\frac{\omega}{2}(ix+jy) + 2\sin\frac{\omega}{2}\cos\frac{\omega}{2}(jx-iy) - \sin^2\frac{\omega}{2}(ix+jy) \right\} \\ &= T^2 \{(ix+jy)\cos\omega + (jx - iy)\sin\omega\} \\ &= T^2 (\cos\omega + k\sin\omega)(ix + jy). \end{aligned}$$

---

[1] Leipzig 1897; 2nd printing, 1914.   [2] Fifth edition, Leipzig 1881.

If we now multiply both sides by the right-hand factor $(-i)$, we obtain
$$x' + ky' = T^2 (\cos\omega + k\sin\omega)(x + ky),$$
which is precisely the rule for multiplying two ordinary complex numbers, and which can be interpreted as a rotation through an angle ω, together with an expansion in the ratio T^2, except that we have used the letter k in place of the usual letter i to denote the imaginary unit $\sqrt{-1}$.

Let us now return to three-dimensional space, and let us modify the formula (1) so that it shall represent a pure rotation without an expansion. To do so, we must replace x', y', z' by $x' \cdot T^2, y' \cdot T^2, z' \cdot T^2$, that is, we must replace q' by $q' \cdot T^2$. If we notice that $p^{-1} = 1/p = p/T^2$, we may write the formula for a pure rotation in the form

(6) $\qquad ix' + jy' + kz' = p \cdot (ix + jy + kz) \cdot p^{-1}.$

There is no loss of generality if we assume that p is a quaternion whose tensor is unity, that is,
$$p = \cos\frac{\omega}{2} + \sin\frac{\omega}{2}(i\xi + j\eta + k\zeta), \quad \text{where} \quad \xi^2 + \eta^2 + \zeta^2 = 1,$$
whence we see that (6) results from (5) if T is set equal to unity. The formula was first stated in this form by Cayley in 1845[1].

We may express the composition of two rotations in a particularly simple form, precisely as we did above for four-dimensional space. Given a second rotation
$$ix'' + jy'' + kz'' = p'(ix' + jy' + kz')p'^{-1},$$
where
$$p' = \cos\frac{\omega'}{2} + \sin\frac{\omega'}{2}(i\xi' + j\eta' + k\zeta')$$
the direction cosines of the axis of rotation being ξ', η', ζ', and the angle of rotation being ω', we may write
$$ix'' + jy'' + kz'' = p' \cdot p \cdot (ix + jy + kz) \cdot p^{-1} \cdot p'^{-1}$$
as the equation for the resultant rotation. Hence the direction cosines of the axis or rotation, ξ'', η'', ζ'', and the angle of rotation, ω'', for the resultant rotation, are given by the equation
$$p'' = \cos\frac{\omega''}{2} + \sin\frac{\omega''}{2}(i\xi'' + j\eta'' + k\zeta'') = p' \cdot p.$$

We have therefore found a brief and simple expression for the composition of two rotations about the origin, whereas the ordinary formulas for expressing the resultant rotation appear rather complicated. Since any quaternion may be expressed as the product of a real number

---

[1] *On certain results relating to quaternions*, Collected Mathematical Papers, vol. 1 (1889), p. 123. According to Cayley's own statement (vol. 1, p. 586), however, Hamilton had discovered the same formula independently.

(its tensor) and the versor of a rotation, we have also found a simple geometric interpretation of quaternion multiplication as the composition of the rotations. The fact that quaternion multiplication is not commutative then corresponds to the well known fact that the order of two rotations about a point cannot be interchanged, in general, without changing the result.

If you desire to make a study of the historical development of the representations and applications of quaternions which we have discussed, I would recommend to you an extremely valuable report on dynamics written by Cayley himself: *Report on the progress of the solution of certain special problems of dynamics*[1].

I shall close with certain general remarks on the value and the dissemination of quaternions. For such a purpose, one should distinguish between the general quaternion calculus and the simple rule for quaternion multiplication. The latter, at least, is certainly of very great usefulness, as appears sufficiently from the preceding discussion. The general quaternion calculus, on the other hand, as Hamilton conceived it, embraced addition, multiplication, and division of quaternions, carried to an arbitrary number of steps. Thus Hamilton studied the algebra of quaternions; and, since he investigated also infinite processes, he may be said to have created a quaternion theory of functions. Since the commutative law does not hold, such a theory takes on a totally different aspect from the theory of ordinary complex variables. It is just to say, however, that these general and far-reaching ideas of Hamilton have not justified themselves, for there have not arisen any vital relationships and interdependencies with other branches of mathematics and its applications. For this reason, the general theory has aroused little general interest.

It is in mathematics, however, as it is in other human affairs: there are those whose views are calmly objective; but there are always some who form regrettable personal prejudices. Thus the theory of quaternions has enthusiastic supporters and bitter opponents. The supporters, who are to be found chiefly in England and in America, adopted in 1907 the modern plan by founding an "Association for the Promotion of the Study of Quaternions". This organization was established as a thoroughly international institution by the Japanese mathematician Kimura, who had studied in America. Sir Robert Ball was for some time its president. They foresaw great possible developments of mathematics to be secured through intensive study of quaternions. On the other hand, there are those who refuse to listen to anything about quaternions, and who go so far as to refuse to consider the very useful idea of quaternion mul-

---

[1] Report of the British Association for the Advancement of Science, 1862; reprinted in Cayley's Collected Mathematical Papers, Cambridge, vol. 4 (1891), pp. 552 ff.

tiplication. According to the view of such persons, all computation with quaternions amounts to nothing but computation with the four components; the units and the multiplication table appear to them to be superfluous luxuries. Between these two extremes, there are many who hold that we should always distinguish carefully between scalars and vectors.

4. Complex Numbers in School Instruction

I shall now leave the theory of quaternions and close this chapter with some remarks about the role which these concepts play in the curriculum of the schools. No one would ever think of bringing up quaternions in a secondary school, but *the common complex numbers $x + iy$ always come up for discussion.* Perhaps it will be more interesting if, instead of telling you at length how it is done and how it ought to be done, I exhibit to you, by means of three books from different periods, *how instruction has developed historically.*

I put before you, first, a book by Kästner who had a leading position in Göttingen in the second half of the eighteenth century. In those days one still studied, at the university, those elementary mathematical things which later, in the thirties of the nineteenth century, went over to the schools. Accordingly, Kästner also gave lectures on elementary mathematics, which were heard by large numbers of non-mathematical students. His book, which formed the basis of these lectures, was called *Mathematische Anfangsgründe**. The portion which interests us here is the second division of the third part: *Anfangsgründe der Analysis endlicher Größen***[1]. The treatment of imaginary quantities begins there on page 20 in something like the following words: "Whoever demands the extraction of an even root of a 'denied' quantity (one said 'denied', then, instead of 'negative'), demands an impossibility, for there is no 'denied' quantity which would be such a power". This is, in fact, quite correct. But on page 34 one finds: "Such roots *are called* impossible or imaginary", and, without much investigation as to justification, one proceeds quietly to operate with them as with ordinary numbers, notwithstanding their existence has just been disputed—as though, so to speak, the meaningless became suddenly usable through receiving a name. You recognize here a reflex of Leibniz's point of view, according to which, imaginary numbers were really something quite foolish but they led, nevertheless, in some incomprehensible way, to useful results.

Kästner was, moreover, a stimulating writer; he achieved quite a place in the literature as a coiner of epigrams. To cite only one of many examples, he expatiates, in the introduction of this book mentioned

---

[1] Third edition. Göttingen 1794.
* *Elements of Mathematics.*
** *Elemements of Analysis of Finite Quantities.*

above, on the *origin of the word algebra*, which, indeed, as the article "al" indicates, comes from the Arabic. According to Kästner, an algebraist is a man who "makes" fractions "whole", who, that is, treats rational functions and reduces them to a common denominator, etc. It is said to have referred, originally, to the practice of a surgeon in mending broken bones. Kästner then cites Don Quixote, who went to an algebraist to get his broken ribs set. Of course, I shall leave undecided, whether Cervantes really adopted this form of expression or whether this is only a lampoon.

The second work which I put before you is more recent, by a whole series of years, and comes from the Berlin professor M. Ohm: *Versuch eines vollständig konsequenten Systems der Mathematik*[*1]; a book with purpose similar to that of Kästner and at one time widely used. But Ohm is much nearer the modern point of view, in that he speaks clearly of the *principle of the extension of the number system*. He says, for example, that, just like negative numbers, so $\sqrt{-1}$ must be added to the real numbers as a *new thing*. But even his book lacks a geometric interpretation, since it appeared before the epoch-making publication by Gauss (1831).

Finally, I lay before you, out of the long list of modern school books, one that is widely used: *Bardeys Aufgabensammlung*[2]. The *principle of extension* comes to the fore here, and, in due course, the *geometric interpretation* is explained. This may be taken as the general position of school instruction today, even if, at isolated places, the development has remained at the earlier level. The point of view adopted in this book seems to me to yield the treatment best adapted to the schools. Withhout tiring the pupil with a systematic development, and without, of course, going into logically abstract explanations, one should explain *complex numbers as an extension of the familiar number concept*, and should avoid any touch of mystery. Above all, one should accustom the pupil, at once, to the *graphic geometric interpretation in the complex plane!*

With this, we come to the end of the first main part of the course, which was dedicated to arithmetic. Before going over to similar discussions of algebra and analysis, I should like to insert a somewhat extended historical appendix in order to throw new light upon the general conduct of instruction at present, and upon those features of it which we would improve.

---

[1] Nine volumes. Berlin 1828. Vol. I: *Arithmetik und Algebra*, p. 276.
[*] *An Attempt to Construct a Consistent System of Mathematics.*
[[2] See also the *Reformausgabe* of Bardeys *Aufgabensammlung*, revised by W. Lietzmann and P. Zülke. Oberstufe. Verlag Teubner. Leipzig.]—See also Fine, H., *The Number-System in Algebra*. Heath. Fine, H., College Algebra. Ginn.

Concerning the Modern Development and the General Structure of Mathematics

Let me proceed from the remark that, in the *history of the development* of mathematics up to the present time, we can distinguish clearly *two different processes of growth*, which now change places, now run side by side independent of one another, now finally mingle. It is difficult to put into vivid language the difference which I have in mind, because none of the current divisions fits the case. You will, however, understand from a concrete example, namely, if I show how one would compile the *elementary chapters of the system of analysis* in the sense of each of these two processes of development.

If we follow the one process, which we will call briefly *Plan A*, the following system presents itself, the one which is most widespread in the schools and in elementary textbooks.

1. At the head stands the *formal theory of equations*, that is to say, the *operating with rational integral functions* and the handling of the cases in which *algebraic equations can be solved by radicals*.

2. The *systematic pursuit of the idea of power and its inverses* yields *logarithms*, which prove to be so useful in numerical calculation.

3. Whereas (up to this point) the analytic development is kept quite separate from geometry, one now borrows from this field, which yields the *definitions of a second kind of transcendental functions, the trigonometric functions*, the further theory of which is built up as a *new separate subject*.

4. Then follows the so called "algebraic analysis", which teaches *the development of the simplest functions into infinite series*. One considers the *general binomial*, the *logarithm* and its inverse, the *exponential function*, together with *the trigonometric functions*. Similarly, the *general theory of infinite series and of operations with them* belongs here. It is here that the *surprising relations between the elementary transcendentals* appear, in particular the famous *Euler formula*

$$e^{ix} = \cos x + i \sin x.$$

Such relations seem the more remarkable because the functions which occur in them had been originally defined in entirely separate fields.

5. The consistent continuation, beyond the schools, of this structure, is the *Weierstrass theory of functions of a complex variable*, which begins with the properties of *power series*.

Let us now set over against this, in condensed form the *second process of development*, which I shall call *Plan B*. Here the controlling *thought is that of analytic geometry*, which seeks a *fusion of the perception of number with that of space*.

1. We begin with the *graphical representation of the simplest functions*, of polynomials, and rational functions of one variable. The point in

which the curves so obtained meet the axis of abscissas put in evidence the *zeros of the polynomials*, and this leads naturally to the *theory of the approximate numerical solution of equations*.

2. The *geometric picture of the curve supplies naturally the intuitive source both for the idea of the differential quotient and that of the integral*. One is led to the former by the *slope of the curve*, to the latter by the *area which is bounded by the curve and the axis of abscissas*.

3. In all those cases in which the *integration process* (or the process of quadrature, in the proper sense of that word) cannot be carried out explicitly with rational and algebraic functions, the process itself gives *rise to new functions*, which are thus introduced in a thoroughly natural and uniform manner. Thus, the *quadrature of the hyperbola* defines the *logarithm*

$$\int_1^x \frac{dx}{x} = \log x,$$

while the *quadrature of the circle* can easily be reduced to the integral

$$\int_0^x \frac{dx}{\sqrt{1-x^2}} = \arcsin x,$$

that is, to the *inverses of the trigonometric functions*. You know that the same line of thought, pursued farther, leads to new classes of functions of higher order, in particular to *elliptic functions*.

4. The *development into infinite power series of the functions thus introduced* is obtained by means of a *uniform principle*, namely *Taylor's theorem*.

5. This method carried higher, yields the *Cauchy-Riemann theory of analytic functions of a complex variable*, which is built upon the *Cauchy-Riemann differential equations* and the *Cauchy integral theorem*. If we try to put the *result of this survey* into definite words, we might say that *Plan A is based upon a more particularistic conception of science which divides the total field into a series of mutually separated parts and attempts to develop each part for itself, with a minimum of resources and with all possible avoidance of borrowing from neighboring fields*. Its ideal is to *crystallize out each of the partial fields into a logically closed system*. On the contrary, *the supporter of Plan B lays the chief stress upon the organic combination of the partial fields, and upon the stimulation which these exert one upon another. He prefers, therefore, the methods which open for him an understanding of several fields under a uniform point of view*. His ideal is the *comprehension of the sum total of mathematical science as a great connected whole*.

One cannot well be in doubt as to which of these two methods has more life in it, as to which would grip the pupil more, in so far as he is not endowed with a specific abstract mathematical gift. In order to bring this home, think only of the *example of the functions e^x and $\sin x$*, about which we shall later have much to say along just this line! In Plan A, which the schools, unfortunately, follow almost exclusively both functions come up in thoroughly heterogeneous fashion: e^x or, as the case may be, the *logarithm*, is introduced as a *convenient aid in numerical calculation*, but sin x appears in the *geometry of the triangle*. How can one understand, thus, that the two are so simply connected, and, more, that the two appear again and again in the most widely differing fields which have not the least to do, either with the technique of numerical calculation or with geometry, and always of their own accord, as the natural expression of the laws that govern the subject under discussion? How far these possibilities of application go is shown by the names *compound interest law* or *law of organic growth*, which have been applied to e^x, and likewise by the fact that sin x plays a central role wherever one has to do with *vibrations*. But in Plan B these *connections make their appearance quite intelligibly, and in accord with the significance of the functions, which is emphasized from the start*. The functions e^x and sin x arise here, indeed, from the same source, the quadrature of simple curves, and one is soon led from there, as we shall see later on, to the *differential equations of simplest type*

$$\frac{de^x}{dx} = e^x, \qquad \frac{d^2 \sin x}{dx^2} = -\sin x,$$

respectively, which lie naturally at the basis of all those applications.

For a complete understanding of the development of mathematics we must, however, think of still a *third Plan C*, which, along side of and within the processes of development A and B, often plays an important role. It has to do with a method which one denotes by the word *algorithm*, derived from a mutilated form of the name of an Arabian mathematician. *All ordered formal calculation* is, at bottom, algorithmic, in particular, the *calculation with letters* is an algorithm. We have repeatedly emphasized what an important part in the development of the science has been played by the algorithmic process, as a *quasi-independent, onward-driving force, inherent in the formulas*, operating apart from the intention and insight of the mathematician, at the time, often indeed in opposition to them. In the beginnings of the infinitesmal calculus, as we shall see later on, the algorithm has often forced new notions and operations, even before one could justify their admissibility. Even at higher levels of the development, these algorithmic considerations can be, and actually have been, very fruitful, so that one can justly call them the *groundwork of mathematical development*. We must then completely ignore history, if, as is sometimes done today, we cast these circumstances contemptuously aside as mere "formal" developments.

Let me now follow more carefully through the history of mathematics the contrast of these different directions of work, confining myself of course to the most important features of the development. The *thoroughgoing difference between A and B, within the whole field of mathematics*, will appear here more clearly than it did above, where our thoughts were directed only to analysis.

With the *ancient Greeks* we find a *sharp separation between pure and applied mathematics*, which goes back to Plato and Aristotle. Above all, the well known *Euclidean structure of geometry* belongs to pure mathematics. In the applied field they developed, especially, *numerical calculation*, the so called *logistics* (λόγος = general number, see p. 32). To be sure, the logistics was not highly regarded, and you know that this prejudice has, to a considerable extent, maintained itself to this day —mainly, it is true, with only those persons who themselves cannot calculate numerically. The slight esteem for logistics may have been due in particular to its having been developed in connection with *trigonometry* and the needs of *practical surveying*, which to some does not seem sufficiently aristocratic. In spite of this fact, it may have been raised somewhat in general esteem by its application in astronomy, which, although related to geodesy, always has been considered one of the most aristocratic fields. You see, even with these few remarks, that the Greek cultivation of science, with its sharp separation of the different fields, each of which was represented with its rigid logical articulation, *belonged entirely in the plan of development A*. Nevertheless the *Greeks were not entire strangers to reflections in the sense of Plan B*, and these may have served them for heuristic purposes, and for a first communication of their discoveries, even if the form *A* appeared to them indispensable for the final presentation. This is indicated quite pointedly in the recently discovered *manuscript of Archimedes*[1], in which he exhibits his calculations of volume through mechanical considerations, in a thoroughly modern, pleasing way, which has nothing to do with the rigid Euclidean system.

Besides the Greeks, in ancient times, the *Hindus*, especially, played a mathematical role as *creators of our modern system of numerals*, and later the *Arabs, as its transmitters*. The *first beginnings of operating with letters* were made also by the Hindus. These advances belong obviously to the *algorithmic course of development C*.

Coming now to *modern times*, we can, first of all, *date the mathematical renaissance from about 1500*, which produced an entire *series of great discoveries*. As an example, I mention the *formal solution of the cubic equation* (Cardan's formula), which was contained in the *"Ars Magna" of Cardano*, published in 1545, in Nürnberg. This was a most significant work, which holds the germs of the modern algebra, reaching out beyond

---

[1] Cf. Heiberg und Zeuthen, *Eine neue Schrift des Archimedes*. Leipzig 1907. Reprint from Bibliotheca Mathematica. Third series, vol. VIII. See also HEATH, T. L., *The Works of Archimedes*. Cambridge University Press.

the scheme of ancient mathematics. To be sure, this work is not Cardano's own, for he is said to have taken from other authors not merely his famous formula but other things as well.

After 1550 *trigonometric calculation* was in the foreground. *The first great trigonometric tables* appeared in response to the needs of *astronomy*, in connection with which I will mention only the name of Copernicus. *From about 1600 on*, the *invention of logarithms* continued this development. The first logarithmic tables, which a Scotchman Napier (or Nepér) compiled, contained, in fact, only the logarithms of trigonometric functions. Thus we see, during these hundred years, a path of development which corresponds to the *Plan A*.

We come now, in the *seventeenth century*, to the *modern era proper*, in which the *Plan B comes distinctly into the foreground*. In *1637* appeared the *analytic geometry of Descartes*, which supplies the fundamental connection between number and space for all that follows. A reprint[1] makes this work conveniently accessible. Now come, in close sequence with this, the *two great problems of the seventeenth century*, the *problem of the tangent*, and *the problem of quadrature*, in other words, the *problems of differentiating and integrating*. For the development of differential and integral calculus, in a proper sense, there was lacking only the knowledge that these two problems are closely connected, *that one is the inverse of the other*. A recognition of this fact was the *principal item in the great advance* which was made at the end of the seventeenth century.

But before this, in the course of the century, the *theory of infinite series, in particular, of power series*, made its appearance, and not, indeed, as an independent subject, in the sense of the algebraic analysis of today, but in *closest connection with the problem of quadrature*. Nicolaus Mercator (the German name "Kaufmann" latinized; 1620—1687), not the inventor of the Mercator projection, was a pioneer here. He had the keen idea of converting the fraction $1/(1+x)$ into a series, by dividing out, and of *integrating this series term by term*, in order to get the *series development for* $\log(1+x)$:

$$\log(1+x) = \int_0^x \frac{dx}{1+x} = \int_0^x (1 - x + x^2 - + \cdots) dx = x - \frac{x^2}{2} + \frac{x^3}{3} - + \cdots$$

That is the substance of his procedure, although he did not, of course, use our simple symbols $\int$, $dx, \ldots$, but rather a much more clumsy form of expression. In the sixties, Isaac Newton (1643—1727) took over this process, to apply it to the *series for the general binomial*, which he had set up. In this process he drew his *conclusions by analogy*, basing

---

[1] Descartes, R., *La Géométrie*. Nouvelle édition. Paris 1886. Translation by Smith, D. E., and Latham, M. L., 1925. Open Court.

them on the known simplest cases, without having a rigorous proof and without knowing the limits within which the series development was valid. We observe here, again, the operation of the *algorithmic process C*. By applying the binomial series to $\frac{1}{\sqrt{1-x^2}} = (1-x^2)^{-1/2}$ and using Mercator's process, he gets the *series for* $\int_0^x \frac{dx}{\sqrt{1-x^2}} = \arc \sin x$. By a very skillful *inversion of this series*, and also of the one for log x, he finds the series for sin x and for e^x. The conclusion of this chain of discoveries is due to Brook Taylor (1685—1731) who, in 1715, published his *general principle for developing functions into power series*.

As is indicated above, the *origin of infinitesimal calculus, at the end of the seventeenth century*, was due to G. W. Leibniz (1646—1716) and Newton. The fundamental idea with Newton is the *notion of flowing*. Both variables x, y, are tought of as functions, $\varphi(t)$, $\psi(t)$, of the time t; and as the time "flows", they flow also. Newton, accordingly, calls the variable *fluens* and designates as *fluxion* $\dot{x}$, $\dot{y}$, that which we call differential coefficient. You see how everything here is *based firmly on intuition*.

It was much the same with the representation of Leibniz, whose first publication appeared in 1684. He himself declares that his greatest discovery was the *principle of continuity in all natural phenomena*, that "Natura non facit saltum". He bases his mathematical developments upon this concept, another example of the *Plan B*. However, the *influence of the algorithm C is very strong*, also, with Leibniz. We get from him the algorithmically valuable symbols dy/dx and $\int f(x)\,dx$.

The sum total, however, of this cursory view is that the *great discoveries of the seventeenth century belong primarily to the plan of development B*.

In the eighteenth century, this period of discovery continues at first *in the same direction*. The most distinguished names to be mentioned here are L. Euler (1707—1783) and J. L. Lagrange (1736—1813). Thus the *theory of differential equations, in the most general sense, including the calculus of variations*, were developed, and *analytical geometry and analytical mechanics were extended*. Everywhere there was a gratifying advance, just as in geography, after the discovery of America, the new countries were first traversed and explored in all directions. But just as there was, as yet, no thought of exact surveys, just as at first one had entirely false notions as to the location of these new places (Columbus, indeed, thought at first, that he had reached Eastern Asia!), just so, in the newly conquered region of mathematics, that of infinitesimal calculus, one was, at first, far removed from a reliable logical orientation. Indeed one even cherished illusions concerning the relation of the calculus to the older familiar fields, in that one looked upon infinitesimal calculus as something *mystical* that in no way submitted to a logical analysis.

Just how untrustworthy the ground was on which the theory stood, became manifest only when it was attempted to prepare textbooks which should present the new subject in an intelligible way. Then it became evident that the method of procedure B was no longer adequate, and it was Euler who first abandoned it. He had, to be sure, no serious doubts concerning infinitesimal calculus, but he thought that it caused too many difficulties and misgivings for the beginner. For this pedagogical reason he thought it advisable to give a preparatory course, such as he provided in his text book *Introductio in analysin infinitorum* (1748), and which we call today *algebraic analysis*. To this he relegated, in particular, the *theory of infinite series and other infinite processes*, which he then afterwards used as a foundation in constructing the infinitesimal calculus.

Lagrange took a much more radical course, nearly fifty years later, in his *Thèorie des fonctions analytiques, in 1797*. He could satisfy his scruples as to the current foundations of infinitesimal calculus only by discarding it entirely, as a general branch of knowledge, and by considering it as an aggregate of formal rules applying to certain special classes of functions. Indeed, he considers *exclusively such functions as can be expressed by means of power series:*

$$f(x) = a_0 + a_1 x + a_2 x^2 + a_3 x^3 + \cdots.$$

He calls these *analytic functions*, meaning thereby functions which appear in analysis and with which one can reasonably hope to do something. The *differential quotient of such a function, f(x), is then defined, purely formally, by means of a second power series*, as we shall see later. Differential and integral calculus was concerned, then, with the mutual relations of power series. This restriction to formal consideration obviated, for a time, of course, a number of difficulties.

As you see, *the turn which Euler gave, and still more, the entire method of Lagrange, belongs strictly to the direction A, in that the perceptual genetic development is replaced by a rigorous closed circle of reasoning*. These two investigators have *had a profound influence upon instruction in the secondary schools*, and when the schools today study infinite series, or solve equations by means of power series according to the so called *method of indeterminate coefficients*, but decline to take up differential and integral calculus proper, they are *exhibiting precisely the after effect of Euler's "introductio" and of Lagrange's thought*.

The *nineteenth century*, to which we come now, begins primarily with a *more secure foundation of higher analysis, by means of criteria of convergence*, about which one had hitherto thought but little. The eighteenth century was the "blissful" period, during which one did not distinguish between good and bad, convergent and divergent. Even in Euler's *Introductio*, divergent and convergent series appear peaceably side by side. But, at the beginning of the new century Gauss (1777—1855) and Abel (1802—1829) made the first rigorous investigations regarding convergence; and in the twenties Cauchy (1789—1857) developed, in lectures and in books, the *first rigorous founding of infinitesimal calculus in the modern sense*. He not only gives an exact *definition of the differential quotient, and of the integral, by means of the limit of a finite quotient and of a finite sum, respectively*, as had previously been done, at times; but, by means of the *mean-value theorem* he erects upon this, for the first time, a *consistent structure for the infinitesimal calculus*. We shall come back to this fully later on. These theories also partake of the nature of *Plan A*, since they work over the field in a logical sytematic way, quite apart from other branches of knowledge. *Meanwhile they had no influence upon the schools*, although they were thoroughly adapted to dispel the old prejudice against differential and integral calculus.

I shall now emphasize only a very little of the *further development of the nineteenth century*. In the first place, I shall speak of a few advances which lie in the *direction B: modern geometry, mathematical physics*, along with *theory of functions of a complex variable, according to Cauchy and Riemann*. The leaders, in the first working over of these three great fileds, were *the French*. This is the place to say a word, also, about the *style of mathematical presentation*. In Euclid, one finds everything according to the scheme "hypothesis, conclusion, proof", to which is added, sometimes, the "discussion", i. e., the determination of the limits which the considerations are valid. The belief is widespread that mathematics always moves thus four steps at a time. But just in the period of which we are speaking, there arose, especially among the French, a *new art form in mathematical presentation*, which might be called *artistically articulated deduction*. The works of Monge or, to mention a more recent book, the *Traité d'Analyse*, by Picard, read just like a well written gripping novel. This is the *style which fits the method of thought B*, whereas the *Euclidean presentation is related, in essence, to the method A*.

Of Germans who achieved distinction in these fields I should mention Jacobi (1804—1851), Riemann (1826—1866), and, coming to a somewhat later time, Clebsch (1833—1872), and the Norwegian Lie (1842—1899). These all belong essentially to the *direction B*, except that occasionally *an algorithmic touch* is noticeable with them.

From the middle of the century on, the *method of thought A* comes again to the front with Weierstrass (1815—1897). His activity, as teacher in Berlin, began in 1856. I have already instanced *Weierstrass function theory* as an example of *A*. The *more recent investigations concerning the axioms of geometry* belong, likewise, to the *type A*. One is concerned here with studies entirely in the Euclidean direction, which approach it, also, in the manner of presentation.

With this I bring our brief historical résumé to an end. Many points of view which could only be alluded to here will be brought up later for more complete discussion. As a summary, we might say that, *in the history of mathematics during the last centuries, both of our chief methods of investigation were of importance; that each of them, and sometimes the two in succession, have resulted in important advances of the science.* It is certain that mathematics will be able to advance uniformally in all directions, only if *neither of the two methods of investigation is neglected.* May each mathematician work in the direction which appeals to him most strongly.

Instruction in the secondary schools, however, as I have already indicated, has long been *under the one-sided control of the Plan A.* Any movement toward reform of mathematical teaching must, therefore, press for *more emphasis upon direction B.* In this connection I am thinking, above all, of an impregnation with the *genetic method of teaching,* of a stronger emphasis upon *space perception,* as such, and, particularly, of giving *prominence to the notion of function,* under *fusion of space perception and number perception!* It is my aim that these lectures shall serve this tendency, especially since these elementary mathematical books to which we are in the habit of going for advice, e g., those of Weber-Wellstein, Tropfke, M. Simon, represent the direction *A* almost exclusively. I called your attention, in the introduction, to this one-sidedness.

And now, gentlemen, enough of these diversions; let us pass to the next main subdivision of this course.

Part II
Algebra

Let me commence by mentioning a *few textbooks of algebra*, in order to introduce you somewhat to a very extensive literature. I suggest, first, Serret's *Cours d'algèbre*[1] which was much used in Germany, formerly, and had great merit. Now, however, we have two great widely used German textbooks: H. Weber's *Lehrbuch der Algebra*[2] and E. Netto's *Vorlesungen über Algebra*[3], each in two volumes; both treat with great fullness the most difficult parts of algebra and are well adapted for extensive special study; they seem to me to be too comprehensive for the average needs of prospective teachers and also too expensive. More fitting in the latter respect is the handy *Vorlesungen über Algebra*[4] by G. Bauer, which hardly goes beyond what the teacher should master[5]. On the *practical side*, for the numerical solution of equations, this book is supplemented by the little book *Praxis der Gleichungen* by C. Runge[6], which I can highly reccomend.

Turning now to the narrower subject, let me remark that I *cannot*, in the limits of this course of lectures, *give a systematic presentation of algebra*; I can give, rather, only a *one sided selection*, and it will be most fitting if I emphasize those things which are, unfortunately, neglected elsewhere, and which are calculated nevertheless to throw light upon school instruction. All of my algebraic developments will group themselves about *one* point, namely, about the *application to the solution of equations of graphical and, generally speaking, of geometrically perceptual methods*. This field alone is a very extensive and widely related chapter of algebra. Even from it, it is obviously possible to select only the most

---

[1] Third edition. Paris 1866 [sixth edition, 1910].
[2] Second edition. Braunschweig 1898/99. [New revision by R. Fricke. Vol. I. 1924.]
[3] Leipzig 1896/99. See also: Chrystal, *Textbook in Algebra* (2 volumes). Macmillan. Bôcher, M., *Introduction to Higher Algebra*. Macmillan.
[4 Second edition. Leipzig 1910.]
[5] See also: Netto, E., *Elementare Algebra*, akademische Vorlesungen für Studierende der ersten Semester. [Second edition. Leipzig 1913, and H. Weber, *Lehrbuch der Algebra*. Small edition in one volume. Second printing. Braunschweig 1921.] See also: Fine, H., *College Algebra*. Ginn. Hall und Knight, *Higher Algebra*. Macmilian.
[6 Second edition. Leipzig 1921.] See also: v. Sanden, H., *Practical Mathematical Analysis*. Dutton & Co.

important and interesting things; in doing this we shall come into organic relation with the most widely differing fields, so that we shall be *studying mathematics quite in the spirit of our system B*. In the first place, we shall treat equations in real unknowns in order that we may follow, later, with the consideration of complex quantities.

I. Real Equations with Real Unknowns

1. Equations with one parameter

We begin with a very simple case, which is susceptible of geometric treatment, namely with a real algebraic equation for the unknown x, in which a parameter λ appears:

$$f(x, \lambda) = 0.$$

We shall obtain a geometric representation most simply if we replace λ by a second variable y and think of

$$f(x, y) = 0$$

as a *curve in the xy plane* (see Fig. 19). *The points of intersection of this curve with the line $y = \lambda$, parallel to the x-axis, give the real roots of the equation $f(x, \lambda) = 0$.* When we have drawn the curve approximately, as we can easily do if f is not too complicated, we can see at a glance by displacing the parallel as λ varies, how the number of real roots changes. This plan is especially effective when f *is linear in* λ, i.e. with equations of the form

$$\varphi(x) - \lambda \psi(x) = 0$$

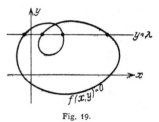

Fig. 19.

If φ and ψ are rational, the curve $y = \varphi(x)/\psi(x)$ will also be rational, and is easy to draw. In these cases one can often use this method to advantage in calculating approximately the roots of equations.

As an *example* consider *the quadratic equation*

$$x^2 + ax - \lambda = 0.$$

The curve $y = x^2 + ax$ is a *parabola*, and one can see at once for what values of λ the equations has two, one, or no real roots according as the horizontal line cuts the parabola in two, one, or no points (see Fig. 20). It seems to me that the presentation of such a simple and obvious construction would be very appropriate in the upper school classes.

As a second example let us take the *cubic equation*

$$x^3 + ax^2 + bx - \lambda = 0,$$

which gives us the *cubical parabola* $y = x^3 + ax^2 + bx$, whose appearance is different according to the values of a, b. In Fig. 21, it is assumed that $x^2 + ax + b = 0$ has two real roots. It is easy to see how the parallels group themselves into those which intersect the curve in one

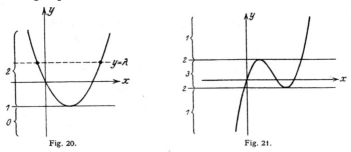

Fig. 20. Fig. 21.

point and those which meet it in three; there can be two limiting positions which yield double roots.

2. Equations with two parameters

When several parameters, let us say two, appear in an equation, more skill is required to handle the problem graphically, but the results are more extensive and interesting. We shall limit ourselves to the case where the *two parameters λ, μ appear linearly*, and we shall write t for the unknown in the equation. The problem is to determine the real roots of the equation

(1) $$\varphi(t) + \lambda \cdot \chi(t) + \mu \cdot \psi(t) = 0,$$

where φ, χ, ψ are polynomials in t.

If x, y are ordinary rectangular point-coordinates, every straight line in the xy plane will be given by an equation of the form

(2) $$y + ux + v = 0.$$

We may call u, v *the coordinates of the straight line*. Then $(-u)$ is the trigonometric tangent of the angle which the line makes with the

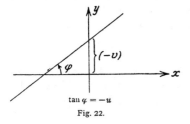

Fig. 22.

x-axis, and $(-v)$ is the y-intercept (see Fig. 22). Let us think of points and lines as of equal importance; and let us give equal attention to point coordinates and line coordinates. This will be especially important later on. Then we may say that the equation $y + ux + v = 0$ *indicates the united position of the line* (u, v) *and of the point* (x, y), i. e., that the point lies on the given line, and the line goes through the given point.

In order now to interpret the equation (1) geometrically, let us identify it with (2). This can be done in *two essentially different ways* which we shall consider, separably.

A. Let us consider the equations

(3a) $$y = \frac{\varphi(t)}{\psi(t)}, \quad x = \frac{\chi(t)}{\psi(t)},$$

(3b) $$u = \lambda, \quad v = \mu.$$

If t is variable, the equations (3a) represent, a *well determined rational curve of the xy plane*, which is called the *normal curve of equation* (1). Since every point on it corresponds to a definite value of t, a certain *scale of* values of t is defined upon it. By means of (3a) we can calculate as many points as we please; and hence we can draw the normal curve, with its scale, as accurately as we please, say on millimeter paper. For every definite pair of values of λ and μ (3b) represents a *straight line* of the plane. From what has been said, it follow that (1) shows, that the point t of the normal curve lies upon this straight line. Thus *we obtain all the real roots of* (1) *if we find all the real intersections of the normal curve with this line and read off their parameter values on the curve scale*. The normal curve is determined, once for all, by the form of equation (1), regardless of the special values which the parameters λ, μ may have. For every equation with definite λ, μ there is, then, one straight line which represents it, in the manner described above, so that, in general, all the straight lines in the plane come into play, whereas before (pp. 87—88) only horizontal lines were used.

As an illustration, let us take the *quadratic equation*
$$t^2 + \lambda t + \mu = 0.$$

The normal curve here is given by the equations

$$y = t^2, \; x = t \quad \text{or} \quad y = x^2,$$

i.e., the normal curve is the parabola shown in Fig. 23, with the scale there indicated.

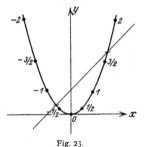

Fig. 23.

We can at once read off the real roots of our equation as the intersections with the line $y + \lambda x + \mu = 0$. In particular, the figure shows that the two roots of the equation $t^2 - t - 1 = 0$ lie between $\frac{3}{2}$ and 2, and between $-\frac{1}{2}$ and -1, respectively. The essential advantage of this method, over that given on pp. 87—88, is that we can now *solve all quadratic equations with one and the same parabola*, if we make use of all the straight lines in the plane. Thus, if we wish to solve, approximately, a considerable number of equations, one can apply this method very effectively.

In a similar way one can treat the totality of *cubic equations*, all of which can, by a linear transformation, be thrown into the *reduced form*

$$t^3 + \lambda t + \mu = 0.$$

The normal curve here is the cubical parabola

$$y = t^3, \quad x = t \quad \text{or} \quad y = x^3$$

sketched in Fig. 24. This method also seems to me to be usable in the schools. The pupils would certainly derive pleasure from drawing such curves.

B. *The second method of interpreting* (1) is got from the first by applying the *principle of duality*, i. e., by interchanging point and line coordinates. To that end, let us write the terms of (2) in reverse order:

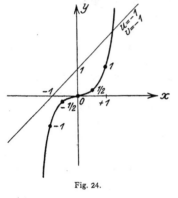

Fig. 24.

$$v + xu + y = 0$$

and identify it, in this form, with (1) by setting

(4a) $\quad v = \dfrac{\varphi(t)}{\psi(t)}, \quad u = \dfrac{\chi(t)}{\psi(t)},$

(4b) $\quad x = \lambda, \quad y = \mu.$

If t is variable, the equation (4a) represents a family of straight lines which will envelope a definite curve, the *normal curve of* (1), in the new interpretation. It is a *rational class curve*, since it is represented, in line coordinates, by rational functions of a parameter. Every tangent, and hence the corresponding point of tangency, is determined by a definite value of t, so that one gets again a *scale on the normal curve*. By drawing a sufficient number of tangents according to (4a), we may draw both curve and scale with any desired degree of exactness. Each parameter-pair λ, μ determines, by virtue of (4b), a point in the xy plane, through which, by virtue of (1), the tangent t of the normal curve (4a) must pass. *We obtain, therefore, all the real roots of* (1) *by reading off the parameter-values t belonging to all the tangents to the normal curve which go through the point $x = \lambda, y = \mu$.* As before, the normal curve is completely determined by the *form* of equation (1). Every equation of this form will be represented, for given values of the parameters λ, μ, by a certain point in the plane, or, if we wish, by its position with respect to the curve.

Let us illustrate by means of the same examples as before. Corresponding to the *quadratic* equation

$$t^2 + \lambda t + \mu = 0$$

the normal curve will be the envelope of the straight lines

$$v = t^2, \quad u = t.$$

This envelope, again, is a *parabola* with its vertex at the origin. The graph, drawn on fine cross section paper exhibits immediately the real roots of $t^2 + \lambda t + \mu = 0$ as parameters t of the tangents drawn to the parabola from the point λ, μ (see Fig. 25).

For the *cubic equation*

$$t^3 + \lambda t + \mu = 0$$

the normal curve

$$v = t^3, \quad u = t$$

will be a curve of the *third class* with a cusp at the origin, shown in Fig. 26.

We can present this method somewhat differently. If we examine the so-called *trinomial equation*

$$t^m + \lambda t^n + \mu = 0,$$

we may represent the *system of tangents to the normal curve* by means of the parameter equation

$$f(t) = t^m + x t^n + y = 0.$$

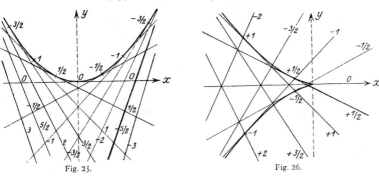

Fig. 25.   Fig. 26.

The equation of the normal curve in point coordinates may be found, as usual, by eliminating t between the last equation and the equation obtained by differentiation with respect to t:

$$f'(t) = m t^{m-1} + n x t^{n-1} = 0$$

for the normal curve, as the envelope of the system of straight lines, is the locus of the intersection of each of these lines with the neighboring line (for t and $t + dt$). If, instead of eliminating t, we express x and y as functions of t from these two equations, we find

(5a) $$x = -\frac{m}{n} t^{m-n}, \quad y = \frac{m-n}{n} t^m,$$

which are the *point equations of the normal curve*.

As normal curves for the quadratic and the cubic equations which were selected above as examples, one finds in this way, respectively,

$$x = -2t, \quad y = t^2$$
$$x = -3t^2, \quad y = 2t^3.$$

These are the curves which are sketched in Figs. 25 and 26.

Let me emphasize the fact that this method is put to practical use by *C. Runge*, in his lectures and exercises, and that it has proved itself *especially appropriate for the actual solution of equations*. We might profitably use one or the other of these graphical methods in school instruction.

If we now compare with each other the two methods which we have developed, we find that, for at least one definite and very important purpose, the *second offers a distinct advantage*, namely, *when one seeks a visible representation of all the equations of a definite type which have a given number of real roots*. Such totalities are represented, according to the first method, by *systems of straight lines*; according to the second, however, by *fields of points*. But because of the peculiar nature of our geometric perception, or of our habit, the latter are essentially easier to grasp than are the former.

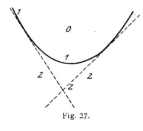

Fig. 27.

I shall show at once, by means of the *example of the quadratic equation*, what can be done in this direction (see Fig. 27). From all points outside of the parabola two tangents can be drawn to the curve; from points within, none. *Hence these two regions represent the manifolds of all equations with two roots and with no roots, respectively*. For all *points of the parabola itself* there is only a single tangent, which can be counted twice. *The normal curve itself is, then, in the general case, the locus of those points whose coordinates λ, μ yield equations with two equal roots*, so that we may call it the *discriminant curve*.

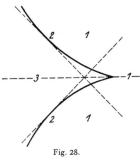

Fig. 28.

In the case of the *cubic equation*, we see that from a point inside the angle of the normal curve one can draw *three* tangents to the curve. This is obvious for points on the median line, because of symmetry; and the number cannot change when the point varies, provided it does not cross the curve. If the point (x, y) moves to the curve, two of the tangents coincide; if it moves into the region outside the curve, both of these tangents become imaginary and there remains but one real tangent. *Accordingly, the region inside the angle of the normal curve represents the totality of cubic equation with three different real roots; the region outside, equations with only one real root; while to the points on the curve itself correspond the equations with one simple and one double real root.* Finally, a triple tangent goes through the *cusp*, corresponding to the *single equation $t^3 = 0$, with a single triple root*. Figure 28 makes this obvious at a glance.

The pictures become much more interesting and show more, if, as is customary in algebra, we *impose definite restrictions upon the roots*, in particular, if we *inquire about all the real roots lying within a given interval* $t_1 \leq t \leq t_2$. As you know, the general answer to this question is furnished by *Sturm's theorem*. We can, however, easily complete our drawing so that it will give a satisfying graphical solution of this general question also. For this purpose we simply *add to the normal curve the tangents to it determined by the parameter values* t_1, t_2 and consider the division of the plane into fields which these tangents bring about.

To carry through these considerations for the quadratic equation, we must *determine the number of tangents which touch the parabolic arc between* t_1 *and* t_2. Through every point of the triangle (see Fig. 29) bounded by the parabolic arc and these two tangents there are obviously two tangents. If the point crosses either of the tangents t_1, t_2, one of the tangents through it will touch the parabola beyond the arc (t_1, t_2), and so will be lost for our purpose. Tangents from points which lie within the two crescent shaped areas bounded by the parabola and the tangents t_1, t_2 touch the parabola outside

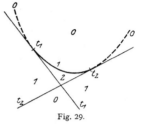

Fig. 29.

the arc $(t_1 t_2)$; and from points within the parabola there are no real tangents at all. The two parabolic arcs $t \leq t_1$ and $t \geq t_2$ are thus of no significance in effecting the desired subdivision of the plane. There remain, then, only those lines in the figure which are drawn full; these, together with the numbers assigned to them, *give at a glance exact information as to the manifolds of quadratic equations which have* 2, 1, *or* 0 *real roots between* t_1 *and* t_2.

We may proceed similarly with the *cubic equation* (see Fig. 30). Let us take, say, $t_1 > 0$, $t_2 < 0$. Again we draw the tangents with these parameter values and examine the subdivisions of the plane brought about by them and the arc of the normal curve which lies

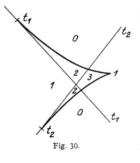

Fig. 30.

between t_1 and t_2. Through every point in the four-cornered region at the cusp there will be three real tangents which touch between t_1 and t_2. If point crosses either of the tangents t_1, t_2, there is a loss of one tangent of this character. When it crosses the normal curve two are lost. From these considerations we obtain the picture, shown in Fig. 30, of the *regions of the plane which correspond to equations with three, two, one, or no roots lying between* t_1 *and* t_2. In order to see the great usefulness of the *graphical method*, one need only make a single attempt to picture abstractly this classification of cubic equations, without making any appeal whatever to space perception; it will require a disproportionately great amount of time. And the proof, which here becomes evident by a glance at the picture, will not be at all easy.

Now as to the *relation of this geometric method to the well known algebraic criteria of Sturm, Cartesius, and Budan-Fourier* I remark, merely, that the geometric method includes them all, for equations of the types which we have considered. You will find these relations carried out more fully in my article[1] *"Geometrisches zur Abzählung der Wurzeln algebraischer Gleichungen"* and in *W. Dycks "Katalog mathematischer Modelle"*[2]. I am glad to take this occasion to refer you to this catalog. It was published on the occasion of the exposition, in Munich, in 1893, by the German Mathematical Society, and remains today the best means of orientation in the field of mathematical models.

3. Equations with three parameters λ, μ, ν

Finally, I shall also show you that one can apply analogous considerations to equations with three parameters. We shall need to use *space of three dimensions instead of the plane*. It will suffice if I consider the *special equation of four terms*

$$t^p + \lambda t^m + \mu t^n + \nu = 0.$$

The method of procedure can be applied immediately to equations of other forms.

In addition to this equation, we shall use the condition, from space geometry, that a point (x, y, z) and a plane with the plane coordinates (u, v, w) shall be "in united position", i.e., that the plane (u, v, w) shall contain the point (x, y, z). This condition is

(2) $$z + ux + vy + w = 0$$

or

(3) $$w + xu + yv + z = 0.$$

We now identify this equation, written in the one form or the other, with (1) and we obtain, exactly as before, two mutually dual interpretations.

Let us then set

(2a) $$z = t^p, \quad x = t^m, \quad y = t^n.$$

These equations determine a certain *space curve*, the *normal curve of the four-term equation* (1), *together with a scale of the values t*. Then we

---

[1] Reprinted in Klein, F., Gesammelte Mathematische Abhandlungen, vol. II, pp. 198—208.]

[2] A catalogue of mathematical and mathematical-physical models, apparatus, and instruments (Munich, 1892), also a supplement to this (Munich, 1893).

consider the plane which is determined by the coefficients λ, μ, ν, of (1):

(2b) $\qquad u = \lambda, \qquad v = \mu, \qquad w = \nu.$

Then equation (1) says that the *real roots of the proposed equation are identical with the parameter values t of the real intersections of the normal curve* (2a) *with the plane* (2b).

If we choose the method dual to the preceding, we must put

(3a) $\quad w = t^p, \quad u = t^m, \quad v = t^n.$

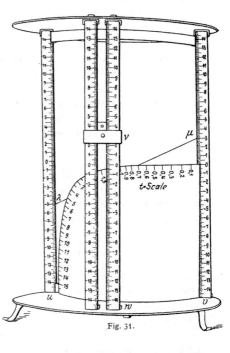

Fig. 31.

These equations represent, for variable t, a *simple infinity of planes*, which we can look upon as the *osculating planes of a definite space curve associated*, as before, with a scale of parameter values t. This will be a normal "*class curve*", being expressed in plane coordinates, in distinction from the previous normal "order curve", which was given in point coordinates. If we now consider, in conjunction with the first curve, also the point

(3b) $\quad x = \lambda, \quad y = \mu, \quad z = \nu,$

it follows that *the real roots of* (1) *are identical with the parameter values t of those osculating planes of the normal class curve* (3a) *which pass through the point* (3b).

Let us next illustrate these two interpretations by *concrete examples*. We have, in our collection, models for both of them, which I shall now put before you.

The *first method* was used by R. *Mehmke*, in Stuttgart, in the *construction of an apparatus for the numerical solution of equations*. His model is a brass frame (see Fig. 31) in which you will notice three vertical rods carrying scales, and into which one can fit curved templates, or stencils, of the normal curves of equations of degree three, four, or five, (after these have been reduced to four terms). Note, however, that while our exposition presupposed the ordinary rectangular coordinate system, Mehmke has so *determined his coordinate system that the appropriate plane coordinates*, i. e., the coefficients u, v, w of the equation of the plane (2), *are precisely the intercepts which this plane makes on the scales of the three vertical rods* and which one can read off there. In order, now, to make possible the fixation of a definite plane $u = \lambda$, $v = \mu$, $w = \nu$, a peep-hole is provided on the w rod, which one sets at the reading v of that scale, while one joins by a stretched string the readings, of the scales on the u and v rods, respectively. The rays joining the peep-hole with this string make our plane, and *by looking through the peep-hole one can observe directly the intersections of this plane with the normal curve as the apparent intersections of the string with the template.* Their parameter values, *the desired roots of the equation, are read at the same time on the scale of the normal curve, which is affixed to the template.* The practical usableness of this apparatus depends, of course, upon the carefulness of its mechanical construction, but the limited power of accommodation of the human eye would, at best, make it very doubtful.

For the *second method,* a model was prepared by *Hartenstein* in connection with his state examination. It has to do with the so-called *reduced form of the equation of degree four,* that is,

$$t^4 + \lambda t^2 + \mu t + \nu = 0,$$

to which every biquadratic equation can be reduced. I shall present this method in a form somewhat different from the one I used for the two-parameter equation (p. 91). In the present case we have to consider a simple infinity of planes whose plane coordinates are given in (3 a) and whose point equations would be written as follows:

(4) $\qquad f(t) \equiv t^4 + xt^2 + yt + z = 0.$

The envelope of these planes is the system of the straight lines in which each plane $f(t) = 0$ meets the neighboring plane $f(t + dt) = 0$, i.e., the *developable surface whose equation is obtained by eliminating t between $f(t) = 0$ and $f'(t) = 0$.* But in order to obtain the *normal curve* we must seek the *osculating configuration of the system of planes,* i.e., the *locus of the points of intersection of three successive planes.* This locus is, as you know, the *cuspidal edge of that developable surface and its coordinates are found, as functions of t, from the three equations $f(t) = 0$, $f'(t) = 0$, $f''(t) = 0$.* In our case these three equations are:

$$t^4 + xt^2 + yt + z = 0$$
$$4t^3 + x \cdot 2t + y \quad = 0$$
$$12t^2 + x \cdot 2 \quad\quad = 0,$$

and one finds from them:

(5) $\qquad x = -6t^2, \quad y = 8t^3, \quad z = -3t^4.$

These expressions represent the *point equation of the normal class curve* of (4) whose *plane equation,* by (3 a), may be written in the form

(6) $\qquad w = t^4, \quad u = t^2, \quad v = t.$

Both forms are of degree four in t. Hence *the normal curve is both of order four and of class four.*

In order to study it more in detail, let us consider *a few simple surfaces* which pass through it. In the first place, the expressions (5) satisfy identically in t the equation

$$z + \frac{x^2}{12} = 0,$$

Hence our normal curve lies upon a *parabolic cylinder of order two whose generators are parallel to the y-axis*. Likewise, we have the relation

$$\frac{y^2}{8} + \frac{x^3}{27} = 0,$$

so that this *cubic cylinder, whose generators are parallel to the z-axis,* also goes through our normal curve. *Moreover, the normal curve is the finite intersection of these two cylinders.* With these facts in mind, one can form an approximate picture of the course of the normal curve. *Is is a skew curve, symmetric to the x z plane, having a cusp at the origin* (see Fig. 32).

Again the *quadric surface*

$$\frac{x \cdot z}{6} - \frac{3y^2}{64} = 0$$

goes through our normal curve; for, by (5), this equation is also satisfied identically in t. From

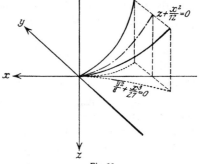

Fig. 32.

it, and the equation of the cubic cylinder, we find another linear combination which represents an especially important *surface of the third degree* passing through the normal curve:

$$\frac{xz}{6} - \frac{y^2}{16} - \frac{x^3}{216} = 0.$$

Let us now consider the *developable surface* whose cuspidal edge is the normal curve, and which we can define as the *totality of the tangents* to the normal curve. The tangent at the point t to any space curve

$$x = \varphi(t), \qquad y = \psi(t), \qquad z = \chi(t)$$

is given by the equations

$$x = \varphi(t) + \varrho \varphi'(t), \qquad y = \psi(t) + \varrho \cdot \psi'(t), \qquad z = \chi(t) + \varrho \chi'(t).$$

in which ϱ is a parameter. For the direction cosines of the tangents to the curve are to each other as the derivatives of the coordinates with respect to t. If t is thought of as variable, we have in these equations, with two parameters t, ϱ, the representation of the developable surface.

All this follows from well known theorems of space geometry. For our curve (5) we get, in particular, the following *equations* for the *developable surface*. If we call the coordinates of its points (X, Y, Z) to distinguish them from the coordinates of the curve, the equations of the developable are

(7) $$\begin{cases} X = -6(t^2 + 2\varrho t) \\ Y = 8(t^3 + 3\varrho t^2) \\ Z = -3(t^4 + 4\varrho t^3). \end{cases}$$

Now this surface is the basis of the Hartenstein model, its straight lines being represented by stretched threads (see Fig. 33).

The parameter representation offers the best starting point for the discussion and the actual construction of the surface. Indeed, it is only from force of habit that we inquire about the *equation of the surface* itself. We can obtain it by eliminating ϱ and t from (7). I shall give you the simplest procedure for this without giving the details of the inner meaning of the several steps. From (7) we form the combination

$$Z + \frac{X^2}{12} = 12\varrho^2 t^2,$$

$$\frac{X \cdot Z}{6} - \frac{Y^2}{16} - \frac{X^3}{216} = 8\varrho^3 t^3,$$

both of which vanish on the curve itself (for $\varrho = 0$). If we equate these to zero, we obtain two of the surfaces mentioned above which pass through the curve. Eliminating the product ϱt from these equations, we find the *equation of the developable surface*

$$\left(Z + \frac{X^2}{12}\right)^3 - 27\left(\frac{X \cdot Z}{6} - \frac{Y^2}{16} - \frac{X^3}{216}\right)^2 = 0.$$

The surface is thus of order six; but it is composed of the plane at infinity and a surface of order five.

As to the *meaning of this formula*, I make the following remark for those who are acquainted with the subject. The expressions in the two parentheses are the *invariants of the biquadratic equation*

$$t^4 + Xt^2 + Yt + Z = 0,$$

with which we started. These play an important role in the theory of elliptic functions and they are designated there, in general, by g_2 and g_3. The left side of the equation of our surface, $\Delta = g_2^3 - 27 g_3^2$, is, as you know, the *discriminant of the biquadratic equation*, which indicates, by its vanishing, the presence of a repeated root. *Our developable surface is therefore the discriminant surface of the biquadratic equation*, i. e., the totality of the points for which it has a double root.

After these theoretical explanations, the construction of a thread model for our surface offers no essential difficulty. By means of the parameter equations (7) we may determine, say, the points in which those tangents which we wish to represent intersect certain fixed planes. We then stretch threads between these planes, which are made out of wood or cardboard. But it requires long trial and great skill to make the model really beautiful and usable, and to bring out the entire interesting course of the surface and of its cuspidal edge, as in the model before us. The sketch on page 99 (see Fig. 33) shows the surface with its straight lines; AOB is the cuspidal edge [see the figure p. 97[1]].

You notice on the model a *double curve* (COD) *along which two sheets of the surface intersect*. This curve is simply the following parabola of the XZ plane:

$$Y = 0, \quad Z - \frac{X^2}{4} = 0.$$

Only one half (CO) *of this parabola, namely that for* $X < 0$, appears, however, as the *intersection of real sheets*, while *the other half lies isolated* in space. This phenomenon is by no means surprising to those who are accustomed to illustrate the theory of algebraic surfaces by concrete geometric representations. It is a common thing, there, for *real branches of double curves* to appear both as *intersections of real sheets* and also in part *isolated*. In the latter case we regard them as *real intersections of imaginary sheets* of the surface. The corresponding phenomenon in the plane is more generally known. In that case, in addition to the ordinary double points of algebraic curves, which appear as intersections of real branches of the curve, there are also the apparently isolated double points, which may be regarded as the intersections of imaginary branches.

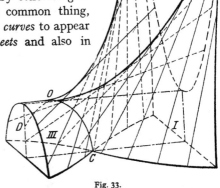

Fig. 33.

Let us now make clear in detail, what this surface with its cuspidal edge, the normal curve, can do for us. We think of the normal curve with its associated scale, or, better, we affix to each tangent its parameter value t, which also belongs to the point of tangency. If, now, someone gives us a biquadratic equation with definite coefficients (x, y, z), we need only to pass through the corresponding point (x, y, z) the osculating plane to the normal curve, or, what would be the same thing, the tangent

---

[1] The Hartenstein string model was put upon the market by the firm of M. Schilling in Leipzig. A dissertation by R. Hartenstein entitled: *Die Discriminantenfläche der Gleichung vierten Grades* goes with the model Leipzig, Schilling, 1909.

plane to the discriminant surface, to obtain the real roots as the parameter values of the points of contact with the curve, or the parameter values of the corresponding tangents, as the case may be. Since the osculating plane cuts the curve where it touches it, every point of contact of an osculating plane with the curve is projected from the point (x, y, z) as an apparent point of inflexion on the curve, and conversely. *Consequently, the real roots of the biquadratic equation are, finally, the parameter values t of the apparent inflexion points of the normal curve, viewed from the point (x, y, z) in space.*

Now it is, of course, quite difficult for the unpractised eye to determine with certainty from the model either the planes of osculation or the apparent inflexions of the curve. But the model exhibits with immediate clearness the next important thing, *the classification of all biquadratic equations according to the number of their real roots*. Let us see, by an abstract examination of equations, just what cases one might expect. If $\alpha, \beta, \gamma, \delta$ are the four roots of the real biquadratic equation (4), then $\alpha + \beta + \gamma + \delta = 0$, because of the vanishing of the coefficient of t^3. So far as the reality of the roots is concerned, the following principal cases are possible:

I. *Four real roots.*
II. *Two real, and two conjugate complex roots.*
III. *No real, and two pairs of conjugate complex roots.*

If, now, two equations of the type I are proposed, with roots $\alpha, \beta, \gamma, \delta$ and $\alpha', \beta', \gamma', \delta'$, respectively, then one certainly could transform $\alpha, \beta, \gamma, \delta$ continuously into $\alpha', \beta', \gamma', \delta'$, respectively, through systems of values whose sum is always zero. At the same time, the one equation would transform continuously into the other, through equations always of the same type, i.e., all equations of type I make up a connected continuum, and the same is true for the other two types. *Our model must therefore exhibit space partitioned into three connected parts such that the points in each part correspond to equations of one type.*

Let us now consider the *transition cases between these three sorts*. Type I goes over into II through equations which have *two different real roots and one double (i.e. two coincident) real root*, which we shall indicate symbolically by $2 + (2)$; similarly we have between II and III the transition case of *one real double root and two complex roots*, which may be indicated by (2). *To both of these sorts there must correspond, in our model, regions of the discriminant surface*, which, indeed, pictures all equations with coincident roots. Considerations similar to those above would show *that to each type there must correspond a connected region of this surface*. Now, again, these two groups, $2 + (2)$ and (2), go over into each other by means of cases with *two real double roots*, symbolically: $(2) + (2)$; the points for which *two* pairs of roots move thus into coincidence must belong simultaneously to *two* sheets of the discriminant surface, that is, to the *non isolated branch of the double curve*. *Accordingly, the discriminant surface falls into two parts, separated by a branch of the double curve;* one of these parts, $2 + (2)$, separates the space regions I and II, the other, (2), the space regions II and III. In order to see, now, how the normal curve lies, we notice that, because of its property as a cuspidal edge, *three tangent planes must merge into one (the osculating plane)* at each point on it, so that we have the case of a triple and a simple real root: $1 + (3)$. This can happen only when one of the simple roots becomes equal to the double root. Consequently, *the cuspidal edge must lie entirely on the first part*, $2 + (2)$, *of the surface*. In the *cusp* of the cuspidal edge ($x = y = z = 0$) we have a quadruple real root, which can arise from the case $(2) + (2)$ through the coincidence of the two double roots. *In fact, the cusp, O, of the cuspidal edge lies also on the double curve.* Finally, *as to the isolated branch of the double curve*, it lies entirely in the space region III and is characterized by the fact that on it the *two pairs of conjugate complex roots merge into one complex double root*. Both double roots are, of course, conjugate to each other.

You can recognize on our model all of the possible cases enumerated above. In the sketch (Fig. 33, p. 99), the interior of the surface to the right of the double curve is region I, to the left, region III; the exterior is region II. You will be able easily to become fully oriented by means of the following tabulation, which exhibits the number and the multiplicity of the real roots which correspond to the points of the several space, surface, and curvilinear regions. In this scheme, the digits not in parentheses denote the number of simple real roots, the others, as before, denote the multiplicity of repeated roots:

	I.	II.	III.
Region:	4	2	0
Discrim. surface:	$2 + (2)$	(2)	
Normal curve:	$1 + (3)$		
Double curve:		$(2) + (2)$	(2 imag. double roots).
Cusp:		(4)	

II. Equations in the field of complex quantities

We shall now remove the restriction to real quantities and shall operate in the field of complex quantities. Of course, we shall endeavor again only to emphasize those things which are susceptible of geometric representation to an extent greater than one finds elsewhere. Let us begin at once with the most important theorem of algebra.

A. The fundamental theorem of algebra

This is, as you know, the theorem *that every algebraic equation of degree n in the field of complex numbers has, in general, n roots, or, more* *accurately, that every polynomial $f(z)$, of degree n, can be separated into n linear factors.*

All proofs of this theorem make fundamental use of the *geometric interpretation* of the complex quantity $x + iy$ in the xy plane. I shall give you the *train of thought of Gauss' first proof (1799)*, which can be presented quite graphically. To be sure, the original exposition of Gauss was somewhat different from mine.

Given the polynomial

$$f(z) = z^n + a_1 z^{n-1} + \ldots + a_n,$$

we may write

$$f(x + iy) = u(x, y) + i \cdot v(x, y),$$

where u, v are real polynomials in the two real variables x, y. The leading thought of Gauss' proof lies now in considering the two curves

$$u(x, x) = 0 \quad \text{and} \quad v(x, y) = 0$$

in the xy plane, and in showing that they must have one point, at least, in common. For this point one would then have $f(x + iy) = 0$, that is, *the existence of a first "root" of the equation $f = 0$ would be proved.* For this purpose, it turns out to be sufficient, to investigate the *behaviour of both curves at infinity*, i.e., at a distance from the origin which is arbitrarily great.

If r, the absolute value of z, is very large, we may neglect the lower powers of z in $f(z)$, in comparison with z^n. If we introduce polar coordinates r, φ into the xy plane, i.e., if we set

$$z = r(\cos \varphi + i \sin \varphi),$$

we have, by De Moivre's formula

$$z^n = r^n (\cos n\varphi + i \sin n\varphi).$$

This expression is approached asymptotically by $f(z)$, as z increases in absolute value. It follows at once that u and v approach, respectively, asymptotically the functions

$$r^n \cos n\varphi, \quad r^n \sin n\varphi:$$

Consequently the ultimate course of the curves $u = 0$, $v = 0$, at infinity, respectively, will be given approximately by the equations

$$\cos n\varphi = 0, \quad \sin n\varphi = 0.$$

Now the curve $\sin n\varphi = 0$ consists of the n straight lines which go through the origin and make with the x-axis the angles 0, π/n, $2\pi/n, \ldots, (n-1)\pi/n$, whereas $\cos n\varphi = 0$ consists of the n rays through the origin which bisect these angles (Fig. 34 is drawn for $n = 3$). In the central part of the figure, the true curves $u = 0$, $v = 0$ can, of course, be essentially different from these straight lines; but they must approach the straight lines asymptotically as the lines recede from the

The Fundamental Theorem of Algebra. 103

origin. We *can indicate their course schematically by retaining the straight lines outside of a large circle and replacing them by anything we please, inside the circle* (see Fig. 35). But no matter what the behavior of the

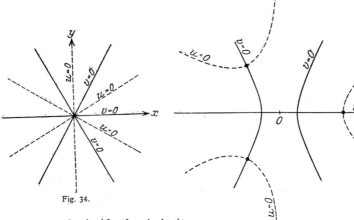

Fig. 34.

curves may be inside the circle, it is certain that, if one makes the circle about the origin sufficiently large, the branches u, v, outside the circle, must alternate, from which *it is graphically clear that these branches must cross one another inside the circle.* In fact, we can give a rigorous[1] proof of this assertion, — and this is the substance of Gauss' proof—if we use the continuity properties of the curves. The preceding argument, however, gives the essentials of the train of thought. If *one* such root has been found, we can divide out a linear factor, and we can then

Fig. 35.

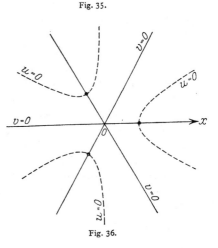

Fig. 36.

---

[1] It should be said here that Gauss does not dispense entirely with geometric considerations. The arithmetization of the proof which he contemplated in his dissertation was first given by A. Ostrowski (Göttinger Nachrichten, 1920, or vol. VIII of the materials for a scientific biography of Gauss, 1920). It is of historical interest that the first proof of the fundamental theorem was by D'Alembert. To be sure, there was an error in his proof, to which Gauss called attention. D'Alembert, namely, failed to distinguish between the upper limit of a function and its maximum, and he made use of the assumption, which in general is false, that a function of a complex variable actually attains its upper limit when this limit exists.

repeat the reasoning for the other polynomial factor of degree $(n-1)$. Continuing in this way, *we may finally break up $f(z)$ into n linear factors*, i.e., *we may prove the existence of n zeros*.

This method of reasoning will be much clearer if you *carry through the construction for special cases*. A simple example would be

$$f(z) = z^3 - 1 = 0.$$

In this case we obviously have

$$u = r^3 \cos 3\varphi - 1, \qquad v = r^3 \sin 3\varphi,$$

so that $v = 0$ consists simply of three straight lines, while $u = 0$ has three hyperbola-like branches. Figure 36 shows the three intersections of the two curves, which give the three roots of our equation. I recommend strongly that you work through other and more complicated examples.

These brief remarks about the fundamental theorem will suffice here, since I am not giving a course of lectures on algebra. Let me close by pointing out that the *significance of the admission of complex numbers into algebra* lies in the fact that it permits a general statement of the fundamental theorem. With the restriction to real quantities one can only say that the equation of degree n has n roots, or fewer, or perhaps none at all.

B. Equations with a complex parameter

The rest of the time which I have set aside for algebra I shall devote to the *discussion, by graphical methods, of all the roots (including the complex ones) of complex equations*, as was done earlier for the real roots of real equations. We shall limit ourselves, however, to equations with one complex parameter and we shall assume, furthermore, that this occurs *only linearly*. The study of a *simple conformal representation* will then give us all that is required.

Let $z = x + iy$ *be the unknown*, and $w = u + iv$ *the parameter*. Then the type of the equation to be considered has the form

(1) $$\varphi(z) - w \cdot \psi(z) = 0$$

where φ, ψ, are polynomials in z. Let n be the highest power of z that occurs. According to the fundamental theorem, this equation has for each definite value of w exactly n roots z which, in general, are different. Conversely, however, it follows from (1) that

(2) $$w = \frac{\varphi(z)}{\psi(z)},$$

i.e., *w is a single-valued rational function of z*, and it is said to be of *degree n*. If we should use, as geometric equivalent of equation (1), simply the conformal representation which this function sets up between the z-plane and the w-plane, the many-valuedness of z as function of w would be visually disturbing. We may help ourselves here, as is always the case in function theory, by *thinking of the w-plane as consisting of n sheets, one over another, which are united in an appropriate manner, by means of branch cuts, into an n leaved Riemann surface.* Such surfaces are familiar to you all from the theory of algebraic functions. *Then our function establishes, between the points of the n-leaved Riemann's surface in the w-plane and the points of the simple z-plane, a one-to-one relation which is, in general, conformal.*

Before we begin a detailed study of this representation, it will be helpful if we set up certain conventions which will do away with the exceptional rôle played by infinite values of w and z, a role not justified by the nature of the case, and which will enable us to state theorems in general form. Inasmuch as these conventions are not so widely employed as they should be, you will permit me to say a word or two more about them than I otherwise should. We cannot be satisfied here when one speaks merely symbolically of an *infinitely distant point of the complex plane*, since such a conception gives no adequate concrete image, so that one must have recourse to special considerations or stipulations, in order to find out what corresponds, for an infinitely distant point, to a definite property of a finite point. *But we can secure all that is desired, if we replace the Gaussian plane, as picture of the complex numbers, once for all, by the Riemannian sphere.* For this purpose, we think simply of a sphere of diameter one, tangent to the xy plane, its south pole S being at the origin, and from its north pole N we project the plane stereographically upon the sphere (see Fig. 37). To every point $Q = (x, y)$ of the plane there corresponds uniquely the second intersection P of the ray NQ

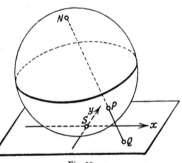

Fig. 37.

with the sphere; and, conversely, to every point P of the sphere, with the exception of N itself, there corresponds a unique point Q with definite coordinate (x, y). *Hence we can consider P as representing the number $x + iy$.* Now if P approaches the north pole N, in any manner, Q moves to infinity; conversely, if Q recedes to infinity in any manner, the corresponding point P approaches the single definite point N. It seems natural, then, to look upon this point N, which does not correspond to any finite complex number, *as the unique representative of all infinitely large $x + iy$, i.e., as the concrete picture of the infinitely* *distant point of the plane, which is otherwise introduced only symbolically, and to affix to it outright the mark* ∞. In this way we bring about, in the geometric picture, *complete equality between all finite points and the infinitely distant point*.

In order to return now to the geometric interpretation of the algebraic relation (1), we shall *replace the w plane also by a w-sphere*. Then our function will be represented by a *mapping of the z-sphere upon the w-sphere*, and, just as in the case of the mapping of the two planes, this is also *conformal*, since the stereographic mapping of the plane upon the sphere is, according to a well known theorem, conformal. To a single position on the w-sphere, there will then correspond, in general, n different positions on the z-sphere. In order to get a *one-to-one* relation we imagine, again, n sheets on the w-sphere, lying one above another, and united, in appropriate manner, by means of branch cuts, so as to form *an n-leaved Riemann surface over the w-sphere*. This picture presents no greater difficulty that that of the Riemann surface over the plane. *Thus, finally, the algebraic equation* (1) *is interpreted as a one-to-one relation, conformal in general, between the Riemann surface over the w-sphere and the simple surface of the z-sphere*. This interpretation obviously takes into account, also, infinite values of z and w which may correspond to each other or to finite values.

In order to make the greatest possible use this geometric device, we must take *a corresponding step in algebra*, one which shall do away with the exceptional role which infinity plays in the formulas, and this step is the *introduction of homogeneous coordinates*. We set, namely,

$$z = \frac{z_1}{z_2}$$

and consider z_1, z_2 as *two independent complex variables, both of which remain finite, and which cannot both vanish simultaneously*. Each definite value of z will then be given by infinitely many systems of values (cz_1, cz_2), where c is an arbitrary constant factor. We shall look upon all such systems of values (cz_1, cz_2) which differ only by such a factor, as the same *"position" in the field of the two homogeneous variables*. Conversely, for every such position there will be a definite value of z, with one exception: to the position (z_1 arbitrary, $z_2 = 0$) there will correspond no finite z; but if one approaches it from other positions, the corresponding z becomes infinite. *This one position is thus to be looked upon as the arithmetic equivalent of the one infinitely distant point of the z-plane or, as the case may be, of the z-sphere, and as carrying the mark* $z = \infty$.

In the same way, of course, we put also $w = w_1/w_2$. We shall now set up the *"homogeneous"* equation between the *"homogeneous"* variables z_1,

z_2 and w_1, w_2, which corresponds to equation (2). Multiplying by z_2^n in order to clear of fractions, we may write the equation in the form

(3) $$\frac{w_1}{w_2} = \frac{z_2^n \varphi\left(\frac{z_1}{z_2}\right)}{z_2^n \psi\left(\frac{z_1}{z_2}\right)} = \frac{\overline{\varphi}(z_1, z_2)}{\overline{\psi}(z_1, z_2)}.$$

In this equation, $\overline{\varphi}(z_1, z_2)$ and $\overline{\psi}(z_1, z_2)$ *are rational integral functions of z_1 and z_2*, since $\varphi(z)$ and $\psi(z)$ contain at most the nth power of $z = z_1/z_2$. Moreover they are *homogeneous polynomials (forms) of dimension n*. For each term z^i of $\varphi(z)$ or $\psi(z)$ is transformed into the term $z_2^n (z_1/z_2)^i = z_2^{n-i} z_1^i$, of dimension n, by clearing of fractions.

We come now to the detailed study of the functional dependence which our equation (1) *or, as the case may be,* (3) *establishes between z and w.* We shall apply consistently our two new aids, mapping upon the complex sphere and homogeneous coordinates. We shall have solved this problem when we can form a complete picture of the conformal relation between the z-sphere and the Riemann surface over the w-sphere.

First of all we must inquire as to the *nature and the position of the branch points of the Riemann surface*. I remind you here that a μ-fold branch point is one in which $\mu + 1$ leaves are connected. Since w is a single-valued function of z, we know the branch points when we know the points of the z sphere which correspond to them, which I am in the habit of calling the *critical or noteworthy points of the z-sphere*. To each of these there corresponds a certain *multiplicity* equal to that of the corresponding branch point. I shall now give, without detailed proof, the theorems which make possible the determination of these points. I assume that the rather simple functiontheoretic facts which enter into consideration here are in general familiar to you, though they may not be in the homogeneous form which I prefer to use. I shall illustrate in concrete graphical form the abstract considerations which I shall present to you, in this connection, by a series of examples.

A little calculation is necessary in order to obtain the *analogue, in homogeneous coordinates,* of the differential coefficient dw/dz. Differentiating equation (3) and omitting the bars over φ and ψ, we obtain

(3') $$\frac{w_2 dw_1 - w_1 dw_2}{w_2^2} = \frac{\psi d\varphi - \varphi d\psi}{\psi^2}.$$

We have also
$$d\varphi = \varphi_1 dz_1 + \varphi_2 dz_2,$$
$$d\psi = \psi_1 dz_1 + \psi_2 dz_2,$$
where
$$\varphi_1 = \frac{\partial \varphi(z_1, z_2)}{\partial z_1}, \qquad \varphi_2 = \frac{\partial \varphi(z_1, z_2)}{\partial z_2},$$
$$\psi_1 = \frac{\partial \psi(z_1, z_2)}{\partial z_1}, \qquad \psi_2 = \frac{\partial \psi(z_1, z_2)}{\partial z_2}.$$

On the other hand, from Euler's theorem for homogeneous functions of degree n, we have

$$\varphi_1 \cdot z_1 + \varphi_2 \cdot z_2 = n \cdot \varphi$$
$$\psi_1 \cdot z_1 + \psi_2 \cdot z_2 = n \cdot \psi ;$$

consequently the numerator on the right side of (3') may be written in the form

$$\psi d\varphi - \varphi d\psi = \begin{vmatrix} d\varphi, & d\psi \\ \varphi, & \psi \end{vmatrix} = \frac{1}{n} \begin{vmatrix} \varphi_1 dz_1 + \varphi_2 dz_2, & \psi_1 dz_1 + \psi_2 dz_2 \\ \varphi_1 z_1 + \varphi_2 z_2, & \psi_1 z_1 + \psi_2 z_2 \end{vmatrix}.$$

This expression, by the multiplication theorem for determinants, becomes

$$= \frac{1}{n} \begin{vmatrix} \varphi_1, & \varphi_2 \\ \psi_1, & \psi_2 \end{vmatrix} \cdot \begin{vmatrix} dz_1, & dz_2 \\ z_1, & z_2 \end{vmatrix}.$$

Thus (3') goes over into the equation

$$\frac{w_2 dw_1 - w_1 dw_2}{w_2^2} = \frac{z_2 dz_1 - z_1 dz_2}{n \cdot \psi^2} (\varphi_1 \psi_2 - \psi_1 \varphi_2).$$

This constitutes the *basal formula of the homogeneous theory of our equation*, and the *functional determinant* $\varphi_1 \psi_2 - \varphi_2 \psi_1$, *of the forms* φ, ψ appears as a crucial expression for all that follows. Except for it and for the factor $z_2^2/(n\psi^2)$, one has on the right the differential of $z = z_1/z_2$, on the left that of $w = w_1/w_2$. Since for finite z and w the critical points are given by $dw/dz = 0$, as is well known, the following theorem appears *plausible*, but I shall here omit the proof. *Each μ-fold zero of the functional determinant is a critical point of multiplicity μ*, i.e., there corresponds to it a μ-fold branch point of the Riemann surface over the w-sphere. The chief advantage of this rule, as compared with those which are otherwise given, lies in the fact that it contains in one statement both finite and infinite values of z and w. It enables us also to make a precise statement concerning the *number of remarkable points*. The four derivatives, namely, are forms of dimension $n - 1$, and the functional determinant is therefore a form of dimension $2n - 2$. Such a polynomial always has $2n - 2$ zeros, if one takes into account their multiplicity. Thus, if $\alpha_1, \alpha_2, \ldots, \alpha_\nu$ *are the remarkable points of the z-sphere* (i.e., if $\varphi_1 \psi_2 - \varphi_2 \psi_1 = 0$ *for* $z_1 : z_2 = \alpha_1, \alpha_2, \ldots, \alpha_\nu$) *and if* $\mu_1, \mu_2, \ldots \mu_\nu$, *are their respective multiplicities, then their sum is*

$$\mu_1 + \mu_2 + \cdots + \mu_\nu = 2n - 2.$$

By virtue of the conformal mapping, to these points there correspond the ν branch points

$$a_1, a_2, \ldots, a_\nu$$

on the Riemann surface over the w-sphere, which must necessarily lie separated on the surface, and about which $\mu_1 + 1, \mu_2 + 1, \ldots, \mu_\nu + 1$ leaves, respectively, must be cyclically connected. It should be noted, however, that different ones of these branch points may lie over the same position on the w sphere, since $w = \varphi(z)/\psi(z)$ for $z = \alpha_1, \alpha_2, \ldots, \alpha_\nu$ may give the same value for w more than once. Over such a point, there would be two or more separate series of leaves, each series being in itself connected. Every such position on the w sphere is called a *branch position*; we shall denote them, in order, by $A, B, C, \ldots$. It should be noted that their number can be smaller than ν.

The statements thus far made furnish only a hazy picture of the Riemann surface. We shall now *build it up so that it can be more readily visualized. For this purpose, let us draw on the w sphere through the branch positions $A, B, C, \ldots$ an arbitrary closed curve $\mathfrak{C}$ without double points and of the simplest possible form* (see Fig. 38), *and distinguish the two spherical caps thus formed as the upper cap and the lower cap.* In all of the examples which we shall discuss later the points $A, B, C, \ldots$ will all be real and we shall then naturally select *as the curve $\mathfrak{C}$ the meridian great circle of real numbers*, so that each of our two partial regions will be a hemisphere.

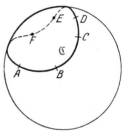

Fig. 38.

Returning to the general case we see that each pair of leaves of the Riemann surface which are connected, intersect along a *branch cut* which joins two branch points. As you know, the Riemann surface remains unchanged in essence if we move these cuts, leaving the end points fixed, that is, if we think of the same leaves as being connected along other curves, provided these join the same branch points. It is in just this variability that the great generality and also the great difficulty of the idea of the Riemann surface lies. In order to give the surface a definite form, which shall be susceptible of concrete visualization, *we move all the branch cuts so that all of them lie over the curve $\mathfrak{C}$, which passes through all the branch points*. It may be that several branch cuts lie over the same part of $\mathfrak{C}$, and none at all over other parts.

Now let us cut this entire complex of leaves, i.e., each individual leaf, along the curve $\mathfrak{C}$. Since we had already moved all the branch cuts into position over $\mathfrak{C}$, the incision just made passes along all of them, so that *our Riemann surface separates into $2n$ "half-leaves" entirely free from branches, n of them over each of the two spherical caps.* If we think of the half-leaves corresponding to the upper cap as being shaded, and those corresponding to the lower as not shaded, we can distinguish briefly, n shaded and n unshaded half-leaves. We can now describe *the original Riemann surface as follows.* On it *each shaded half-leaf meets only unshaded half-leaves, those with which it is connected along segments of the curve $\mathfrak{C}$ lying over $AB, BC, \ldots$; and, similarly, each unshaded half-leaf* is connected along such segments of $\mathfrak{C}$ only to shaded half-leaves. However, more than two half-leaves may meet only at a branch point; and in fact around any μ-fold branch point, $\mu + 1$ shaded half-leaves would alternate with $\mu + 1$ unshades ones.

Since the mapping by means of our function $w(z)$ of the z sphere upon the Riemann surface over the w sphere is a one-to-one correspondence, we can immediately transfer to the z sphere the above conditions of connectivity. Because of continuity, the $2n$ half-leaves of the Riemann surface must correspond to $2n$ connected z regions, which we may call the shaded and the unshaded half-regions. These will be separated from one another by the n images of each of the segments $AB, BC, \ldots$ of the curve $\mathfrak{C}$ which the n-valued function $z(w)$ represents upon the z sphere. *Each shaded half-region meets only shaded half-regions along these image-curves, and each unshaded half-region meets only shaded ones. It is only in a μ-fold critical point that more than two half-regions can meet. At such a point $\mu + 1$ shaded and $\mu + 1$ unshaded half-regions come together.*

This division of the z sphere into partial regions will help us to follow in detail the course of the function $z(w)$ for a few simple characteristic examples. I shall begin with the simplest one possible.

1. The "pure" equation

We shall call the well known equation

(1) $$z^n = w$$

a pure equation. Its solution is given formally by introducing the radical $z = \sqrt[n]{w}$. This gives us no information, however, regarding the functional relation between z and w. We shall proceed according to the general plan by introducing the homogeneous variables

$$\frac{w_1}{w_2} = \frac{z_1^n}{z_2^n},$$

and we shall consider the functional determinant of the numerator and denominator of the right side

$$\begin{vmatrix} n z_1^{n-1}, & 0 \\ 0, & n z_2^{n-1} \end{vmatrix} = n^2 z_1^{n-1} \cdot z_2^{n-1}.$$

This expression obviously has the $(n - 1)$ *fold zeros* $z_1 = 0$ and $z_2 = 0$, or (in non-homogeneous form) $z = 0$ and $z = \infty$. These are the only critical points and they are of total multiplicity $2n - 2$. By our general theorem, therefore, *the only branch points of the Riemann surface over the w sphere* are at the positions $w = 0$ and $w = \infty$. By the equation $w = z^n$ these correspond to the two points $z = 0$ and $z = \infty$. Each of these two points has the *multiplicity* $n - 1$, so that n leaves are cyclically connected at each of them. Let us now mark on the w sphere *the meridian of real numbers as the curve* ℭ and let us cut all the leaves of the Riemann surface along this meridian, after having appropriately displaced all of the branch cuts. Of the $2n$ hemispheres into which the surface separates we think of those over the *rear* half of the w sphere, that is, those which correspond to w values with *positive imaginary parts*, as shaded. Upon the meridian itself, we shall distinguish between the *half meridian of positive real numbers* (drawn full in Fig. 39) and that of the *negative real numbers (dotted)*.

Fig. 39.

Now we must examine the *mappings of this meridian* ℭ *curve upon the z sphere*, where they bring about the characteristic division into half-regions. Upon the positive half meridian $w = r$, where r ranges through positive real values from 0 to ∞; for these values we have by a well known formula of complex numbers,

$$z = \sqrt[n]{w} = \left|\sqrt[n]{r}\right|\left(\cos\frac{2k\pi}{n} + i\sin\frac{2k\pi}{n}\right), \quad \text{where} \quad k = 0, 1, \ldots, n - 1.$$

For the different values of k, this expression gives those n *half-meridians of the z sphere which make with the half-meridian of positive real numbers the angles* $0, 2\pi/n, 4\pi/n, \ldots, 2(n-1)\pi/n$. Thus these curves corres-

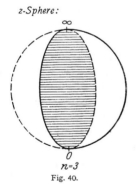

Fig. 40.

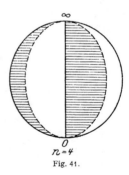

Fig. 41.

pond to the full drawn half of ℭ. On the negative half-meridian of the w sphere we must set $w = -r = r \cdot e^{i\pi}$, where again $0 \leq r \leq \infty$. This gives

$$z = \sqrt[n]{w} = \left|\sqrt[n]{r}\right|\left(\cos\frac{(2k+1)\pi}{n} + i\sin\frac{(2k+1)\pi}{n}\right), \text{where } k = 0, 1, \ldots, n-1.$$

Corresponding to this we have, on the z sphere, those n *half-meridians which have the "longitude"* $\pi/n, 3\pi/n, \ldots, 2(n-1)\pi/n,$ *which thus bisect the angles between the others.* Accordingly, the z sphere is divided into $2n$ congruent *sectors reaching from the north pole to the south pole*, similar to the natural divisions of an orange. This division is exactly in accord with the general theory. In particular, it is only at the remarkable points, the two poles, that more than two half-regions meet. At each of these points $2n$ half-regions meet, corresponding to the multiplicity $n-1$.

As for the *shading of the regions*, we need to fix it for *one* region only. The remainder are then alternately shaded and unshaded. Now note that when we look at the shaded half of the w sphere (the rear) from the point $w = 0$, the full drawn part of the boundary lies to the left, the dotted part to the right. Since we are concerned with a conformal mapping in which angles are not reversed, *each shaded portion of the z sphere, looked at from the correponding point $z = 0$, must have the same property as to position, that is, it must have a full drawn boundary to the left, and a dotted one to the right*. With this we control completely the division of the z sphere into regions. Moreover, one notices a characteristic difference in the distribution of the regions upon two z hemispheres, according as n is even or odd, as can be clearly seen in Figs. 40 and 41 on p. 111 for the first cases $n = 3$, $n = 4$. Let me emphasize how necessary it was to go over to the complex *sphere* in order to get a full understanding of the situation. In the complex z *plane*, one would have a division into angular sectors by straight lines radiating from $z = 0$, and it would not be at all so obvious that $z = \infty$ and $w = \infty$ have equal significance with $z = 0$ and $w = 0$, as critical point and branch point, respectively.

This furnished us with the essentials for exact knowledge of the functional relation between z and w. We need now study only the *conformal mapping of each of the $2n$ spherical sectors upon one or the other of the two w hemispheres*. But I shall not go into the details here. This case, as one of the simplest and most obvious illustrations, will be familiar ground to any one who has had to do with conformal representation. We shall see later (see p. 131) how to deduce from this methods for the numerical calculation of z.

Let us, however, settle here the important question as to the mutual relation *among the various congruent regions of the z sphere*. Speaking more exactly, $w = z^n$ takes on the same value at a point in each one of the n shaded regions. Can the corresponding values of z be expressed in terms of one another? We notice, in fact, that for $z' = z \cdot \varepsilon$ (where ε is any one of the nth roots of unity) $z'^n = z^n$, that is $w = z^n$ *takes the same value at all the n positions*

(2) $$z' = \varepsilon^\nu \cdot z = e^{\frac{2\nu i \pi}{n}} \cdot z \quad (\nu = 0, 1, 2, \ldots, n-1).$$

These n values of z' must therefore be distributed so that just one of them lies in each of the n shaded regions of the z sphere, if z is taken in *one* of the shaded regions and each of them must traverse one of these regions as z traverses its region. The same thing is true of the unshaded regions. Each of the substitutions (2) is represented geometrically by *a rotation of the z sphere through an angle $v \cdot 2\pi/n$ about the vertical axis $0, \infty$*, since, as is well known, multiplication in the complex plane by $e^{2vi\pi/n}$ denotes a rotation through that angle about the origin. Thus *corresponding points of our spherical regions, as well as the regions themselves, go over into one another by means of these n rotations about the vertical axis.*

If, then, we had determined at the start only one shaded partial region of the sphere, this remark would have furnished *all the similar partial regions*. In this we have made use only of the property of the substitutions (2) *that they transform equation* (1) *into itself* (*i.e.*, $z^n = w$ *into* $z'^n = w$) *and that their number is equal to the degree*. In the examples that follow, we shall always be able to give such linear substitutions at the outset, and by means of them to simplify the determination of the division into subregions.

By using the present example I should like to illustrate *an important general notion*, namely, the *notion of irreducibility for equations which contain a parameter w rationally*. We have already discussed irreducibility of equations with rational *numerical* coefficients in connection with the construction of the regular heptagon (p. 51 et seq.). *An equation $f(z, w) = 0$* (e.g., our equation $z^n - w = 0$), *where $f(z, w)$ is a polynomial in z, whose coefficients are rational functions of w, is called reducible with respect to the parameter w, when f can be split into the product of two polynomials of the same sort, in each of which z really appears*

$$f(z, w) = f_1(z, w) \cdot f_2(z, w);$$

otherwise the equation is called irreducible with respect to w. The entire generalization, in comparison with the earlier conception, lies in the fact that the "*domain of rationality*" in which we operate and in which the coefficients of the admissible polynomials are to lie, consists of the *totality of rational functions of the parameter w instead of the totality of rational numbers*, in other words, that we pass from a numbertheoretic to a functiontheoretic conception.

If we illustrate this, for each equation $f(z, w) = 0$, by means of its Riemann surface, we can set up *a simple criterion for reducibility in this new sense*. If the equation, namely, is reducible, every system of the values z, w which satisfies it satisfies either $f_1(z, w) = 0$ or $f_2(z, w) = 0$; now the solutions of $f_1 = 0$ and $f_2 = 0$ are represented by means of their Riemann surfaces, which have nothing to do with each other, and, in particular, are not connected. Thus, *the Riemann surface which belongs to a reducible equation $f(z, w) = 0$ must break down into at least two separates pieces*.

According to this, we can now assert that the equation $z^n - w = 0$ *is certainly irreducible in the function theoretic sense*. For, on its Riemann surface, which we known exactly, all the n leaves are cyclically connected at each of its branch points. Moreover, the entire surface is mapped upon the unpartitioned z sphere. Hence such a breaking down cannot occur.

In connection with this, we can answer one of the popular problems of mathematics which we touched earlier (p. 51), *namely, that of the possibility of dividing an arbitrary angle φ into n equal parts, in particular, for $n = 3$, the possibility of trisecting an angle*. The problem is to give an exact construction *with ruler and compasses* for dividing into three equal parts *any angle φ whatever*. (It is easy, of course, to give a construction for a series of *special values of φ*). I shall give you the train of thought for the proof of the impossibility of trisecting an angle in the sense just mentioned, and I shall ask you to recall, in this connection, the proof of the impossibility of constructing the regular heptagon with ruler and compasses (see p. 51 et seq.). Just as at that time, we shall reduce the problem to that of the solution of an *irreducible cubic equation*, and we shall then show that this equation cannot be solved by a series of square roots; except that, now, the equation will contain *a parameter* (the angle φ), whereas, before, the coefficients were integers. Accordingly, *functiontheoretic irreducibility must replace numbertheoretic* irreducibility.

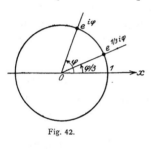

Fig. 42.

In order to set up the equation of the problem let us think of the angle φ as laid off from the positive real half-axis in the w plane (see Fig. 42). Then its free arm will cut the unit circle in the point

$$w = e^{i\varphi} = \cos\varphi + i\sin\varphi.$$

Our problem consists in finding, independently of special values of the parameter φ, a construction, involving a finite number of applications of the ruler and compasses, which shall give the point of intersection with the unit circle of the arm of the angle $\varphi/3$, i. e., the point

$$z = e^{\frac{i\varphi}{3}} = \cos\frac{\varphi}{3} + i\sin\frac{\varphi}{3}.$$

This value of z satisfies the equation:

(3) $$z^3 = \cos\varphi + i\sin\varphi,$$

and the analytic equivalent of our geometric problem consists in *solving this equation* (see p. 51) *by means of a finite number of square roots, one over another, of rational functions of $\sin\varphi$ and $\cos\varphi$*, since these quantities are the coordinates of the point w with which we start the construction

We must show, first, *that the equation* (3) *is irreducible in the function theoretic sense*. To be sure, this equation does not have just the form we assumed while explaining the notion, since, instead of the a *complex* parameter w that enters rationally, we have now two functions cos and sin of a *real* parameter φ, both of which appear rationally. As a natural extension here of our notion, we shall *call the polynomial $z^3 - (\cos\varphi + i\sin\varphi)$ reducible if it can be split into polynomials whose coefficients are likewise rational functions of* co φ *and* sin φ; and we can, as before, assign a criterion for this. If we let φ assume all real values in (3), $w = e^{i\varphi} = \cos\varphi + i\sin\varphi$ will describe the unit circle of the w plane, to which the equation of the w sphere corresponds by stereographic projection. The curve which lies over this, on the Riemann surface of the equation $z^3 = w$, and which describes, in one stroke, all three leaves, is mapped by equation (3) uniquely upon the unit circle of the z sphere. Hence it can be regarded, in a sense as its "*one dimensional Riemann image*". In the same way, we can obviously assign such a Riemann image to every equation of the form $f(z, \cos\varphi, \sin\varphi) = 0$ by taking as many copies of the unit circle with arc length φ as the equation has roots, and joining them according to the connectivity of the roots. It follows, just as before, *that the equation* (3) *can be reducible only when its one-dimensional Riemann image breaks down into separate parts*, and this is obviously not the case. *This proves the function theoretic irreducibility of our equation* (3).

Now, however, the former proof of the theorem, that a cubic equation with rational numerical coefficients is reducible if it can be solved by a series of square roots, can be applied literally to the present case of the function-theoretically irreducible equation (3) (see p. 51 et seq.). We need only to replace "rational numbers" there by "rational functions of cos φ and sin φ". *This proves our assertion that the trisection of an arbitrary angle cannot be accomplished by a finite number of applications of a ruler and compasses*. Hence the endeavors of angle-trisection zealots must always be fruitless!

I pass on now to the treatment of a somewhat more complicated example.

2. The dihedral equation

The equation

(1) $$w = \frac{1}{2}\left(z^n + \frac{1}{z^n}\right).$$

is called the dihedral equation, for reasons that will appear later. Clearing of fractions, we see that its degree is $2n$. Introducing homogeneous variables we get

$$\frac{w_1}{w_2} = \frac{z_1^{2n} + z_2^{2n}}{2 z_1^n \cdot z_2^n},$$

in which, in fact, forms of dimension $2n$ appear in numerator and denominator. The functional determinant of these forms is

$$\begin{vmatrix} 2n z_1^{2n-1}, & 2n z_2^{2n-1} \\ 2n z_1^{n-1} z_2^n, & 2n z_1^n z_2^{n-1} \end{vmatrix} = 4n^2 z_1^{n-1} z_2^{n-1} (z_1^{2n} - z_2^{2n}).$$

It has an $(n-1)$-fold zero at $z_1 = 0$ and at $z_2 = 0$; the other $2n$ zeros are given by

$$z_1^{2n} - z_2^{2n} = 0 \quad \text{or:} \quad \left(\frac{z_1}{z_2}\right)^n = \pm 1.$$

If in addition to the n-th root of unity

$$\varepsilon = e^{\frac{2i\pi}{n}}$$

which we have already used, we introduce also the primitive n-th root of -1:

$$\varepsilon' = e^{\frac{i\pi}{n}},$$

the last $2n$ zeros are given by the equations

$$\frac{z_1}{z_2} = \varepsilon^\nu \quad \text{and} \quad \frac{z_1}{z_2} = \varepsilon' \cdot \varepsilon^\nu, \qquad (\nu = 0, 1, \ldots, n-1).$$

Since the values of $z = z_1/z_2$ corresponding to them all have the absolute value one, they all lie therefore on the equator of the z sphere (corresponding to the unit circle of the z plane), at equal angular spacings of π/n. We have therefore *as critical points on the z sphere*:

(a) *the south pole $z = 0$ and the north pole $z = \infty$, each of multiplicity $n-1$*;

(b) *the $2n$ equatorial points $z = \varepsilon^\nu$, $\varepsilon' \cdot \varepsilon^\nu$, each of multiplicity one.*

The sum of all the multiplicities is $2 \cdot (n-1) + 2n \cdot 1 = 4n - 2$, as is demanded by the general theorem on p. 108 for the degree $2n$. By virtue of equation (1) there will correspond to the remarkable points $z = 0$, $z = \infty$ of the z sphere, the position $w = \infty$ on the w sphere. Moreover, to all the points $z = \varepsilon^\nu$, corresponds the position $w = +1$; and, to all the points $z = \varepsilon' \cdot \varepsilon^\nu$ the position $w = -1$. There are, accordingly, *only three branch points ∞, $+1$, -1 on the w sphere*. These will lie as follows:

$w = \infty$ two branch points of multiplicity $n - 1$;
$w = +1$ n branch points of multiplicity 1;
$w = -1$ n branch points of multiplicity 1.

The $2n$ leaves of the Riemann surface group themselves therefore over the point $w = \infty$ in two separate series, each of n cyclically connected leaves; over $w = +1$ and $w = -1$ in n series, each of two leaves. The disposition of the leaves will become clear when we study the *corresponding subdivision of the z sphere* into half-regions.

The Dihedral Equation.

To this end it will be well, as we remarked above, *to know the linear substitutions which transform equation* (1) *into itself*. As in the case of the pure equation, it is unchanged by the n substitutions

(2a) $\quad z' = \varepsilon^\nu \cdot z \,(\nu = 0, 1, \ldots n-1),\quad$ where $\quad \varepsilon = e^{\frac{2i\pi}{n}}$,

since for these $z'^n = z^n$. Likewise, however, it is unchanged by the n additional substitutions

(2b $\quad\quad\quad\quad z' = \dfrac{\varepsilon^\nu}{z}\,(\nu = 0, 1, \ldots n-1).$

since these only change z^n into $1/z^n$.

We have therefore $2n$ linear substitutions of equation (1) into itself, exactly *as many as its degree indicates*. Thus, if we know for a given value w_0 of w one root z_0 of the equation, we know immediately $2n$ roots

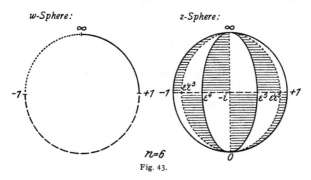

Fig. 43.

$\varepsilon^\nu \cdot z_0$ and $\varepsilon^\nu/z_0\,(\nu = 0, 1, 2, \ldots, n-1)$, in general all different, for which w has the same value w_0, i.e., we know *all* the roots of the equation when we have obtained the n-th root of unity ε.

Let us now *proceed to examine the subdivision of the z sphere corresponding to cuts along the real meridian of the Riemann surface over the w sphere*. In this, as in the previous example, we distinguish on the real meridian of the w sphere the three segments made by the branch points —that from $+1$ to ∞ (drawn full), that from ∞ to -1 (short dotted), and that from -1 to $+1$ (long dotted) (see Fig. 43). *To each of these three segments there correspond on the z sphere $2n$ different curvilinear segments which can be derived from any one of them by means of the $2n$ linear substitutions* (2). It will always suffice, therefore, to find one of them. Moreover all these segments must connect the critical points $z = 0, \infty, \varepsilon^\nu, \varepsilon' \cdot \varepsilon^\nu$, which we therefore mark on the z sphere. Just as in the previous case, their form is of a somewhat different type according as n is even or odd. It will suffice if we exhibit a definite case, say for $n = 6$. Fig. 43 shows the front half of the z sphere in orthogonal projection. One sees, on the equator, from left to right with spacings of

$60°$, $\varepsilon^3 = -1$, $\varepsilon^4, \varepsilon^5, \varepsilon^6 = 1$; and lying midway between the others, $\varepsilon' \cdot \varepsilon^3$, $\varepsilon' \cdot \varepsilon^4 = -i$, and $\varepsilon' \cdot \varepsilon^5$.

Now we shall see that the quadrant $+1 < z < \infty$ of the meridian of real z corresponds to the part of the real w meridian $+1 < w < \infty$ (full drawn). In fact, if we put $z = r$ and let r range through real values from 1 to ∞, then $w = \frac{1}{2}(z^n + 1/z^n) = \frac{1}{2}(r^n + 1/r^n)$ will vary also through real values that are always increasing, from 1 to ∞. We obtain n other full drawn curves on the z sphere, from this one, by means of the n linear substitutions (2a). But, as we saw in the previous example, these substitutions mean rotations of the sphere about the vertical axis $(0, \infty)$ through the angles $2\pi/n$, $4\pi/n$, ..., $2(n-1)\pi/n$. We get in this way the n quarter-meridians from the north pole ∞ to the points ε^ν on the equator. We get an additional full drawn curve if we apply the substitution $z' = 1/z$, which transforms the meridian quadrant from $+1$ to ∞ into the lower real meridianquadrant from $+1$ to 0. If we subject this quadrant to the n rotations (2a), the composition of which with $z' = 1/z$ gives the n substitutions (2b), we obtain, in addition, the n meridian quadrants which join the south pole with the equatorial points ε^ν. We have now in fact the $2n$ full drawn curves which correspond to the full drawn w meridian quadrant. In particular, for $n = 6$, they make up the three entire meridians into which the real meridian is transformed by rotations of $0°, 60°, 120°$.

It is now also obvious that the totality of the values $z = \varepsilon' \cdot r$, where r again ranges through real values from $+1$ to ∞, corresponds to the dotted part of the real w meridian; for the equation (1) yields then:

$$w = \frac{1}{2}\left(\varepsilon'^n r^n + \frac{1}{\varepsilon'^n r^n}\right) = -\frac{1}{2}\left(r^n + \frac{1}{r^n}\right),$$

and this expression actually decreases through real values from -1 to $-\infty$. But $z = \varepsilon' \cdot r$ represents the meridian quadrant from ∞ to the equatorial point ε^ν. If we now apply to it the substitutions (2a), (2b), we find, as before, that to the dotted part of the real w meridian there correspond all the meridian quadrants joining the poles to the equatorial points $\varepsilon' \cdot \varepsilon^\nu$, which thus bisect the angles between the meridian quadrants which we obtained before. In particular, for $n = 6$, they make up the three entire meridians into which the real meridian is transformed by rotations of $30°, 90°, 150°$.

There remain to be found the $2n$ curvilinear segments which correspond to the long-dotted half-meridian $-1 < w < +1$. I shall prove that they are the segments of the equator of the z sphere determined by the points ε^ν and $\varepsilon' \cdot \varepsilon^\nu$. In fact, the equator represents the points of absolute value one and is given therefore by $z = e^{i\varphi}$ where φ is real and ranges from 0 to 2π. Hence we have

$$w = \frac{1}{2}\left(z^n + \frac{1}{z^n}\right) = \frac{1}{2}(e^{ni\varphi} + e^{-ni\varphi}) = \cos n\varphi.$$

The Dihedral Equation.

This expression is always real, and its absolute value is not greater than 1. In fact, it assumes once every value between $+1$ and -1 as φ varies from one multiple of π/n to the next one, i. e., when z traverses one of the segments of which we are speaking.

The curves determined in this manner divide the z sphere into $2 \cdot 2n$ triangular half-regions which are bounded by one curve of each of the three sorts, and each half-region corresponds to a half leaf of the Riemann surface. Several regions can meet only at the critical points, and then in accordance with the table of multiplicities (p. 116), namely, *$2n$ at the north pole, and at the south pole, and $2 \cdot 2$ at each of the points ε^ν and $\varepsilon' \cdot \varepsilon^\nu$.* In order to determine which of these regions are to be shaded, we notice that when w traverses, in order, the full-drwan, the long-dotted, and the short dotted parts of the real w meridian, the rear half of the w sphere lies at its left. Since the mapping is conformal with preservation of angles, we should shade those half-regions whose boundaries follow one another in this same sense, and we should leave the others unshaded.

We have now obtained a complete geometric picture of the mutual dependence between z and w which is set up by our equation. We might follow it out in greater detail by examining more closely the conformal mapping of the single triangular regions upon the w hemisphere, but we shall forego this. *I shall describe only,* and *briefly, the case $n = 6$, to which I have already given special attention.* The z sphere is then divided into twelve shaded and twelve unshaded triangles of which six of each sort are visible in Fig. 44. Six of each sort meet at each pole, and two of each sort at each of twelve equidistant points of the equator. Each triangle is mapped conformally upon a w half-leaf of the same sort. Of the half-leaves of the Riemann surface, six of each sort are connected at the branch position ∞, and two of each sort at each of the branch positions ± 1, corresponding to the grouping of the half-regions on the z sphere.

We may obtain a convenient picture of the division of the z sphere, and one which is especially valuable because of its analogy with pictures soon to come, as follows. If we join the n equidistant points on the equator (e. g., the ε^ν) with one another in order by straight lines, and also join each of them to the two poles, one obtains a *double pyramid, with $2n$ faces, inscribed in the sphere* (in Fig. 44, twelve faces). If we now project, from the center, the subdivision of the z sphere upon this double pyramid, every pyramid face is divided into a shaded and an unshaded half by the altitude of that face dropped from the pole. If we represent the division of the z sphere, and consequently our function, by means of this double pyramid, the latter will render a service quite analogous to that which we shall get in the coming examples from the *regular polyhedra.* We obtain a *complete analogy if we think of the double pyramid as collapsed into its base,* and consider the *double regular n-gon* (hexagon) which results whose two faces (upper and lower) are divided each into $2n$ triangles by the straight lines which join the center with the vertices and the middle points of the sides (see Fig. 45). *I have been in the habit of calling this figure a dihedron and of classing it with the five regular polyhedra which have been studied since Plato's time.* It fulfills, in fact, all the conditions by means of which a regular polyhedron is usually defined, since its faces (the two faces of the n-gon) are congruent regular polygons, and since it has congruent edges (the sides of the n-gon) and congruent vertices (the vertices of the n-gon). The only difference is that it does not bound a proper solid body but encloses the volume zero. Thus the theorem of Plato, that there are

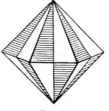
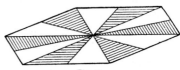

Fig. 44.   Fig. 45.

only five regular solids, is correct only when one includes in the definition the requirement of a *proper* solid, which is always tacitly assumed in the proof.

If we start with the dihedron, we obtain our subdivision of the z sphere by projecting upon that sphere not only its vertices but also the centers of its edges and its faces, the projecting rays for the latter being perpendicular to the plane of the dihedron. *Thus the dihedron can also be looked upon as representing the functional relation which our equation sets up between w and z.* Hence the brief name which we have already used, dihedral equation, is appropriate.

In addition, we shall now consider those equations which, as already intimated, are closely related to the platonic regular solids.

3. The tetrahedral, the octahedral, and the icosahedral equations.

We shall see that the last two could, with equal right, be called the hexahedral and the dodecahedral equations, so that all five regular bodies will have been covered. We shall follow here a route that is the reverse of the one we followed in the preceding example. *Starting from the regular body, we shall first deduce a division of the sphere into regions, and we shall then set up the appropriate algebraic equation, for which that figure* is the proper *geometric interpretation.* I shall have to confine myself frequently to suggestions, however, and I therefore refer you at once to my book: *Vorlesungen über das Ikosaeder und die Auf-*

lösung der Gleichungen vom fünften Grade[1], in which you will find a systematic presentation of the entire extensive theory with its numerous relations to allied fields.

Moreover, I shall give a parallel treatment of all three cases and I shall begin by *deducing the subdivision of the sphere for the tetrahedron*.

1. *The tetrahedron* (see Fig. 46). We divide each of the four equilateral face-triangles of the tetrahdron, by means of the three altitudes into *six partial triangles*. These are congruent in two groups of three each, while any two non-congruent ones are symmetric. We obtain thus a *division of the entire surface of the tetrahedron into twenty-four triangles, which fall into two groups, each containing twelve congruent triangles, while any triangle of one group is symmetric to every triangle of the other group*. We shall shade the

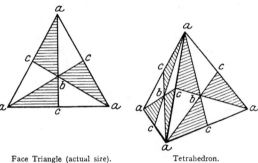

Face Triangle (actual size).   Tetrahedron.

Fig. 46.

triangles of one group. Among the vertices of these twenty-four triangles we can distinguish *three sorts*, such that each triangle has one vertex of each sort:

a) *the four vertices of the initial tetrahedron, at each of which three shaded and three unshaded triangles meet;*

b) *the four centers of gravity of the faces,* which determine again another regular tetrahedron (the co-tetrahedron); *at each of these, three triangles of each kind meet;*

c) *the six middle points of the edges,* which determine a *regular octahedron; at each of these, two triangles of each kind meet.*

If from the center of gravity of the tetrahedron we project this subdivision into triangles upon the circumscribed sphere, the latter will be *subdivided into* 2 · 12 *triangles, which are bounded by arcs of great circles and are mutually congruent or symmetric.* About each vertex of the sort a), b), c), there will be respectively 6, 6, 4 equal angles, and since the sum of the angles about a point on a sphere is 2π, *each of the spherical triangles will have an angle $\pi/3$ at a vertex of the sort a or b and an angle $\pi/2$ at a vertex of the sort c.*

It is a *characteristic property of this division of the sphere that it, as well as the tetrahedron itself, is transformed into itself by a number*

---

[1] Leipzig 1884; referred to hereafter as "*Ikosaeder*". Translation into English by G. C. Morrice: *Lectures on the Icosahedron by Klein.* Revised Edition, 1911, Kegan Paul & Co.

of rotations of the sphere about its center. This will be clear to you in detail if you examine a model of the tetrahedron with its divisions, like the one in our collection. For the lecture, it will suffice if I indicate the *number of possible rotations* (whereby the position of rest is included as the *identical rotation*. If we select a definite vertex of the original tetrahedron, we can, by means of a rotation, transform it into every vertex of the tetrahedron (including itself), which gives four possibilities. If we keep this vertex fixed, however, in any one of these four positions, we can still transform the tetrahedron. This gives *altogether $4 \cdot 3 = 12$ rotations* which transform the tetrahedron, or the corresponding triangular division of the circumscribed sphere, into itself. By means of these rotations we can transform a preassigned shaded (or unshaded) triangle into every other shaded (or unshaded) triangle, and the particular rotation is determined when that second triangle is chosen. These *twelve rotations* form obviously what one calls *a group G_{12}* of twelve operations, i. e., if we performs two of them in succession, the result is one of the twelve rotations.

If we think of this sphere as the z sphere, each of these twelve rotations will be represented by a *linear transformations* of z, and *the twelve linear transformations which arise in this manner will transform into itself the equation which corresponds to the tetrahedron.* For purposes of comparison, I remark that *one can interpret the $2n$ linear substitutions of the dihedral equation as the totality of the rotations of the dihedron into itself.*

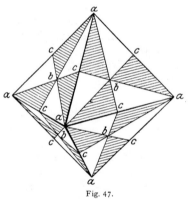

Fig. 47.

2. We shall now treat the *octahedron* similarly (see Fig. 47) and we may be somewhat briefer. We divide each of the faces, just as before, into *six partial triangles* and obtain a *division of the entire surface of the octahedron into twenty-four congruent shaded triangles, and twenty-four unshaded triangles which are congruent among themselves but symmetric to the other twenty-four.* We can again distinguish *three sorts of vertices*:

a) *the six vertices of the octahedron*, at each of which four triangles of each kind meet;

b) *the eight centers of gravity of the faces*, which form the *vertices of a cube*; at each of these, three triangles of each kind meet;

c) the twelve mid-points of the edges, at each of which two triangles of each kind meet.

If we pass now to the *circumscribed sphere*, by means of central projection, we obtain a division into 2 · 24 spherical triangles which are either congruent or symmetric, and each of which has an angle $\pi/4$ at the vertex a, $\pi/3$ at the vertex b, and $\pi/2$ at the vertex c. Since the vertices b form a cube, it is easy to see that *one would have obtained the same division on the sphere if one had started with a cube and had projected its vertices, and the centers of its faces and edges, upon the sphere*. In other words, we do not need to give special attention to the cube.

Just as in the previous case, it is easy to see *that the octahedron, as well as this division of the sphere, is transformed into itself by twenty-four rotations which form a group G_{24}; again each rotation is determined in that it transforms a preassigned shaded triangle into another definite shaded triangle.*

3. We come finally to the *icosahedron* (see Fig. 48). Here, also, we start with the same subdivision of each of the twenty-four triangular *faces and obtain altogether sixty shaded and sixty unshaded partial triangles.* The *three sorts of vertices* are:

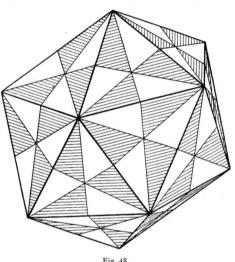

Fig. 48.

a) *the twelve vertices of the icosahedron, at each of which five triangles of each kind meet;*

b) *the twenty centers of gravity of the faces*, which are the *vertices of a regular dodecahedron; at each of them three triangles of each kind meet*;

c) *the thirty mid-points of the edges, at each of which two triangles of each sort meet.*

When this is carried over to the sphere each spherical triangle has at the vertices a, b, c the angles $\pi/5$, $\pi/3$, $\pi/2$, respectively. From the property of the vertices b one can conclude, as before, *that the same division of the sphere would have resulted if one had considered the dodecahedron.*

Finally, the icosahedron, as well as the corresponding division of the sphere, is transformed into itself by a group G_{60} of sixty rotations of the sphere about its center. These rotations, as well as those for the octahedron, will become clear to you upon examination of a model.

124   Algebra: Equations in the Field of Complex Quantities.

Let me make a list of *the angles of the spherical triangles* which have appeared in the three cases which we have considered, to which I shall add the dihedron also; they are

$$\begin{aligned}
\text{Dihedron:} & \quad \pi/2, \quad \pi/2, \quad \pi/n; \\
\text{Tetrahedron:} & \quad \pi/3, \quad \pi/3, \quad \pi/2; \\
\text{Octahedron:} & \quad \pi/4, \quad \pi/3, \quad \pi/2; \\
\text{Icosahedron:} & \quad \pi/5, \quad \pi/3, \quad \pi/2.
\end{aligned}$$

As a variation of a joke of Kummer's I might suggest that the student of natural science would at once conclude from this, that there were additional subdivisions of the sphere, having analogous properties, and with angles such as $\pi/6$, $\pi/3$, $\pi/2$; $\pi/7$, $\pi/3$, $\pi/2$. The mathematician, to be sure, does not risk making such inferences by analogy, and his cautiousness justifies itself here, for *the series of possible spherical subdivisions of this sort ends*, in fact, *with our list*. Of course this is connected with the fact *that there are no more regular polyhedrons*. We can see the ultimate reason in *a property of whole numbers*, which does not admit a reduction to simpler reasons. It appears, namely, that the angles of each of our triangles must be aliquot parts of π, say π/m, π/n, π/r, such that the denominators satisfy the inequality

$$1/m + 1/n + 1/r > 1.$$

This inequality has the property of existing only for the integral solutions given above. Moreover, we can understand it readily, since it only expresses the fact that the sum of the angles of a spherical triangle exceeds π.

I should like to mention that, as some of you doubtless know, an *appropriate generalization* of the theory does carry one byeond these apparently too narrow bounds: *The theory of automorphic functions* involves subdividing the sphere into *infinitely many triangles* whose angle sum is less than or equal to π.

4. Continuation: Setting up the Normal Equation.

We come now to the second part of our problem, *to set up that equation of the form*

(1) $$\varphi(z) - w\psi(z) = 0, \quad \text{or} \quad w = \frac{\varphi(z)}{\psi(z)},$$

which belongs to a definite one of our three spherical subdivisions, that is, which maps the two hemispheres of the w sphere upon the $2 \cdot 12$, or the $2 \cdot 24$, or the $2 \cdot 60$ partial triangles of the z sphere. To each value of w there must correspond then, in general, 12, 24, 60 values, respectively, of z, each one in a partial triangle of the right kind. Hence the desired equation must have the *degree 12, 24, 60* in the three cases respectively, for which we shall write N in general. Now each partial region touches three critical points; hence there must be, in every case, *three branch positions* on the w sphere. We assign these, as is customary, to $w = 0$, $1, \infty$; and we choose again the *meridian of real numbers* as the *section curve* $\mathfrak{E}$ through these three points, whose three segments shall correspond to the boundaries of the z triangles.

We shall assume (see Fig. 49) that in each of the three cases *the centers of gravity of the faces* (vertices b in the former notation) *correspond to the point $w = 0$, the mid-point of the edges* (vertices c) *to the point $w = 1$, and the vertices of the polyhedron* (vertices a) *to the point $w = \infty$*. The sides of the triangles will then correspond to the three segments of the w meridian in the manner indicated by the mapping, and the shaded triangles will correspond to the rear w hemisphere, the unshaded to the front w hemisphere. By virtue of these correspondences, the equation (1) is to effect a unique mapping of the z sphere upon an N-leaved Riemann surface over the w sphere with branch points at $0, 1, \infty$.

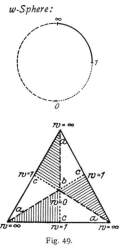

Fig. 49.

We might deduce, a priori, a proof for the existence of this equation by means of general functiontheoretic theorems. However, I prefer not to presuppose the knowledge which this method would require, but to *construct* the various equations *empirically*. This method will give us perhaps a more vivid perception of the individual cases.

Let us think of equation (1) written in homogeneous variables

$$\frac{w_1}{w_2} = \frac{\Phi_N(z_1, z_2)}{\Psi_N(z_1, z_2)},$$

where Φ_N, Ψ_N are homogeneous polynomials of dimension N in z_1, z_2 ($N = 12, 24,$ or 60). In this form of the equation, the positions $w_1 = 0$, $w_2 = 0$ (i.e., $w = 0, \infty$) on the w sphere seem to be favored more than the third branch position $w = 1$ (in homogeneous form, $w_1 - w_2 = 0$). Since, however, the three branch positions are, for our purpose, of equal importance, it is expedient to consider also the following form of the equation:

$$\frac{w_1 - w_2}{w_2} = \frac{X_N(z_1, z_2)}{\Psi_N(z_1, z_2)},$$

where $X_N = \Phi_N - \Psi_N$ denotes also a form of dimension N. Both forms are embraced in the continued proportion

(2) $\qquad w_1 : (w_1 - w_2) : w_2 = \Phi_N(z_1, z_2) : X_N(z_1, z_2) : \Psi_N(z_1, z_2).$

This furnishes us with a completely homogeneous form of equation (1) which gives the same consideration to all the branch points.

Our problem now is *to set up the forms* Φ_N, X_N, Ψ_N. For this purpose, we shall bring them into relation to our subdivision of the z sphere. From equation (2) we see that the form $\Phi_N(z_1, z_2) = 0$ for $w_1 = 0$, i.e., *that $w = 0$ corresponds to the N zeros of Φ_N on the z sphere.* On the other hand, *the centers of gravity of the faces of the polyhedron* (vertices b in the subdivision), of which there are $N/3$ in every case, must, according to our assumptions, correspond to the branch position $w = 0$. But every one of these centers must be *a triple root* of our equation, since in each of them there meet three shaded and three unshaded triangles of the z sphere. *Thus these points, each with multiplicity three, supply all the positions which correspond to $w = 0$, and consequently all the zeros of Φ_N.* Hence Φ_N has only triple zeros and must, therefore, be *the third power* of a form $\varphi_N(z_1, z_2)$ of degree $N/3$:

$$\Phi_N = [\varphi_{N/3}(z_1, z_2)]^3.$$

In the same way, it follows that the zeros of $X_N = 0$ correspond to the position $w = 1$ (i.e., $w_1 - w_2 = 0$), and that these are identical with the $N/2$ midpoints, each counted twice, of the edges of the polyhedron (vertices c of our subdivision). Consequently X_N must be the square of a form of dimension $N/2$:

$$X_N = [\chi_{N/2}(z_1, z_2)]^2.$$

Finally the zeros of Ψ_N are to correspond to the point $w = \infty$, so that they must be identical with the vertices of the polyhedron (vertices a of the subdivision); but at these vertices 3, 4, or 5 triangles meet, in the several cases, so that we get

$$\Psi_N = [\psi_{N/\nu}(z_1, z_2)]^\nu, \quad \text{where} \quad \nu = 3, 4 \text{ or } 5.$$

Our equation (2) *must then necessarily have the form*

(3) $\qquad w_1 : (w_1 - w_2) : w_2 = \varphi(z_1, z_2)^3 : \chi(z_1, z_2)^2 : \psi(z_1, z_2)^\nu,$

where the degrees and powers of φ, χ, ψ, and the values of the degree N of the equation are exhibited in the following table:

Tetrahedron: φ_4^3, χ_6^2, ψ_4^3; $N = 12$.
Octahedron: φ_8^3, χ_{12}^2, ψ_6^4; $N = 24$.
Icosahedron: φ_{20}^3, χ_{30}^2, ψ_{12}^5; $N = 60$.

I shall now show briefly *that the dihedral equation which we discussed, fits also into the scheme* (3). We need only to recall that in that case we chose $-1, +1, \infty$ as the branch positions on the w sphere instead of $0, +1, \infty$ which we selected later. We shall, then, obtain actual analogy with (3) only if we throw the dihedral equation into the form

$$(w_1 + w_2) : (w_1 - w_2) : w_2 = \Phi : X : \Psi.$$

Now from the dihedral equation (p. 115) which we used:
$$\frac{w_1}{w_2} = \frac{z_1^{2n} + z_2^{2n}}{2 \cdot z_1^n z_2^n},$$
we get by simple reduction
$$(w_1 + w_2) : (w_1 - w_2) : w_2 = (z_1^{2n} + z_2^{2n} + 2z_1^n z_2^n) : (z_1^{2n} + z_2^{2n} - 2z_1^n z_2^n) : (2z_1^n z_2^n)$$
$$= (z_1^n + z_2^n)^2 : (z_1^n - z_2^n)^2 : 2(z_1 z_2)^n.$$
Thus we can, in fact, add to the above table:

$$\text{Dihedron:} \quad \varphi_n^2, \quad \chi_n^2, \quad \psi_2^n; \quad N = 2n.$$

The critical points together with their multiplicities which can at once be read off from this form of the equation are in full agreement with those which we found above (see p. 116).

We come now to the *actual setting up of the forms* φ, χ, ψ *in the three new cases. I shall give details here only for the octahedron,* for which the relations turn out to be the simplest. But even here I shall, at times, give only suggestions or results, in order to remain within the confines of a brief survey. For those who desire more, there is easily accessible the detailed exposition in my book on the icosahedron. For the sake of simplicity we think of the octahedron as so inscribed in the z sphere *that the six vertices fall on* (see Fig. 50):
$$z = 0, \infty, +1, +i, -1, -i.$$

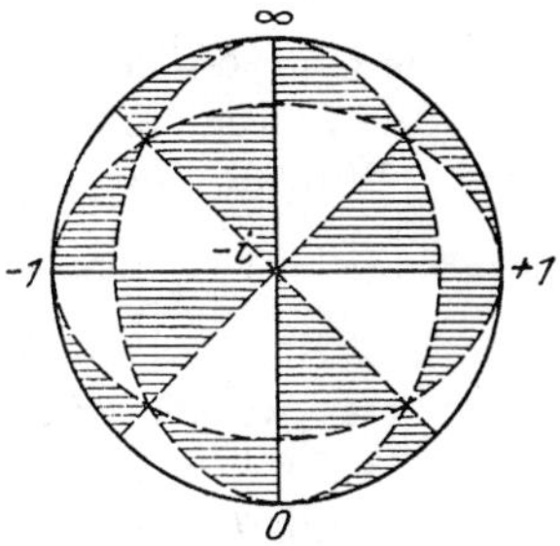

Fig. 50.

It will then be a simple matter to give the *twenty-four linear substitutions of* z which represent the rotations of the octahedron, i.e., which permute these six points. We begin with the four rotations in which the vertices 0 and ∞ remain fixed

(4a) $$z' = i^k \cdot z, \qquad (k = 0, 1, 2, 3).$$

Then we can interchange the points $0, \infty$ by means of the substitution $z' = 1/z$ (i.e., a rotation through $180°$ about the horizontal axis $(+1, -1)$ which transforms every point of the octahedron into another one. If we now apply the four rotations (4a), we get four new substitutions:

(4b) $$z' = \frac{i^k}{z}. \qquad (k = 0, 1, 2, 3)$$

In the same way, we now throw in succession the four remaining vertices $z = 1, i, -1, -i$ to ∞ by means of the substitutions
$$z' = \frac{z+1}{z-1}, \quad \frac{z+i}{z-i}, \quad \frac{z-1}{z+1}, \quad \frac{z-i}{z+i},$$
which obviously permute the six vertices of the octahedron, and again

apply, each time, the four rotations (4a). Thus we get $4 \cdot 4 = 16$ additional substitutions for the octahedron

(4c) $\begin{cases} z' = i^k \dfrac{z+1}{z-1}, & z' = i^k \dfrac{z-1}{z+1}, \\ z' = i^k \dfrac{z+i}{z-i}, & z' = i^k \dfrac{z-i}{z+i}. \end{cases}$ $(k = 0, 1, 2, 3)$

We have therefore found the desired twenty-four substitutions, and we can easily show, by calculation, that *they really permute the six vertices of the octahedron and that they form a group* G_{24}, i. e., that the successive application of any two of them gives again one of the substitutions in (4).

I shall now construct *the form* ψ_6 which vanishes in each of the vertices of the octahedron. The point $z = 0$ gives the factor z_1, the point $z = \infty$ the factor z_2; the form $z_1^4 - z_2^4$ has a simple zero at each of the points $\pm 1, \pm i$, so that we obtain finally

(5a) $$\psi_6 = z_1 \cdot z_2 (z_1^4 - z_2^4).$$

It is more difficult to construct the forms φ_8 and χ_{12} which have as zeros the centers of gravity of the faces and the midpoints of the edges. Without deducing them, I may state that they are[1]

(5b) $\begin{cases} \varphi_8 = z_1^8 + 14 z_1^4 z_2^4 + z_2^8, \\ \chi_{12} = z_1^{12} - 33 z_1^8 z_2^4 - 33 z_1^4 z_2^8 + z_2^{12}. \end{cases}$

It goes without saying that there is an undetermined constant multiplier in each of these three forms. If φ_8, ψ_6, χ_{12} stand for the normal forms (5), we must insert, in the octahedral equation (3), two undetermined constants c_1, c_2, and we must write

$$w_1 : (w_1 - w_2) : w_2 = \varphi_8^3 : c_1 \chi_{12}^2 : c_2 \psi_6^4.$$

The constants c are now to be so determined that these two equations give actually only *one* equation between z and w. This is possible when and only when

$$\varphi_8^3 - c_2 \psi_6^4 = c_1 \chi_{12}^2$$

is an identity in z_1 and z_2. Now this relation can be satisfied by definite constants c_1 and c_2. A brief calculation shows that the identity

$$\varphi_8^3 - 108 \psi_6^4 = \chi_{12}^2$$

must hold, so that the octahedral equation (3) becomes:

(6) $$w_1 : (w_1 - w_2) : w_2 = \varphi_8^3 : \chi_{12}^2 : 108 \psi_6^4.$$

This equation surely maps *the points* $0, 1, \infty$ *respectively upon the centers of gravity of the faces, the midpoints of the edges, and the vertices of the octahedron, with the proper multiplicity*, because the forms φ, χ, ψ were so constructed. Furthermore, the twenty-four octahedron substi-

---

[1] See *Ikosaeder*, p. 54.

tutions (4) transform it into itself, for they transform the zeros of each of the forms φ, χ, ψ into themselves and at the same time change each of the forms by a multiplicative factor. And calculation shows that these factors cancel when the quotients are formed.

It only remains to show that equation (6) really maps *each shaded or unshaded triangle of the z sphere conformally upon the rear or front w hemisphere*. We know that the points $0, 1, \infty$ of the real w meridian correspond to the three vertices of each of the triangles; but the equation has, moreover, twenty-four roots z for each value of w. Since these must distribute themselves among the twenty-four triangles, w can take a given value but once, at most, within a triangle. If we could only show that *w remains real on the three sides of a triangle*, we could then easily show that there is a one-to-one mapping of each side upon a segment of the real w meridian, and also a *similar mapping of the entire interior of the triangle upon the corresponding hemisphere, one which is conformal without reversal of angles*. You will be able to make these deductions yourselves by making use of the continuity and the analytic character of the function $w(z)$. I shall indicate the only noteworthy step of the proof, that of showing the reality of w upon the sides of the triangle.

It is more convenient to prove this by showing *that w is real upon all the great circles that arise in the octahedral subdivision*. These are, first, the three mutually perpendicular circles which pass each through four of the six vertices of the octahedron (*principal circles*; full drawn in Fig. 50, p. 127), and, second, the six circles, corresponding to the altitudes of the faces, which bisect the angles of the principal circles (*auxiliary circles*; long dotted in Fig. 50). By means of the octahedron substitutions, one can transform every principal circle into any other and every auxiliary circle into any other. Hence it will suffice to show *that the function w is real at every point on one principal and one auxiliary circle*, since it must take the same values on the other circles. Now the meridian of real numbers z is one of the principal circles. By (6), the values on this circle are

$$w = \frac{w_1}{w_2} = \frac{\varphi_8^3}{108\,\psi_6^4},$$

which are, of course, real, since φ and ψ are real polynomials in z_1 and z_2. Of the auxiliary circles let us select the one through 0 and ∞ which makes an angle of $45°$ with the real meridian and on which z takes the values $z = e^{\frac{i\pi}{4}} \cdot r$, where r ranges through real values from $-\infty$ to $+\infty$. On this circle $z^4 = e^{i\pi} \cdot r^4 = -r^4$ is real. Since by (5) only the fourth powers of z_1 and z_2 occur in φ_8 and in the fourth power of ψ_6, the last formula shows that w is real.

This concludes the proof: *Equation* (6), *in fact, maps the w hemisphere, or the Riemann surface over it, conformally upon that triangular subdivision* of the z sphere which corresponds to the octahedron, and consequently we have in this case, as completely as in the earlier examples, a geometric control of the dependence which this equation sets up between z and w.

The treatment of the tetrahedron and of the icosahedron proceeds according to the same plan. I shall give only the results. As before, these results are those obtained when the subdivision of the z sphere has the simplest possible position. The *tetrahedral equation*[1] is

$$w_1 : (w_1 - w_2) : w_2 = \{z_1^4 - 2\sqrt{-3}\, z_1^2 z_2^2 + z_2^4\}^3$$
$$: -12\sqrt{-3}\,\{z_1 z_2 (z_1^4 - z_2^4)\}^2$$
$$: \{z_1^4 + 2\sqrt{-3}\, z_1^2 z_2^2 + z_2^4\}^3$$

and the *icosahedral equation*[2] is

$$w_1 : (w_1 - w_2) : w_2 = \{-(z_1^{20} + z_2^{20}) + 228(z_1^{15} z_2^5 - z_1^5 z_2^{15}) - 494 z_1^{10} z_2^{10}\}^3$$
$$: -\{(z_1^{30} + z_2^{30}) + 522(z_1^{25} z_2^5 - z_1^5 z_2^{25}) - 10005(z_1^{20} z_2^{10} + z_1^{10} z_2^{20})\}^2$$
$$: 1728\{z_1 z_2 (z_1^{10} + 11 z_1^5 z_2^5 - z_2^{10})\}^5,$$

i.e., *these equations map the w hemispheres conformally upon the shaded and the unshaded triangles of that subdivision of the z sphere which belongs to the tetrahedron and to the icosahedron respectively.*

5. Concerning the Solution of the Normal Equations

Let us now consider somewhat the *common properties of the equations* which we have been discussing and which we shall call the *normal equations*.

Note, first of all, that *the extremely simple nature of all our normal equations* is due to the fact that *they have exactly the same number of linear substitutions into themselves as is indicated by the degree*, i.e., that all their roots are linear functions of a single one; and, further, that we have, in the divisions of the sphere, a very obvious geometric picture of all the relations that come up for consideration. Just how simple many things appear which are ordinarily quite complicated with equations of such high degree will be evident if I raise a certain question in connection with the icosahedral equation.

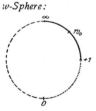

Fig. 51.

Let a *real value w_0* be given, say on the segment $(1, \infty)$ of the real w meridian (see Fig. 51). *Let us inquire about the sixty roots z of the icosahedral equation when $w = w_0$.* Our theory of the mapping tells us at once that *one of them must lie on a side of each of the sixty triangles on the z sphere which arise in the case of the icosahedron* (drawn full in Fig. 49, p. 125). This supplies what one calls, in the theory of

---

[1] See *Ikosaeder*, p. 51, 60. [2] Loc. cit., p. 56, 60.

equations, the *separation of the roots,* usually a laborious task, which must precede the numerical calculation of the roots. *The task is that of assigning separated intervals in each of which but one root lies.* But we can also tell at once *how many of the roots are real.* If we take into account, namely, that the form of the icosahedral equation given above implies such a placing[1] of the icosahedron in the z sphere *that the real meridian contains four vertices of each of the three sorts* a, b, c, then it follows (see Fig. 48, p. 123, and Fig. 49, p. 119) that four full-drawn triangle sides lie on the real meridian, *so that there are just four real roots.* The same is true if w lies in one of the other two segments of the real w meridian, *so that for every real w different from* 0, 1, ∞ *the icosahedral equation has four real and fifty-six imaginary roots; for* $w = 0$, 1, ∞ *there are also four different real roots, but they are repeated.*

I shall now say *something about the actual numerical calculation of the roots of our normal equations.* We have here again the great advantage that we *need to calculate but one root,* because the others follow by linear substitutions. Let me remind you, however, *that the numerical calculation of a root is really a problem of analysis, not of algebra,* since it requires necessarily the application of infinite processes when the root to which one is approximating is irrational, as is the case in general.

I shall go into details only for the *simplest example of all,* the *pure equation*

$$w = z^n.$$

Here I come again *into immediate touch with school mathematics.* For this equation, i. e., the calculation of $\sqrt[n]{w}$, at least for the small values of n and for real values of $w = r$, is treated there also. The method of calculating square and cube root, as you learned it in school, depends, in essence, upon the following procedure. One determines the position which the radicand $w = r$ has in the series of the squares or cubes, respectively, of the natural numbers 1, 2, 3, ... Then, using the decimal notation, one makes the same trial with the tenths of the interval concerned, then with the hundreths, and so on. In this way one can, of course, approximate with any desired degree of closeness.

I should like to apply a *more rational process,* one in which we can admit not only arbitrary integral values of n but also *arbitrary complex values of w.* Since we need to determine *only one* solution of the equation, we shall seek, in particular, that value $z = \sqrt[n]{w}$ which lies *within the angle $2\pi/n$ laid off on the axis of real numbers.* Generalizing the elementary method mentioned above, we begin by dividing this angle into ν equal parts ($\nu = 5$ in Fig. 52), and by drawing circles intersecting the dividing rays by circles which have the origin as common center and

---

[1] See *Ikosaeder,* p. 55.

whose radii are measured by the numbers $r = 1, 2, 3, \ldots$ In this way, after choosing ν, we find all the points

$$z = r \cdot e^{\frac{2i\pi}{n} \cdot \frac{k}{\nu}} \qquad \begin{pmatrix} k = 0, 1, 2, \ldots, \nu - 1 \\ r = 1, 2, 3, \ldots \end{pmatrix}$$

marked within the angular space, and we can at once mark in the w plane the corresponding w values

$$w = z^n = r^n \cdot e^{2i\pi \frac{k}{\nu}}.$$

These will be the corners of a corresponding network (see Fig. 53) covering the *entire* w plane and consisting of circles with radii 1^n, 2^n, 3^n, ... together with rays inclined to the real axis at angles of 0, $2\pi/\nu$,

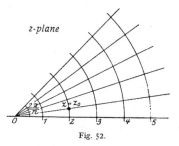

Fig. 52.

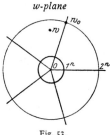

Fig. 53.

$4\pi/\nu, \ldots, (\nu - 1) 2\pi/\nu$. Let the given value of w lie either within or on the contour of one of the meshes of this lattice, and suppose that w_0 is the lattice corner nearest to it. We know a value z_0 of $\sqrt[n]{w_0}$ is a corner of the lattice in the z plane; hence the value we are seeking will be

$$Z = \sqrt[n]{w} = \sqrt[n]{w_0 + (w - w_0)} = \sqrt[n]{w_0} \sqrt[n]{1 + \frac{w - w_0}{w_0}} = z_0 \left(1 + \frac{w - w_0}{w_0}\right)^{\frac{1}{n}}.$$

We expand the right side by the *binomial theorem*, which we may consider known, inasmuch as we are now, in reality, in the domain of analysis

$$Z = z_0 \left\{ 1 + \frac{1}{n} \cdot \frac{w - w_0}{w_0} + \frac{1 - n}{2n^2} \left(\frac{w - w_0}{w_0} \right)^2 + \cdots \right\}.$$

We can decide at once as to the convergence of this series if we look upon it as the *Taylor's development of the analytic function* $\sqrt[n]{w}$ and apply the theorem that it converges within the circle which has w_0 as center and which passes through the nearest singular point. Since $\sqrt[n]{w}$ has only 0 and ∞ as singular points, *our development will converge if, and only if, w lies within that circle about w_0 which passes through the origin*, and we can always bring this about by starting, in the z plane, with a similar lattice which may have smaller meshes, if necessary. But in order that the convergence should be good, i.e., in order that the series *should be adapted to numerical calculation* $(w - w_0)/w_0$ *must be sufficiently small*. This can always be effected by a further reduction of the lattice. This is really a very usable method for the actual calculation of numerical roots.

Now is it worthy of remark that the numerical solution of the remaining normal equations of the regular solids is not essentially more difficult, but I shall omit the proof. If we apply, namely, the same method to our normal equations, starting from the mapping upon the w sphere of two neighboring triangles, there will appear, in place of the binomial series, certain other series that are well known in analysis and are well adapted to practical use, called *the hypergeometric series*. In the year 1877 I set up[1] this series numerically.

6. Uniformization of the Normal Irrationalities by Means of Transcendental Functions

I shall now discuss *another method of solving our normal equations* which is characterized by the *systematic employment of transcendental functions*. Instead of proceeding, in each individual case, with series developments in the neighborhod of a known solution, we try to represent, once for all, *the whole set of number pairs* (w, z) *which satisfy the equation, as single-valued analytic functions of an auxiliary variable: or, as we say, to uniformize the irrationalities defined by the equation*. If we can succeed by using only functions which can easily be tabulated, or of which one already has, perhaps, numerical tables, one can obtain the *numerical solution of the equation without farther calculation*. I am the more willing to discuss this connection with transcendental functions because it sometimes plays a part in *school instruction*, where it still often has a hazy, almost mysterious, aspect. The reason for this is that one is still clinging to traditional imperfect conceptions, although the modern theory of functions of a complex variable has provided perfect clearness.

I shall apply these general suggestions first to the *pure equation. Even in the schools, we always use logarithms in calculating the positive solution of* $z^n = r$, *for real positive values of* r. We write the equation in the form $z = e^{\log r/n}$, where $\log r$ stands for the positive principal value. The logarithmic tables supply first $\log r$, and then, conversely, z is the number that corresponds to $\log r/n$. Moreover, we ordinarily use 10 as base instead of e. This solution can be extended immediately to complex values. We satisfy the equation

$$z^n = w,$$

---

[1] *Weitere Untersuchungen über das Ikosaeder*, Mathematische Annalen, vol. 12, p. 515. See also Klein, F., *Gesammelte Mathematische Abhandlungen*, vol. 2, p. 331 et seq.]

by putting x equal to the general complex logarithm, $\log w$, after which we obtain w and z actually as single-valued analytic functions of x:

$$w = e^x, \quad z = e^{\frac{x}{n}}$$

In view of the many-valuedness of $x = \log w$, which we shall study later in detail, one obtains here for the same w precisely n values of z. *We call x the uniformizing variable.*

Since the tables contain only the *real logarithms of real numbers*, we are apparently unable to read off immediately the value of the given solution. But by the aid of a simple property of logarithms, we can *reduce the calculation to the use of trigonometric tables which are accessible to everybody*. If we put

$$w = u + iv = \left|\sqrt{u^2 + v^2}\right| \left(\frac{u}{|\sqrt{u^2+v^2}|} + i \frac{v}{|\sqrt{u^2+v^2}|} \right),$$

then the first factor, as a positive real number, has a real logarithm, the second, as a number of absolute value 1, *a pure imaginary logarithm* $i\varphi$ (i.e., the second factor is equal to $e^{i\varphi}$), and we obtain φ from the equation

(a) $$\frac{u}{\sqrt{u^2+v^2}} = \cos\varphi, \quad \frac{v}{\sqrt{u^2+v^2}} = \sin\varphi.$$

This gives $x = \log w = \log\left|\sqrt{u^2+v^2}\right| + i\varphi$, and the root of the equation is therefore

$$z = e^{\frac{x}{n}} = e^{\frac{1}{n}\log|\sqrt{u^2+v^2}|} \cdot e^{\frac{1}{n}i\varphi},$$

i.e., we have

(b) $$z = \sqrt[n]{u+iv} = e^{\frac{1}{n}\log|\sqrt{u^2+v^2}|}\left(\cos\frac{\varphi}{n} + i\sin\frac{\varphi}{n}\right).$$

Since φ is determined only to within multiples of 2π, this formula supplies all the n roots. With the aid of ordinary logarithmic and trigonometric tables, we can now get first φ from (a) und then z from (b). We have obtained this *"trigonometric solution" from the logarithms of complex numbers in an entirely natural way.* However, if we assume that these are not known and try to develop this trigonometric solution, as is done in the schools, it must appear as something entirely foreign and unintelligible.

Occasionally it becomes necessary to find roots of numbers that are not real. Thus, in school instruction, such roots must be found in the so called *Cardan's solution of the cubic equation* about which I should like to interpolate here a few remarks. If this equation is given in the reduced form

(1) $$x^3 + px - q = 0.$$

then the formula of Cardan states that its three roots x_1, x_2, x_3 are contained in the expression

(2) $$x = \sqrt[3]{\frac{q}{2} + \sqrt{\frac{q^2}{4} + \frac{p^3}{27}}} + \sqrt[3]{\frac{q}{2} - \sqrt{\frac{q^2}{4} + \frac{p^3}{27}}}.$$

Since every cube root is three valued, this expression has, all told, nine values, in general all different; among these, x_1, x_2, x_3 are determined by the condition *that the product of the two cube roots employed each time is* $-p/3$. If we replace the coefficients p, q in the well known manner by their expressions as symmetric functions of x_1, x_2, x_3, and if we note that the coefficient of x^2 vanishes, that is, $x_1 + x_2 + x_3 = 0$, we get

$$\frac{q^2}{4} + \frac{p^3}{27} = -\frac{(x_1 - x_2)^2 (x_2 - x_3)^2 (x_3 - x_1)^2}{108},$$

that is, the radicand of the square root is, to within a negative factor, the discriminant of the equation. This shows at once that *it is negative when all three roots are real, but positive when one root is real and the other two conjugate imaginary.* It is precisely in the apparently simplest case of the cubic equation, namely when all the roots are real, that the formula of Cardan requires the extraction of the square root of a negative number, and hence of the cube root of an imaginary number.

This passage through the complex must have seemed something quite impossible to the mediaeval algebraists at a time when one was still far removed from a theory of complex numbers, 250 years before Gauss gave his interpretation of them in the plane! One talked of the "*Casus irreducibilis*" of the cubic equation and said that the Cardan formula failed here to give a reasonable usable solution. When it was discovered later that it was possible, precisely in this case, to establish a simple relation between the cubic equation and the trisection of an angle, and to get in this way a real "trigonometric solution" in place of the defective Cardan formula, it was believed that something new had been discovered which had no connection with the old formula. Unfortunately this is the position taken occasionally even today in elementary instruction.

In opposition to this view, I should like to insist here emphatically that this trigonometric solution is nothing else than the application, in calculating the roots of complex radicands, of the process which we have just discussed. It is obtained therefore in a perfectly natural way in this case, where the cube root has a complex radicand, if we transform the Cardan formula, for numerical calculation, in the same convenient way that one pursues in school for the case of the real radicand. In fact, let us suppose

$$\frac{q^2}{4} + \frac{p^3}{27} < 0,$$

where p must be negative if q is real. If we then write the first cube root in (2) in the form

$$\sqrt[3]{\frac{q}{2}+i\left|\sqrt{-\frac{q^2}{4}-\frac{p^3}{27}}\right|}.$$

We note that its absolute value value (as positive cube root of the value $\sqrt{-p^3/27}$ of the radicand) is equal to $|\sqrt{-p/3}|$; but since the product of this by the second cube root is equal to $-p/3$, that second cube root must be the conjugate complex of this, and the sum of the two, i.e., the solution of the cubic equation, is simply twice the real part, that is,

$$x_1, x_2, x_3 = 2\Re\left(\sqrt[3]{\frac{q}{2}+i\left|\sqrt{-\frac{q^2}{4}-\frac{p^3}{27}}\right|}\right).$$

Now let us apply the general procedure of p. 134. We write the radicand of the cube root, after separating out its absolute value, in the form

$$\left|\sqrt{-\frac{p^3}{27}}\right|\left\{\frac{\frac{q}{2}}{\left|\sqrt{-\frac{p^3}{27}}\right|}+i\frac{\left|\sqrt{-\frac{q^2}{4}-\frac{p^3}{27}}\right|}{\left|\sqrt{-\frac{p^3}{27}}\right|}\right\}$$

and determine an angle φ from the equations

$$\cos\varphi = \frac{\frac{q}{2}}{\left|\sqrt{-\frac{p^3}{27}}\right|}, \qquad \sin\varphi = \frac{\left|\sqrt{-\frac{q^2}{4}-\frac{p^3}{27}}\right|}{\left|\sqrt{-\frac{p^3}{27}}\right|}.$$

Then, since the positive cube root of $|\sqrt{-p^3/27}|$ is $|\sqrt{-p/3}|$, our cube root takes the form

$$\left|\sqrt{-\frac{p}{3}}\right|\cdot\left(\cos\frac{\varphi}{3}+i\sin\frac{\varphi}{3}\right),$$

and hence, remembering that φ is determinate only to within multiples of 2π, we obtain

$$x_k = 2\left|\sqrt{-\frac{p}{3}}\right|\cdot\cos\frac{\varphi+2k\pi}{3} \qquad (k=0,1,2).$$

But this is the usual form of the trigonometric solution.

I should like to take this opportunity to make a remark about the expression *"casus irreducibilis"*. "Irreducible" is used here in a sense entirely different from the one in use today and which we shall often use in these lectures. In the sense here used it implies that the solution of the cubic equation cannot be reduced to the cube roots of real numbers. This is not in the least the modern meaning of the word. You see how the unfortunate use of words, together with the general fear of complex numbers, has created at least the possibility for a good deal of misunderstanding in just this field. I hope that my words may serve as a preventive, at least among you.

Let us now inquire briefly about uniformization by means of transcendental functions in the case of the remaining normal irrationalities. In the dihedral equation

$$z^n + \frac{1}{z^n} = 2w$$

we put simply

$$w = \cos \varphi.$$

De Moivre's formula shows that the equation is then satisfied by

$$z = \cos \frac{\varphi}{n} + i \sin \frac{\varphi}{n}.$$

Since all values of $\varphi + 2k\pi$ and of $2k\pi - \varphi$ give the same value of w this formula gives, in fact, for every w, $2n$ values of z, which we can write

$$z = \cos \frac{\varphi + 2k\pi}{n} \pm i \sin \frac{\varphi + 2k\pi}{n}. \qquad (k = 0, 1, 2, \ldots, n-1)$$

In the case of the equations of the octahedron, tetrahedron, and icosahedron these "elementary" transcendental functions do not suffice. However, we can obtain the corresponding solution by means of elliptic modular functions. Although one may not consider this solution as belonging to elementary methematics, I should, nevertheless, like to give, at least, the formulas[1] which relate to the icosahedron. They are, namely, closely related to the solution of the general equation of degree five by means of elliptic functions, to which allusion is always made in textbooks and about which I shall have something to say later by way of explanation. The icosahedral equation had the form (see pp. 130, 126)

$$w = \frac{\varphi_{20}(z)^3}{\psi_{12}(z)^5}.$$

Now we identify w with the absolute invariant J from the theory of elliptic functions and think of J as a function of the period quotient $w = \omega_1/\omega_2$ (in Jacobi's notation iK'/K), i.e., we set

$$w = J(\omega) = \frac{g_2^3(\omega_1, \omega_2)}{\Delta(\omega_1, \omega_2)},$$

where g_2 and Δ are certain transcendental forms of dimension -4 and -12, respectively, in ω_1 and ω_2, which play an important rôle. If we introduce the usual abbreviation of Jacobi

$$q = e^{i\pi\omega} = e^{-\pi\frac{K'}{K}}$$

[1 See *Mathematische Annalen*, vol. 14 (1878/79), p. 111 et seq., or Klein, *Gesammelte Abhandlungen*, vol. 3, p. 13 et seq., also *Ikosaeder*, p. 131.]

the roots z of the icosahedral equation will be given by the following quotients of ϑ functions

$$z = -q^{\frac{3}{5}} \frac{\vartheta_1(2\pi\omega, q^5)}{\vartheta_1(\pi\omega, q^5)}.$$

If we take into account that ω as a function of w, coming from the first equation, is infinitely many-valued, then this formula yields in fact all sixty roots of the icosahedral equation for a given w.

7. Solution in Terms of Radicals

There is one question in the theory of the normal equations which I have not yet touched, namely, whether or not our normal equations yield algebraically anything that is essentially new; and whether or not they can be resolved into one another or, in particular, into a sequence of pure equations. In other words, is it possible to build up the solution z of these equations in terms of w by means of a finite number of radical signs, one above another?

So far as the equations of the dihedron, tetrahedron, and octahedron are concerned, it is easy to show, by means of algebraic theory, that they can be reduced, in fact, to pure equations. It will be sufficient if I give the details here for the dihedral equation only:

$$z^n + \frac{1}{z^n} = 2w.$$

If we set:

$$z^n = \zeta,$$

the equation goes over into

$$\zeta^2 - 2w\zeta + 1 = 0.$$

It follows from this that

$$\zeta = w \pm \sqrt{w^2 - 1},$$

and consequently

$$z = \sqrt[n]{w \pm \sqrt{w^2 - 1}},$$

which is the desired solution by means of radicals.

On the other hand, however, the icosahedral equation does not admit such a solution by means of radicals, so that this equation defines an essentially new algebraic function. I am going to give you a particularly graphic proof of this, which I have recently published (Mathematische Annalen, Vol. 61 [1905]), and which follows from consideration of the familiar functiontheoretic construction of the icosahedral function $z(w)$. For this purpose I shall need the following theorem, due to *Abel*, a proof of which you will find in every treatise on algebra: *If the solution of an algebraic equation can be expressed as a sequence of radicals, then every radical of the sequence can be expressed as a rational function of the n roots of the given equation.*

Let us now apply this theorem to the icosahedral equation. If we assume its root z can be expressed as a sequence of roots of rational functions of the coefficients, i. e., of rational functions of w, then every radical in the sequence is a rational function of the sixty roots:

$$R(z_1, z_2, \ldots, z_{60}).$$

(We shall show that this leads to a contradiction.) In the first place, we can replace this expression by a rational function $R(z)$ of z alone since all the roots can be derived from any one of them by a linear substitution. Let us now convert this $R(z)$ into a function of w by writing for z the sixty-valued icosahedral function $z(w)$, and consider the result. Since every circuit in the w plane which returns z to its initial value must of necessity return $R(z)$ also to its initial value, it follows that $+R[z(w)]$ can have branch points only at the positions $w = 0, 1, \infty$ (where $z(w)$ has branch points), and the number of leaves of the Riemann surface for $R[z(w)]$ which are cyclically connected at each of these positions must be a divisor of the corresponding number belonging to $z(w)$. We know that this number is 3, 2, 5 at the three positions, respectively. Hence every rational function $R(z)$ of an icosahedral root, and consequently every radical which appears in the assumed solution, considered as function of w, can have branch points, if at all, only at $w = 0$, $w = 1$, $w = \infty$. If branching occurs, then there must be three leaves connected at $w = 0$, two at $w = 1$, and five at $w = \infty$, since 3, 2, 5 have no divisor other than 1.

We shall now see that this result leads to a contradiction. To this end let us examine the innermost radical which appears in our hypothetically assumed expression for $z(w)$. Its radicand must be a rational function $P(w)$. We can assume that the index of the radical is a prime number p, since we could otherwise build it up out of radicals with prime indices. Moreover $P(w)$ cannot be the p-th power of a rational function $\varrho(w)$ of w, for if it were, our radical would be superfluous, and we could direct our attention to the next really essential radical.

Let us now see what kind of branchings the function $\sqrt[p]{P(w)}$ can have. For this purpose it will be convenient to write it in the homogeneous form

$$P(w) = \frac{g(w_1, w_2)}{h(w_1, w_2)},$$

where g and h are forms of the same dimension in the variables w_1, w_2 ($w = w_1/w_2$). According to the fundamental theorem of algebra we can separate g and h, into linear factors and write

$$P(w) = \frac{l^\alpha \cdot m^\beta \cdot n^\gamma \ldots}{l'^{\alpha'} \cdot m'^{\beta'} \cdot n'^{\gamma'} \ldots}.$$

where

$$\alpha + \beta + \gamma + \cdots = \alpha' + \beta' + \gamma' + \cdots$$

since the numerator and the denominator are of the same degree. Not all the exponents $\alpha, \beta, \ldots, \alpha', \beta' \ldots$ can be divisible by p, since P would then be a perfect p-th power. On the other hand, $\alpha + \beta + \ldots - \alpha' - \beta' - \ldots$ is equal to zero, and is therefore divisible by p. Consequently at least two of these numbers are not divisible by p. It follows that the zeros of both the corresponding linear factors must be branch points of $\sqrt[p]{P(w)}$, at each of which p leaves are cyclically connected. But herein lies the contradiction of the previous theorem, which, of course, must be equally valid for $\sqrt[p]{P(w)}$. For we enumerated at that time all possible branch points, and we found among them no two at which the same number of leaves were connected. Our assumption is therefore not tenable, and the icosahedral equation cannot be solved by radicals.

This proof depends essentially upon the fact that the numbers 3, 2, 5 which are characteristic for the icosahedron have no common divisor. When such a common divisor appears, as in the case of the numbers 3, 2, 4 of the octahedron, it is at once possible to have rational functions $R[z(w)]$ which exhibit the same kind of branching at two points, e.g., one in which two leaves are connected at 1 and at ∞, and these can then be really represented as roots of a rational function $P(w)$. It is in this way that the solution by means of radicals comes about in the case of the octahedron and tetrahedron (with the numbers 3, 2, 3), and of the dihedron (2, 2, n).

I should like to show you here how slightly the language used in wide mathematical circles keeps pace with knowledge. The word "root" is used today nearly everywhere in two senses: once for the solution of any algebraic equation, and, secondly, in particular, for the solution of a pure equation. The latter use, of course, dates from a time when only pure equations were studied. Today it is, if not actually harmful, at least rather inconvenient. Thus it seems almost a contradiction to say that the "*roots*" of an equation cannot be expressed by means of *radical signs*. But there is another form of expression which has lingered on from the beginnings of algebra and which is a more serious source of misunderstanding, namely, that algebraic equations are said to be "not algebraically solvable", if they cannot be solved in terms of radicals i.e. if they cannot be reduced to pure equations. This use is in immediate contradiction with the modern meaning of the word "algebraic". Today we say that an equation can be solved algebraically when we can reduce it to a chain of simplest algebraic equations in which one controls the dependence of the solutions upon the parameters, the relation of the different roots to one another, etc. as completely as one does in the case of the pure equation. It is not at all necessary that these equations should be *pure* equations. In this sense we may say that the icosahedral equation can be solved algebraically, for our discussion shows that we can construct its theory in a manner that meets all the demands mentioned above. The fact that this equation cannot be solved by radicals lends it special interest by suggesting it as an appropriate normal equation to which one might try to reduce, (i. e., completely solve) still other equations which are in the old sense algebraically unsolvable.

The last remark leads us to the last section of this chapter, in which we shall try to get a general view of such reductions.

8. Reduction of General Equations to Normal Equations

It turns out, namely, that the following reductions are possible:
The general equation of the third degree to the dihedral equation for $n = 3$;
The general equation of the fourth degree to the tetrahedral or to the octahedral equation;
The general equation of the fifth degree to the icosahedral equation.
This result is the most recent triumph of the theory of the regular bodies which have always played such an important rôle since the beginning of mathematical history, and which have a decisive influence in the most widely separated fields of modern mathematics.

In order to show you the meaning of my general assertion I shall go somewhat more into details for the equation of degree three, without, however, fully proving the formulas. We again take the cubic equation in the reduced form

(1) $$x^3 + px - q = 0.$$

Denoting solutions by x_1, x_2, x_3, we try to set up a rational function z of them which undergoes the six linear substitutions of the dihedron for $n = 3$ when we interchange the x_i in all six possible ways. The values that z should take on are

$$z,\ \varepsilon z,\ \varepsilon^2 z,\ \frac{1}{z},\ \frac{\varepsilon}{z},\ \frac{\varepsilon^2}{z} \left(\text{where } \varepsilon = e^{\frac{2i\pi}{3}}\right).$$

It is easily seen that

(2) $$z = \frac{x_1 + \varepsilon x_2 + \varepsilon^2 x_3}{x_1 + \varepsilon^2 x_2 + \varepsilon x_3}$$

satisfies these conditions. The dihedral function $z^3 + 1/z^3$ of this quantity must remain unaltered by all the interchanges of the x_k, since the six linear substitutions of the z leave it unchanged. Hence, by a well known theorem of algebra, it must be a rational function of the coefficients of (1). A calculation shows that

(3) $$z^3 + \frac{1}{z^3} = -27\frac{q^2}{p^3} - 2.$$

Conversely, if we solve this dihedral equation, and if z is one of its roots, we can express the three values x_1, x_2, x_3 rationally in terms of

z, p, and q by means of (2) and the well known relations

$$x_1 + x_2 + x_3 = 0, \qquad x_1 x_2 + x_2 x_3 + x_3 x_1 = p, \qquad x_1 x_2 x_3 = q.$$

Doing this, we find

(4)
$$\begin{cases} x_1 = -\dfrac{3q}{p} \cdot \dfrac{z(1+z)}{1+z^3}, \\ x_2 = -\dfrac{3q}{p} \cdot \dfrac{\varepsilon z(1+\varepsilon z)}{1+z^3}, \\ x_3 = -\dfrac{3q}{p} \cdot \dfrac{\varepsilon^2 z(1+\varepsilon^2 z)}{1+z^3}. \end{cases}$$

Thus, as soon as the dihedral equation (3) has been solved, the formulas (4) give at once the solution of the cubic (1).

In the same way we may reduce the general equations of the fourth and fifth degrees. The equations would be, of course, somewhat longer, but not more difficult in principle. The only new thing would be that the parameter w of the normal equation, which was expressed above rationally in the coefficients of the equation $\left(2w = -27\dfrac{q^2}{p^2} - 2\right)$, would now contain square roots. You will find this theory for the equation of degree five given fully in the second part of my lectures on the icosahedron. Not only are the formulas calculated, but also the essential reasons for the appearance of the equations are explained.

Finally, let me say a word about the relation of this development to the usual presentation of the theory of equations of the third, fourth, and fifth degree. In the first place, we can obtain the usual solutions of the cubic and biquadratic from our formulas by appropriate reductions, if we use the solutions of the equations of the dihedron, octahedron, and tetrahedron in terms of radicals. In the case of equations of degree five, most of the textbooks confine themselves unfortunately to the establishment of the negative result that the equation cannot be solved by radicals, to which is then added the vague hint that the solution is possible by elliptic functions, to be exact one should say elliptic *modular* functions. I take exception to this procedure because it exhibits a one-sided contrast and hinders rather than promotes a real understanding of the situation. In view of the preceding survey, using first algebraic and then analytic language, we may say:

1. *The general equation of the fifth degree cannot be reduced, indeed, to pure equations, but it is possible to reduce it to the icosahedral equation as the simplest normal equation.* This is the real problem of its algebraic solution.

2. *The icosahedral equation, on the other hand, can be solved by elliptic modular functions.* For purposes of numerical calculation, this is the full analog of the solution of pure equations by means of logarithms.

This supplies the complete solution of the problem of the equation of fifth degree. Remember that when the usual road does not lead to success, one should not be content with this determination of impossibility, but should bestir oneself to find a new and more promising route. Mathematical thought, as such, has no end. If someone says, to you that mathematical reasoning cannot be carried beyond a certain point, you may be sure that the really interesting problem begins precisely there.

In conclusion, it might be remarked that these theories do not stop with equations of degree five. On the contrary, one can set up analogous developments for equations of the sixth and higher degrees if one will only make use of the higher-dimensional analogs of the regular bodies. If you are interested in this, you might read my article[1] *Über die Auflösung der allgemeinen Gleichung fünften und sechsten Grades**. In connection with this article the problem was successfully attacked by P. Gordan[2] and A. B. Coble[3]. The investigation is somewhat simplified in the latter memoir[4].

---

[1] Journal für Mathematik, vol. 129 (1905), p. 151; and Mathematische Annalen, vol. 61 (1905), p. 50.

* *Concerning the solution of the general equation of fifth and of sixth degree.*

[2] Mathematische Annalen, vol. 61 (1905), p. 50; and vol. 68 (1910), p. 1.

[3] Mathematische Annalen, vol. 70 (1911), p. 337.

[4] See also Klein, F., Gesammelte Mathematische Abhandlungen, vol. 2, p. 502—503.

Part Three

Analysis

During this second half of the semester we shall select certain chapters in analysis which are important from our standpoint and we shall discuss them as we did arithmetic and algebra. The most important thing for us to discuss will be the elementary transcendental functions, i. e. logarithmic and exponential functions and trigonometric functions, since they play an important part in school instruction. Let us begin with the first.

I. Logarithmic and Exponential Functions

Let me recall briefly the familiar curriculum of the school, and the continuation of it to the point at which the so called algebraic analysis begins.

1. Systematic Account of Algebraic Analysis

One starts with powers of the form $a = b^c$, where the exponent c is a positive integer, and extends the notion step by step for negative integral values of c, then for fractional values of c, and finally, if circumstances warrant it, to irrational values of c. In this process the concept of *root* appears as that of a particular power. Without going into the details of involution, I will only recall the rule for multiplication

$$b^c \cdot b^{c'} = b^{c+c'},$$

which reduces the multiplication of two numbers to the addition of exponents. The possibility of this reduction, which, as you know, is fundamental for logarithmic calculation, lies in the fact that the fundamental laws for multiplication and addition are so largely identical, that both operations, namely, are commutative as well associative.

The operation inverse to that of raising to a power yields the logarithm. The quantity c is called the *logarithm* of a to the base b:

$$c = \log_{(b)} a.$$

At this point a number of essential difficulties appear which are usually passed over without any attempt at explanation. For this reason

I shall try to be especially clear at this point. For the sake of convenience we shall write x and y instead of a and c, inasmuch as we wish to study the mutual dependence of these two numbers. Our fundamental equations then become

$$x = b^y, \qquad y = \log x. \quad (b)$$

Let us first of all notice that b is always assumed to be positive. If b were negative, x would be alternately positive and negative for integral values of y, and would even include imaginary values for fractional values of y, so that the totality of number pairs (x, y) would not give a continuous curve. But even with $b > 0$ one cannot get along without making stipulations that appear to be quite arbitrary. For if y is rational, say $y = m/n$, where m and n are integers prime to each other, $x = b^{m/n}$ is, as you know, defined to be $\sqrt[n]{b^m}$ and it has accordingly n values, of which, for even values of n, we should have two to deal with even if we confined ourselves to real numbers. It is customary to stipulate that x shall always be the positive root, the so-called *principal* root.

If you will permit me to use, somewhat prematurely, the familiar graph of the logarithm $y = \log x$ (Fig. 54), you will see that neither the above stipulation nor its suitableness is by any means self-evident. If y traverses the dense set of rational values, the corresponding points whose abscissas are the positive principal values $x = b^y$ constitute a dense set on our curve. If, now, when the denominator n of y is even, we should mark the points which correspond to

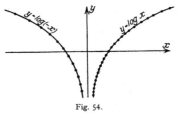

Fig. 54.

negative values of x, we have a set of points which would be, one might say, only *half so dense*, but nevertheless dense on the curve which is the reflection in the y axis of our curve $[y = \log(-x)]$. If we now admit all real, including irrational, values of y, it is certainly not immediately clear why the principal values which we have been marking on the right now constitute a continuous curve and whether or not the set of negative values which we have marked on the left is similarly raised to a continuum. We shall see later that this can be made clear only with the profounder resources of function theory, an aid which is not at the command of the elementary student. For this reason, one does not attempt in the schools to give a complete exposition. One adopts rather an authoritative convention, which is quite convincing to the pupils, namely that one must take $b > 0$ and must select the positive principal values of x, that everything else is prohibited. Then the theorem follows, of course, that the logarithm is a single-valued function defined only for a positive argument.

Once the theory is carried to this point, the logarithmic tables are put into the hands of the pupil and he must learn to use them in practical calculation. There may still be some schools — in my school days this was the rule — where little or nothing is said as to how these tables are made. That was despicable utilitarianism which is scornful of every higher principle of instruction, and which we must surely and severly condemn. Today, however, the calculation of logarithms is probably discussed in the majority of cases, and in many schools indeed the theory of natural logarithms and the development into series is taught for this purpose.

As for the first of these, the base of the system of natural logarithms is, as you know, the number

$$e = \lim_{n=\infty}\left(1 + \frac{1}{n}\right)^n = 2.7182818\ldots.$$

This definition of e is usually, in imitation of the French models, placed at the very beginning in the great text books of analysis, and entirely unmotivated, whereby the really valuable element is missed, the one which mediates the understanding, namely, an explanation why precisely this remarkable limit is used as base and why the resulting logarithms are called natural. Likewise the development into series is often introduced with equal abruptness. There is a formal assumption of the development

$$\log(1+x) = a_0 + a_1 x + a_2 x^2 + \cdots,$$

the coefficients $a_0, a_1, \ldots$, are calculated by means of the known properties of logarithms, and perhaps the convergence is shown for $|x| < 1$. But again there is no explanation as to why one would ever even suspect the possibility of a series development in the case of a function of such arbitrary composition as is the logarithm according to the school definition.

2. The Historical Development of the Theory

If we wish to find all the fundamental connections whose absence we have noted, and to ascertain the deeper reasons why those apparently arbitrary conventions must lead to a reasonable result, in short, if we wish really to press forward to a full understanding of the theory of logarithms, it will be best to follow the historical development in its broad outlines. You will see that it by no means corresponds to the practice mentioned above, but rather that this practice is, so to speak, a projection of that development from a most unfavorable standpoint.

We shall mention first a German mathematician of the sixteenth century, the Swabian, Michael Stifel, whose *Arithmetica Integra* appeared in Nürnberg in 1544. This was the time of the first beginnings of our present algebra, a year before the appearance, also in Nürnberg, of the book by Cardanus, which we have mentioned. I can show you this book, as well as most of those which I shall mention later, thanks to our unusually complete university library. You will find that it uses, for the first time, operations with powers where the exponents are any rational numbers, and, in particular, emphasizes the rule for multiplication. Indeed, Stifel gives, in a sense, the very first logarithmic table (see p. 250) which, to be sure, is quite rudimentary. It contains only the integers from -3 to 6 as exponents of 2, along with the corresponding powers $\frac{1}{8}$ to 64. Stifel appears to have appreciated the significance of the development of which we have here the beginning. He declares, namely, that one might devote an entire book to these remarkable number relations.

But in order to make logarithms really available for practical calculation Stifel lacked still an important device, namely, decimal fractions; and it was only when these became common property, after 1600, that the possibility arose of constructing real logarithmic tables. The first tables were due to the Scotchman Napier (or Neper), who lived 1550—1617. They appeared in 1614, in Edinburgh, under the title *Mirifici logarithmorum canonis descriptio*, and the enthusiasm which they aroused is evidenced by the verses with which different authors in their prefaces sang the virtues of logarithms. However, Napier's method for calculating logarithms was not published until 1619, after his death, as *Mirifici logarithmorum canonis constructio*[1].

The Swiss, Jobst Bürgi (1552—1632), had calculated a table independently of Napier, which did not appear, however, until 1620, in Prag, under the title *Arithmetische und geometrische Progresstabuln*. We, in Göttingen, should have a peculiar interest in Bürgi, as one of our countrymen, since he lived for a long time in Cassel. In general, Cassel, particularly the old observatory there, has been of importance for the development of arithmetic, astronomy, and of optics prior to the discovery of infinitesimal calculus, just as Hannover became important later as the home of Leibniz. Thus our immediate neighborhood was historically significant for our science long before this university was founded.

It is very instructive to follow the train of thought of Napier and Bürgi. Both start from values of $x = b^y$ for integral values of y and seek an arrangement whereby the numbers x shall be as close together as possible. Their object was to find for every number x, as nearly as possible, a logarithm y. This is achieved today, in school, by considering fractional values of y, as we saw before. But Napier and Bürgi, with the

---

[1] Lugduni 1620. There is a later edition in phototype. (Paris 1895.)

intuition of genius, avoided the difficulties which thus present themselves by grasping the thing by the smooth handle. They had, namely, the simple and happy thought of choosing the base b close to one, when, in fact, the successive integral powers of b are close to one another. Bürgi takes
$$b = 1.0001,$$
while Napier selects a value less than one, but still closer to it:
$$b = 1 - 0.0000001 = 0.9999999.$$

The reason for this departure by Napier from the method of today is that he had in mind the application to trigonometric calculation, where one has to do primarily with logarithms of proper fractions (sine and cosine) and these are negative for $b > 1$ but positive for $b < 1$. But with both investigators the chief thing was that they made use only of integral powers of this b and so avoided, completely, the many valuedness which embarrassed us above.

Let us now calculate, in the system of Bürgi, the powers for two neighboring exponents, y and $y + 1$:
$$x = (1.0001)^y, \qquad x + \Delta x = (1.0001)^{y+1}.$$

By subtraction, then, we have
$$\Delta x = (1.0001)^y (1.0001 - 1) = \frac{x}{10^4}$$
or, writing Δy for the differences, 1, of the values of the exponent:

(1a) $$\frac{\Delta y}{\Delta x} = \frac{10^4}{x}.$$

We have thus obtained a difference equation for the Bürgi logarithms, one which Bürgi himself used directly in the calculation of his tables. After he had determined the x corresponding to a y he obtained the following x belonging to $y + 1$ by the addition of $x/10^4$. In the same way it follows that the logarithms of Napier satisfy the difference equation

(1b) $$\frac{\Delta y}{\Delta x} = -\frac{10^7}{x}.$$

In order to see the close relationship between the two systems, we need only write for y on the one hand $y/10^4$, on the other hand $y/10^7$, i.e., we need only displace the decimal point in the logarithm. If we denote the new numbers so obtained simply by y, we shall have in each case a series of numbers which satisfy the difference equation

(2) $$\frac{\Delta y}{\Delta x} = \frac{1}{x}$$

and in which the values of y proceed by steps of 0.0001 in the one case and of -0.0000001 in the other.

If, for the sake of convenience, we now make use of the graph of the continuous exponential curve (we ought really to obtain it as the result of our discussion) we shall have a tangible representation of the points which correspond to the number series of Napier and of Bürgi. These points will be the corners of a stairway inscribed in one of the two exponential curves

(3) $\quad x = (1.0001)^{10\,000\,y}$, and $\quad x = (0.9999999)^{10\,000\,000\,y}$,

respectively, where the risers have the constant value $\Delta y = 0.0001$ and $\Delta y = 0.0000001$ in the two systems, respectively (see Fig. 55).

We can get another geometric interpretation in which we do not need to presuppose the exponential curve, which will rather point out the natural way to obtain that curve, if we replace the difference equation (2) by a summation equation, that is, if we integrate it, in a sense:

(4) $\quad \eta = \sum \dfrac{\Delta \xi}{\xi}.$

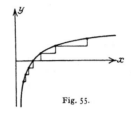

Fig. 55.

During this summation ξ increases discontinuously, from unity on, by such steps that the corresponding $\Delta \eta = \Delta \xi / \xi$ is always constant and equal to 10^{-4} and 10^{-7} respectively, so that $\Delta \xi = \xi/10^4$ and $-\xi/10^7$, in the two cases. With the last step ξ attains the value x. Once can easily give geometric expression to this procedure. For this purpose let us draw the hyperbola $\eta = 1/\xi$ in an $\xi\eta$ plane (see Fig. 56) and, beginning at $\xi = 1$, construct successively on the ξ axis the points that are given by the law of progression $\Delta \xi = \xi/10^4$ (confining ourselves to the Bürgi formulation). The rectangle of altitude $1/\xi$ erected upon each of the intervals so obtained will have the constant area

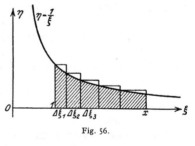

Fig. 56.

$\Delta \xi \cdot 1/\xi = 1/10^4$. The Bürgi logarithm will then be, according to (4), the 10^4-fold sum of all these rectangles inscribed in the hyperbola and lying between 1 and x. A similar result is obtained for the logarithm of Napier.

Proceeding from this last representation, one is led immediately to the natural logarithm if, instead of the sum of the rectangles, one takes the area under the hyperbola itself between the ordinates $\xi = 1$ and $\xi = x$ (shaded in the figure). This finds expression in the well-known formula

$$\log x = \int_1^x d\xi/\xi.$$

This was, in fact, the historical way, and the decisive step was taken about 1650, when analytic geometry had become the common possession of mathematicians and the infinitesimal calculus was achieving the quadrature of known curves.

If we desire to use this definition of the natural logarithm as our starting point, we must, of course, convince ourselves that it possesses the fundamental property of replacing the multiplication of numbers by the addition of logarithms; or, in modern terms, we must show that the function

$$f(x) = \int_1^x d\xi/\xi$$

defined thus by means of the area under the hyperbola, has the simple addition theorem

$$f(x_1) + f(x_2) = f(x_1 \cdot x_2).$$

In fact, if we vary x_1 and x_2, then, according to the definition of an integral, the increments of the two sides $dx_1/x_1 + dx_2/x_2$ and $d(x_1 \cdot x_2)/(x_1 \cdot x_2)$ are equal. Consequently $f(x_1) + f(x_2)$ and $f(x_1 \cdot x_2)$ can differ only by a constant, and this turns out to be zero when we put $x_1 = 1$ (since $f(1) = 0$).

If we wish to determine the "base" of the logarithms obtained in this way, we need only notice that the transition from the series of rectangles to the area under the hyperbola can be made by changing the increment $\Delta \xi = \xi/10^4$ to $\Delta \xi = \xi/n$ and allowing n to become infinite. This is the same thing as replacing the Bürgi sequence $x = (1.0001)^{10\,000\,y}$ by $x = (1 + 1/n)^{ny}$, where ny becomes infinite through integral values. According to the general definition of a power, this amounts to saying that x is the y-th power of $(1 + 1/n)^n$. Accordingly it seems plausible to say that the base is $\lim_{n=\infty} (1 + 1/n)^n$, the very limit which is ordinarily assumed at the start as the definition of e. It is interesting to note, moreover, that Bürgi's base $(1.0001)^{10\,000} = 2.718146$ coincides with e to three decimal places.

Let us now examine the historical development of the theory of the logarithm after Napier and Bürgi. First of all I shall make the following statements.

1. Mercator, whom we have already met in these pages (see p. 81) was one of the first to make use of the definition of the logarithm by means of the area of the hyperbola. In his book *Logarithmotechnica* of 1668, as well as in articles in the *Philosophical Transactions* of the London Royal Society in 1667 and 1668, he shows, by means of the same argument which I have just given you in modern terms, that $f(x) = \int_1^x d\xi/\xi$ differs from the common logarithm with the base 10, which was already the base used in calculations, only by a constant factor,

the so called modulus of the system of logarithms. Moreover he had already introduced[1] the name "natural logarithm" or "hyperbolic logarithm". But the greatest achievement of Mercator was the setting up of the power series for the logarithm, which he obtained (essentially, at least) from the integral representation by dividing out and integrating term by term. I mentioned this to you (p. 81) as an epochmaking advance in mathematics.

2. In that same connection, I told you also that Newton had taken up these ideas of Mercator's and had enriched them with two important results, namely, the general binomial theorem and the method for the reversion of series. This last appeared in a work of Newton's youth *De analysi per aequationes numero terminorum infinitas* which appeared late in print but which from 1669 on was distributed in manuscript form[2]. In this[3] Newton derives the exponential series

$$x = 1 + \frac{y}{1!} + \frac{y^2}{2!} + \frac{y^3}{3!} + \cdots$$

for the first time by reverting Mercator's series for $y = \log x$. This yields, as the number whose natural logarithm $y = 1$

$$e = 1 + \frac{1}{1!} + \frac{1}{2!} + \frac{1}{3!} + \cdots,$$

and it is now easy, with the aid of the functional equation for the logarithm, to show that, for every real rational y, x is one of the values of e^y, and in fact the positive value, in the sense of the customary definition of power. We shall go into this more in detail later on. The function $y = \log x$ thus turns out to be precisely what one would call the logarithm of x to the base e, according to the ordinary definition, in which e is defined by means of the series and not as $\lim_{n=\infty} (1 + 1/n)^n$.

3. Brook Taylor could follow a more convenient path in deriving the exponential series, after he had devised the general series-development which bears his name, which appeared in his work *Methodus Incrementorum*[4] and of which we shall have much to say later on. He could then use the relation

$$\frac{d \log x}{dx} = \frac{1}{x},$$

which is implied in the integral definition of the logarithm, infer from it the inverse relation

$$\frac{de^y}{dy} = e^y$$

---

[1] Philosophical Transactions of the Royal Society of London, vol. 3 (1668), p. 761.
[2] Newton, I.: Opuscula, Tome I, op. 1, Lausanne 1744. Appeared first in 1711.
[3] Loc. cit., p. 20.
[4] London, 1715.

and so write down at once the exponential series as a special case of his general series.

We have already seen (p. 82) how this productive period was followed by the period of criticism, I should almost like to say the period of moral despair, in which every effort was directed toward placing the new results upon a sound basis and in separating out what was false. Let us now see what attitude was taken toward the exponential function and the logarithm in the books of Euler and Lagrange, which tended in this new direction.

We shall begin with Euler's *Introductio in analysin infinitorum*[1]. Let me, first of all, praise the extraordinary and admirable analytic skill which Euler shows in all his developments, noting, however, at the same time, that he shows no trace of the rigor which is demanded today.

At the head of his developments Euler places the binomial theorem

$$(1+k)^l = 1 + \frac{l}{1}k + \frac{l(l-1)}{1\cdot 2}k^2 + \frac{l(l-1)(l-2)}{1\cdot 2\cdot 3}k^3 + \cdots,$$

in which the exponent l is assumed to be an integer. Now integral exponents are not considered in the *Introductio*. This development is specialized for the expression

$$\left(1 + \frac{1}{n}\right)^{ny},$$

in which ny is integral. He then allows n to become infinite, applies this limit process to each term of the series, thinks of e as defined by $\lim_{n\to\infty} (1 + 1/n)^n$, and so obtains the exponential series

$$e^y = 1 + y + \frac{y^2}{2!} + \frac{y^3}{3!} + \cdots.$$

To be sure, Euler is not in the least concerned here as to whether or not the individual steps in this process are rigorous, in the modern sense; in particular, whether the sum of the limits of the separate terms of the series is really the limit of the sum of the terms, or not. Now this derivation of the exponential has been, as you know, a model for numerous textbooks on infinitesimal calculus, although, as time went on, the different steps have been more and more elaborated and their legitimacy put to the test of rigor. You will see how influential Euler's work has been for the entire course of these things if you recall that the use of the letter e for that important number is due to him. "Ponamus autem brevitatis gratia pro numero hoc 2.71828... constanter litteram e", as he writes on page 90.

---

[1] Lausanne, 1748, Caput VII, p. 85 et seq. Translation by Maser, Berlin 1885, p. 70. [See also vol. VIII (1923) of Euler's Works, edited by F. Rudio, A. Krazer, and P. Stäckel.]

I might add that Euler immediately follows this with an entirely analogous derivation of the series for the sine and cosine. For this purpose he starts with the development of sin φ in powers of sin (η/n) and lets n become infinite. This is nothing else than a limit process applied to the binomial theorem, as is evident if one obtains the power series in question from De Moivre's formula:

$$\cos\varphi + i\sin\varphi = \left(\cos\frac{\varphi}{n} + i\sin\frac{\varphi}{n}\right)^n = \left(\cos\frac{\varphi}{n}\right)^n \cdot \left(1 + i\,\text{tg}\,\frac{\varphi}{n}\right)^n.$$

Let us now consider Lagrange's *Théorie des fonctions analytiques*[1]. Again it is to be noted that questions of convergence are treated, at most, only incidentally. I have already stated (p 83) that Lagrange considers only those functions that are given by power series, and defines their differential quotients formally by means of the derived power series. Consequently the Taylor's series

$$f(x+h) = f(x) + hf'(x) + \frac{h_2}{2!}f''(x) + \cdots$$

is for him simply the result of a formal reordering of the series for $f(x+h)$ proceeding originally according to powers of $x+h$. Of course, if one wishes then to apply this series to a given function, one ought really to show in advance that this function is *analytic*, i.e., that it can be developed into a power series.

Lagrange begins with the investigation of the function $f(x) = x^n$, for rational n, and determines $f'(x)$ as the coefficient of h in the expansion of $(x+h)^n$, the first two terms of which he thinks of as calculated. Then, by the same law, he obtains at once $f''(x)$, $f'''(x)$, ..., and the binomial expansion of $(x+h)^n$ appears as a special case of Taylor's series for $f(x+h)$. Moreover, let me note expressly that Lagrange does not give special consideration to the case of irrational exponents, but rather looks upon it as obviously settled when he has considered all rational values. It is interesting to contemplate this fact, since it is upon the rigorous justification of precisely this sort of transition that the greatest importance is laid today.

Lagrange uses these results in a similar treatment of the function $f(x) = (1+b)^x$. By recording the binomial series for $(1+b)^{x+h}$ he finds, namely, $f'(x)$ as the coefficient of h, then determines $f''(x)$, $f'''(x)$, ... according to the same law, and forms, finally, the Taylor series for $f(x+h) = (1+b)^{x+h}$. He is then in possession, for $h=0$, of the desired exponential series.

I should like now to finish this brief historical sketch, in which I have, of course, mentioned only names of the very first rank, by indicating what essentially new turns came with the nineteenth century.

---

[1] Paris, 1797, Reprinted in Lagrange, Œuvres, vol. 4. Paris 1881. Compare especially chapter 3, p. 34 et seq.

1. At the head of this list I should place the precise ideas concerning the convergence of infinite series and other infinite processes. Gauss takes precedence here with his *Abhandlung über die hypergeometrische Reihe** in 1812 (*Disquisitiones generales circa seriem infinitam* $1 + [(a \cdot b)/(1 \cdot c)] x + \cdots$)[1]. After him comes Abel with his memoir on the binomial series in 1826 (*Untersuchungen über die Reihe* $1 + (m/1) x + \cdots$[2]), while Cauchy, in the early twenties in his *Cours d'Analyse*[3] undertook, for the first time, a general discussion of the convergence of series. The result of these investigations, for the series which we have under consideration, is that all the earlier developments are sometimes correct, although the rigorous proofs are very complicated. For the detailed consideration of such proofs, in modern form, I refer you again to Burkhardt's *Algebraische Analysis* or to Weber-Wellstein.

2. Although we shall have occasion to talk about it in detail later, I must mention here the final foundation by Cauchy of the infinitesimal calculus. By means of it the theory of the logarithm, which we discussed above as taking its start at the hands of Bürgi and Napier in the seventeenth century, was established with full mathematical exactness.

3. Finally, we must mention the rise of that theory which is indispensable to a complete understanding of the logarithmic and exponential functions, namely, the theory of functions of a complex argument, often called, briefly, function theory. Gauss was the first to have a complete view of the foundations of this theory, even though he published little or nothing concerning it. In a letter to Bessel, dated December 18, 1811, but published much later[4], he sketches and explains with admirable clearness the significance of the integral $\int_1^z dz/z$ in the complex plane, in so far as it is an infinitely many-valued function. The fame of having also created independently the complex function theory and of having made it known to the mathematical world belongs, however, to Cauchy.

The result of these developments, insofar as it concerns our special subject, might be briefly stated as follows: The introduction of the logarithm by means of the quadrature of the hyperbola is the equal in rigor of any other method, whereas it surpasses all others, as we have seen, in simplicity and clearness.

---

* Memoir on the hypergeometric series.

[1] Commentationes societatis regiae Göttingiensis recentiores, vol. 11 (1813), No. 1, pp. 1—46. Werke vol. 3, pp. 123—162. German translation by Simon, Berlin 1888.

[2] Journal für Mathematik, vol. 1 (1826), pp. 311—339. Ostwalds Klassiker No. 71.

[3] Première Partie, Analyse Algébrique. Paris 1821. = Œuvres, 2nd series, vol. 3, Paris, 1897. German translation by Itzigsohn. Berlin 1885.

[4] *Briefwechsel zwischen Gauss und Bessel*, edited by Auwers. Berlin 1880; or Gauss Werke, vol. 8 (1900), p. 90.

3. The Theory of Logarithms in the Schools

It is remarkable that this modern development has passed over the schools without having, for the most part, the slightest effect on the instruction, an evil to which I have often alluded. The teacher manages to get along still with the cumbersome algebraic analysis, in spite of its difficulties and imperfections, and avoids the smooth infinitesimal calculus, although the eighteenth century shyness toward it has long lost all point. The reason for this probably lies in the fact that mathematical instruction in the schools and the onward march of investigation lost all touch with each other after the beginning of the nineteenth century. And this is the more remarkable since the specific training of future teachers of mathematics dates from the early decades of that century. I called attention in the preface to this discontinuity, which was of long standing, and which resisted every reform of the school tradition: In the schools, namely, one cared little whether and how the given theorems were extended at the university and one was therefore satisfied often with definitions which were perhaps sufficient for the present, but which failed to meet later demands. In a word, Euler remained the standard for the schools. And conversely, the university frequently takes little trouble to make connection with what has been given in the schools, but builds up its own system, sometimes dismissing this or that with brief consideration and with the inappropriate remark: "You had this at school".

In view of this, it is interesting to note that those university teachers who give lectures to wider circles, e.g. to students of natural science and technology, have, of their own accord, adopted a method of introducing the logarithm which is quite similar to the one which I am recommending. Let me mention here, in particular, Scheffer's *Lehrbuch der Mathematik für Studierende der Naturwissenschaften und Technik*[*][1]. You will find there in chapters six and seven a very detailed theory of the logarithm and the exponential function, which coincides entirely with our plan and which is followed in chapter eight by a similar theory of the trigonometric functions. I urge you to make the acquaintance of this book. It is very appropriate for teachers, for whom it is designed, in that the material is presented fully, in readable form, and adapted to the comprehension even of the less gifted. Note, too, the great pedagogic skill of Scheffers when he (to cite one example) continually draws attention to the small number of formulas in the theory of logarithms that one needs to know by heart, provided the subject is once understood; for one can then easily look them up when they are needed. In this way he encourages the reader to persevere in face of

---

[*] *Textbook of Mathematics for Students of Natural Science and Technology.*
[1] Leipzig, 1905; fifth ed. 1921.

the great mass of new material. I call your attention also to the fact that although Scheffers takes it for granted that the subject has been studied in school, he nevertheless develops it here in detail, on the assumption that most of what was learned in school has been forgotten. In spite of this, it does not occur to Scheffers to make proposals for a reform of instruction in the schools, as I am doing.

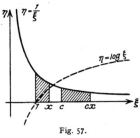

Fig. 57.

I should like to outline briefly once more my plan for introducing the logarithm into the schools in this simple and natural way. The first principle is that the proper source from which to bring in new functions is the quadrature of known curves. This corresponds, as I have shown, not only to the historical situation but also to the procedure in the higher fields of mathematics, e. g., in elliptic functions. Following this principle one would start with the hyperbola $\eta = 1/\xi$ and define the logarithm of x as the area under this curve between the ordinates $\xi = 1$ and $\xi = x$ (see Fig. 57). If the end ordinate is allowed to vary, it is easy to see how the area changes with ξ and hence to draw approximately the curve $\eta = \log \xi$.

In order now to obtain simply the functional equation of the logarithm we can start with the relation

$$\int_1^x \frac{d\xi}{\xi} = \int_c^{cx} \frac{d\xi}{\xi},$$

which is obtained by applying the transformation $c\xi = \xi'$ to the variable of integration. This means that the area between the ordinates 1 and x is the same as that between the ordinates c and cx which are c times as far from the origin. We can make this clear geometrically by observing that the area remains the same when we slide it along the ξ axis under the curve provided we stretch the width in the same ratio as we shrink the height. From this the addition theorem follows at once:

$$\int_1^{x_1} \frac{d\xi}{\xi} + \int_1^{x_2} \frac{d\xi}{\xi} = \int_1^{x_1} \frac{d\xi}{\xi} + \int_{x_1}^{x_1 \cdot x_2} \frac{d\xi}{\xi} = \int_1^{x_1 \cdot x_2} \frac{d\xi}{\xi}.$$

I wish very much that some one would give this plan a practical test in the schools. Just how it should be carried out in detail must, of course, be decided by the experienced school man. In the Meran school curriculum we did not quite venture to propose this as the standard method.

4. The Standpoint of Function Theory

Let us, finally, see how the modern theory of functions disposes of the logarithm. We shall find that all the difficulties which we met in our earlier discussion will be fully cleared away. From now on we shall use, instead of y and x, the complex variables $w = u + iv$ and $z = x + iy$. Then

1. The logarithm is defined by means of the integral

$$(1) \qquad w = \int_1^z \frac{d\zeta}{\zeta},$$

where the path of integration is any curve in the ζ plane joining $\zeta = 1$ to $\zeta = z$.

2. The integral has infinitely many values according as the path of integration encircles the origin $0, 1, 2, \ldots$ times, so that $\log z$ is an infinitely-many-valued function. One definite value, the principal value $[\log z]$, is determined if we slit the plane along the negative real axis and agree that the path of integration shall not cross this cut. It still remains arbitrary, of course, whether we shall choose to reach the negative real values from above or from below. According to the decision on this point the logarithm has $+\pi i$ or $-\pi i$ for its imaginary part. The general value of the logarithm is obtained from the principal value by the addition of an arbitrary multiple of $2i\pi$:

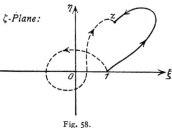

Fig. 58.

$$(2) \qquad \log z = [\log z] + 2k\pi i, \quad (k = 0, \pm 1, \pm 2, \ldots).$$

3. It follows from the integral definition of $w = \log z$ that the inverse function $z = f(w)$ satisfies the differential equation

$$(3) \qquad \frac{df}{dw} = f.$$

From this we can at once write down the power series for f

$$z = f(w) = 1 + \frac{w}{1!} + \frac{w^2}{2!} + \frac{w^3}{3!} + \cdots.$$

Since this series converges for every finite w, we can infer that the inverse function is a single-valued function which can be singular only for $w = \infty$, i.e., that it is an integral transcendental function.

4. The addition theorem for the logarithm is derived from the integral definition, just as for real variables. From it we obtain for the inverse function the equation

$$(4) \qquad f(w_1) \cdot f(w_2) = f(w_1 + w_2).$$

Similarly, it follows from (2) that

$$(5) \qquad f(w + 2k\pi i) = f(w), \quad (k = 0, \pm 1, \pm 2, \ldots)$$

i.e., $f(w)$ is a simply periodic function with the period $2\pi i$.

Analysis: Logarithmic and Exponential Functions.

5. If we put $f(1) = e$, it follows from (4) that for every rational value m/n of w the function $f(w)$ will be one of the n values of $\sqrt[n]{e^m}$, as this expression is usually defined; that is

$$f\left(\frac{m}{n}\right) = \sqrt[n]{e^m} = e^{\frac{m}{n}}.$$

We shall adopt the customary notation, and denote this one value of $f(w)$ by $e^w = e^{m/n}$, so that e^w is a well defined single-valued function, and indeed, the one given by equation (3).

6. What sort of a function, then, shall we understand, in the most general sense, by the power b^w with an arbitrary base b? We must adopt such conventions, of course, that the formal rules for exponents are satisfied. In order then to establish a connection between b^w and the function e^w which we have just defined, let us put b equal to $e^{\log b}$, where $\log b$ has the infinitely many values

$$\log b = [\log b] + 2k\pi i, \qquad (k = 0, \pm 1, \pm 2, \ldots).$$

It follow then that

$$b^w = (e^{\log b})^w = e^{w \cdot \log b} = e^{w[\log b]} \cdot e^{2k\pi i w}, \qquad (k = 0, \pm 1, \pm 2, \ldots),$$

and this expression represents, for the different values of k, infinitely many functions which are completely unconnected. We have thus the remarkable result that the values of the general exponential expression b^w, as these are obtained by the processes of raising to a power and extracting a root, do not belong at all to one coherent analytic function, but to infinitely many different functions of w, each of which is single-valued.

The values of these functions are, to be sure, related to each other in various ways. In particular they are all equal when w is an integer; and there are only a finite number of different ones among them (namely, n) when w is a fraction m/n in its lowest terms. These n values are $e^{(m/n)\log b} \cdot e^{2k\pi i(m/n)}$ for $k = 0, 1, \ldots, n-1$, that is, the n values of $\sqrt[n]{b^m}$, as we should expect.

7. It is only now that we can appreciate the inappropriateness of the traditional method which starts from involution and evolution and expects to arrive at a single-valued exponential function. It finds itself in an outright labyrinth in which it cannot possibly find its way by so called elementary means, especially since it restricts itself to real quantities. You will see this clearly if you will consider the situation when b is negative, with the aid of the illuminating results which we have just obtained. In this connection I merely remind you that we are only now in a position to understand the suitableness of the definition of the principal value ($b > 0$ and $b^{m/n} > 0$; see p. 145) which at the

time seemed arbitrary. It yields the values of one only of our infinitely many functions, namely those of the function
$$[b^w] = e^{w[\log b]}.$$
On the other hand, if n is even, the negative real values of $b^{m/n}$ will constitute a set which is everywhere dense, but they belong to an entirely different one of our infinitely many functions, and cannot possible combine to form a continuous analytic curve.

I should now like to add a few remarks of a more serious nature concerning the function theoretic nature of the logarithm. Since $w = \log z$ suffers an increment of $2\pi i$ every time z makes a circuit about $z = 0$, the corresponding Riemann surface of infinitely many sheets must have at $z = 0$ a branch point of infinitely high order so that each circuit means a passage from one sheet into the next one. If one goes over to the Riemann sphere it is easy to see that $z = \infty$ is another branch point of the same order and that there are no others. We can now make clear what one calls the uniformizing power of the logarithm of which we have already spoken in connection with the solution of certain algebraic equations (see p. 133 et sq.). To fix ideas let us consider a rational power, $z^{m/n}$. By reason of the relation
$$z^{\frac{m}{n}} = e^{\frac{m}{n}\log z}$$
this power will be a single-valued function of $w = \log z$. This is expressed by saying that it is uniformized by means of the logarithm. In order to understand this, let us think of the Riemann surface of $z^{m/n}$ as well as that of the logarithm, both spread over the z plane. This will have n sheets and its branch points will also be at $z = 0$ and $z = \infty$, at each of which all the n sheets will be cyclically connected. If we now think of any closed path in the z plane (see Fig. 59) along which the logarithm returns to its initial value, which implies that its path on the infinitely many sheeted surface is also closed, it is easy to see that the image of this path will likewise be closed when it is mapped upon the n sheeted surface. We infer from this geometric consideration that $z^{m/n}$ will always return to its initial value when $\log z$ does, and hence that it is a singlevalued function of $\log z$. I am the more willing to give this brief explanation because we have here the simplest case of the principle of uniformization, which plays such an important part in modern function theory.

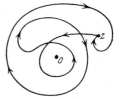

Fig. 59.

We shall now try to make clearer the nature of the functional relation $w = \log z$ by considering the conformal mapping upon the w plane of the z plane and of the Riemann surface spread upon it. In order not to be obliged to go back too far, let us refrain from including the corresponding spheres within the scope of our deliberations, in spite of the fact that it would be preferable to do so. As before, we divide the z plane along the axis of reals into a shaded (upper) and a unshaded (lower) half plane. Each of these must have infinitely many images in the w plane, since $\log z$ is infinitely many valued, and all these images must lie in smooth connection with one another since the inverse function $z = e^w$ is one valued. This means that the w plane is divided into parallel strips of width π separated from one another by parallels to the real axis (see Fig. 60). These strips are to be alternately shaded and left blank (the first one above the real axis is shaded) and they represent, accordingly, alternate conformal maps of the upper and lower z half planes while the separating parallels correspond to the parts of the real z axis. As to the correspondence in detail, I shall remark only that z always approaches 0 when w, within a strip, tends to the left toward infinity, that z becomes infinite when w approaches infinity to the right, and that the inverse function e^w has an essential singularity at $w = \infty$.

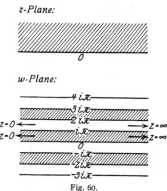

Fig. 60.

I must not omit here to draw attention to the connection between this representation and the theorem of Picard, since that is one of the most interesting theorems of the newer function theory. Let $z(w)$ be an integral transcendental function, that is, a function which has an essential singularity only at $w = \infty$ (e.g. e^w). The question is whether there can be values z, and how many of them, which cannot be taken at any finite value of w, but which are approached as a limit when w becomes infinite in an appropriate way. The theorem of Picard states that a function in the neighborhood of an essential singularity can omit at most two different values; that an integral transcendental function, therefore, can omit, besides $z = \infty$, (which it of necessity omits), at most one other value. e^w is an example of a function which really omits one other value besides ∞, namely $z = 0$. In each of the parallel strips of our division e^w approaches each of these values but it assumes neither of them for any finite value of w. The function $\sin w$ is an example of a function which omits no value except $z = \infty$.

I should like to conclude this discussion by bringing up again a point which we have repeatedly touched and applying to it these geometric aids. I refer to the passage to the limit from the power to the exponential function which is given by the formula

$$e^w = \lim_{n=\infty}\left(1 + \frac{1}{n}\right)^{nw}.$$

If we put $n \cdot w = \nu$ this takes the form

$$e^w = \lim_{\nu = \infty} \left(1 + \frac{w}{\nu}\right)^\nu.$$

Let us, before passing to the limit, consider the function

$$f_\nu(w) = \left(1 + \frac{w}{\nu}\right)^\nu,$$

whose function-theoretic behavior, as a power, is known to us. It has a critical point, at $w = -\nu$ and $w = \infty$, where the base becomes 0 and ∞ respectively, and it maps the f_ν half planes conformally upon sectors of the w plane which have $w = -\nu$ as common vertex and the angular opening π/ν (see Fig. 61). If ν is not an integer this series of sectors can cover the w plane a finite or an infinite number of times, corresponding to the many valuedness of f_ν. If now ν becomes infinite, the vertex, $-\nu$, of the sectors moves off without limit to the left and it is clear that

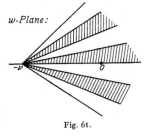

Fig. 61.

these sectors lying to the right of $-\nu$ go over into the parallel strips of the w plane which belong to the limit function e^w. This explains geometrically the limit definition of e^w. One can verify by calculation that the width of the sectors at $w = 0$ goes over into the strip width π of the parallel division.

But a doubt arises here. If ν becomes infinite continuously, it passes through not only integral but also rational and irrational values, for which the f_ν will be many valued and will correspond to many sheeted surfaces. How can these go over into the smooth plane which corresponds to the single-valued function e^w? If, for example, we allow ν to approach infinity only through rational values having

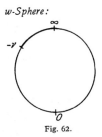

Fig. 62.

n for a denominator each $f_\nu(w)$ will have an n sheeted Riemann surface. In order to follow the limit process, let us, for a moment, consider the w sphere. It is covered for each $f_\nu(w)$ with n sheets which are connected at the branch points $-\nu$ and ∞. Let the branch cut lie along the minor meridian segment joining these points, as shown in Fig. 62. If now ν approaches ∞ the branch points coincide and the branch cut disappears. Thus the bridge is destroyed that supplied the connection between the sheets, there emerge n separate sheets and, corresponding to them, n single-valued functions, of which only one is our e^w. If we now allow ν to vary through all real values, we shall have, in general, surfaces with infinitely many sheets whose connection is broken in the limit. The values on one leaf of each of these surfaces converge toward the single-valued function e^w, which is spread over the smooth sphere, while the sequences of values on the other sheets have, in general, no limit whatever. We thus have a complete explanation of the right complicated and wonderful passage to the limit from the many valued power to the single-valued exponential function.

As a general moral of these last considerations we might say that a complete understanding of such problems is possible only when they are taken into the field of complex numbers. Is this, then, not a sufficient reason for teaching complex function theory in the schools? Max Simon, for one, has in fact supported similar demands. I hardly believe, however, that the average pupils, even in the highest class, can be carried so far, and I think, therefore, that we should abandon those aspects of method as to algebraic analysis in the schools which incline toward such considerations, in favor of the simple and natural way which we have developed above. I am, to be sure, all the more desirous that the teacher shall be in full possession of all the function-theoretic connections that come up here; for the teacher's knowledge should be far greater than that which he presents to his pupils. He must be familiar with the cliffs and the whirlpools in order to guide his pupils safely past them.

After these detailed discussions we can now be briefer in the corresponding consideration of the goniometric functions.

II. The Goniometric Functions

Let me say, before beginning, that the name *goniometric functions* seems preferable to the customary name *trigonometric functions*, since trigonometry is but a particular application of these functions, which are of the greatest importance for mathematics as a whole. Their inverse functions are analogous to the logarithm, while they themselves are analogous to the exponential function. We shall call these inverse functions the *cyclometric functions*.

1. Theory of the Goniometric Functions

As a starting point for our theoretical considerations let me suggest the question as to the most appropriate way of introducing the goniometric functions in the schools. I think that here also it would be best to make use of our general principle of quadrature. The customary procedure, which begins with the measurement of the circular arc, does not seem to me to be so very obvious, and it lacks, above all, the advantage of affording a simple and coherent control both of elementary and advanced fields.

Again I shall make immediate use of analytic geometry. Let us start with the unit circle
$$x^2 + y^2 = 1$$

and consider the sector formed by the radii to the points $A\ (x=1, y=0)$ and $P\ (x, y)$ (see Fig. 63). In order to be in agreement with the usual notation, I shall denote the area of this sector by $\varphi/2$. (Then the arc in the customary notation will be φ.)

I shall define the goniometric functions sine and cosine of φ as the lengths of the coordinates x and y of the limiting point P of the sector $\varphi/2$:

$$x = \cos\varphi, \qquad y = \sin\varphi.$$

The origin of this notation is not clear. The word "sinus" probably arose through an erroneous translation of an Arabic word into Latin. Since we did not start from the arc we cannot well designate the inverse functions, i. e., the double sector, as, a function of the coordinates, by using the customary terms arc sine and arc cosine, but it is natural by analogy to call $\varphi/2$ the "area" of the sine (or cosine) and to write

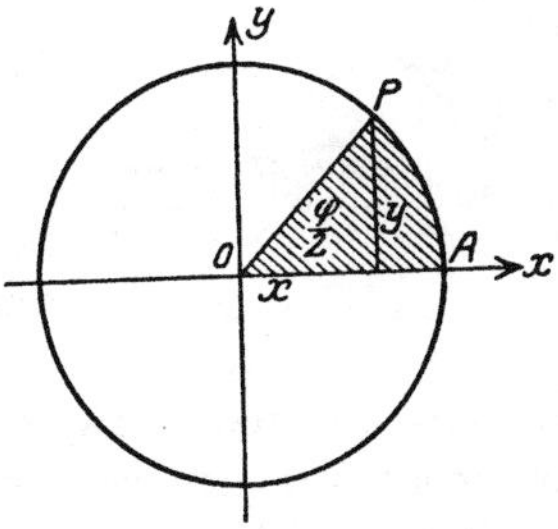

$$\varphi = 2\ \text{area} \sin y = \text{arc} \sin y,$$
$$\varphi = 2\ \text{area} \cos x = \text{arc} \cos x.$$

Fig. 63.

The following notation, used in England and in America is also quite appropriate:

$$\varphi = \cos^{-1} x, \qquad \varphi = \sin^{-1} y.$$

The further goniometric functions:

$$\tan\varphi = \frac{\sin\varphi}{\cos\varphi}, \qquad \text{ctn}\,\varphi = \frac{\cos\varphi}{\sin\varphi}$$

(in the older trigonometry also secant and cosecant) are defined as simple rational combinations of the two fundamental functions. They are introduced only with a view to brevity in practical calculation and have for us no theoretical significance.

If we follow the coordinates of P with increasing φ we can at once obtain qualitatively a representation of the cosine and

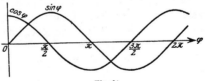

Fig. 64.

sine curves in a rectangular coordinate system. They are the well known wave lines with a certain period 2π (see Fig. 64), where π is defined as the area of the entire unit circle, instead of as usual, the length of the semi-circle.

Let us now compare once more our introduction of the logarithm and the exponential function with these definitions. You will recall that our point of departure was a rectangular hyperbola referred to its asymptotes as axes.

$$\xi \cdot \eta = 1.$$

The semi axis of this hyperbola is $OA = \sqrt{2}$ (see Fig. 65), whereas the circle had the radius 1. Let us now consider the area of the strip between the fixed ordinate AA' ($\xi = 1$) and the variable ordinate PP'. If this is called Φ, we may put $\Phi = \log \xi$, and the coordinates of P are expressed in terms of Φ in the form

$$\xi = e^{\Phi}, \qquad \eta = e^{-\Phi}.$$

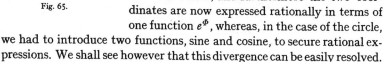

Fig. 65.

You notice a certain analogy with the preceding discussion, but that the analogy fails in two respects. In the first place, Φ is not a sector as it was before, and furthermore the two coordinates are now expressed rationally in terms of one function e^{Φ}, whereas, in the case of the circle, we had to introduce two functions, sine and cosine, to secure rational expressions. We shall see however that this divergence can be easily resolved.

Notice, in the first place, that the area of the triangle $OP'P$, namely $\frac{1}{2} \xi \cdot \eta = \frac{1}{2}$, is independent of the position of P. In particular, then, it is the same as that of $OA'A$. Therefore, if we add the latter triangle to Φ and then subtract the former triangle from this sum, we see that Φ can be defined as the area of a hyperbolic sector lying between a radius vector to the vertex A and one to a variable point P, just as in the case of the circle. There is still a difference in sign. Before, the arc AP, looked at from O, was counterclockwise, whereas now it is clockwise. We can remove this difference by reflecting the hyperbola in OA, i.e., by interchanging ξ and η. We get then as coordinates of P

$$\xi = e^{-\Phi}, \qquad \eta = e^{\Phi}.$$

Finally let us introduce the principal axes in place of the asymptotes as axes of reference, by turning Fig. 65 through 45° (after reflection in OA). If we call the new coordinates (X, Y), the equations of this transformation are

$$X = \frac{\xi + \eta}{\sqrt{2}}, \qquad Y = \frac{-\xi + \eta}{\sqrt{2}}.$$

The equation of the hyperbola then becomes

$$X^2 - Y^2 = 2,$$

and the sector Φ now has precisely the same position that sector $\Phi/2$ had in the circle. The new coordinates of P as functions of Φ may be written in the form

$$X = \frac{e^{\Phi} + e^{-\Phi}}{\sqrt{2}}, \qquad Y = \frac{e^{\Phi} - e^{-\Phi}}{\sqrt{2}}.$$

It remains only to reduce the entire figure in the ratio $1:\sqrt{2}$ in order to make the semi axis of the hyperbola 1 instead of the $\sqrt{2}$, as it was in the case of the circle. Then the sector in question has the area $\varphi/2$, in complete accord with the preceding. If we call the new coordinates (x, y) again, they will be the following functions of Φ

$$x = \frac{e^{\Phi} + e^{-\Phi}}{2},$$

$$y = \frac{e^{\Phi} - e^{-\Phi}}{2},$$

which satisfy the relation

$$x^2 - y^2 = 1,$$

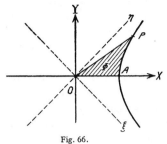

Fig. 66.

which is the equation of a hyperbola. These functions are called hyperbolic cosine and sine and are written in the form

$$x = \cosh \Phi = \frac{e^{\Phi} + e^{-\Phi}}{2}, \qquad y = \sinh \Phi = \frac{e^{\Phi} - e^{\Phi}}{2}.$$

The final result, then, is that if we treat the circle and the rectangular hyperbola, each with semiaxis one, in literally the same way we obtain on the one hand the ordinary goniometric functions, on the other the hyperbolic functions, so that these functions correspond fully to one another.

You know that these functions cosh and sinh can be used to advantage in many cases. Nevertheless we have really taken a step backward here, so far as the treatment of the hyperbola is concerned. Whereas at first, the coordinates (ξ, η) could be rationally expressed in terms of a single function e^{Φ}, it now requires two functions, which are connected by an algebraic relation (the equation of the hyperbola). It is natural, therefore to attempt a converse treatment for the goniometric functions, analogous to the original developments for the hyperbola. This is, in fact, quite easy if one does not object to the use of complex quantities, and it leads to the setting up of a single fundamental function in terms of which $\cos \varphi$ and $\sin \varphi$ can be expressed rationally, just as $\cosh \Phi$ and $\sinh \Phi$ are in terms of e^{Φ}, and which is therefore entitled to play the chief rôle in the theory of the goniometric functions.

To this end we introduce into the equation of the circle $x^2 + y^2 = 1$ (where $x = \cos \varphi$, $y = \sin \varphi$) the new coordinates

$$x - iy = \xi, \qquad x + iy = \eta,$$

which gives

$$\xi \cdot \eta = 1.$$

Analysis: The Goniometric Functions.

The desired central function is now the second coordinate η, just as it was above in the case of the hyperbola. If we denote it by $f(\varphi)$ we have, by virtue of the equations of transformation:

$$\eta = f(\varphi) = \cos\varphi + i\sin\varphi, \qquad \xi = \frac{1}{f(\varphi)} = \cos\varphi - i\sin\varphi.$$

From the last equations we get

$$\cos\varphi = \frac{\xi+\eta}{2} = \frac{f(\varphi)+[f(\varphi)]^{-1}}{2}, \qquad \sin\varphi = \frac{-\xi+\eta}{2i} = \frac{f(\varphi)-[f(\varphi)]^{-1}}{2i},$$

where we have complete analogy with the earlier relations between $\cosh\Phi$, $\sinh\Phi$, and e^Φ. If prominence is thus given, from the start, to the analogy between the circular and the hyperbolic functions, the great discovery of Euler that $f(\varphi) = e^{i\varphi}$ is divested of the mystery that usually attaches to it.

The question now arises whether we cannot effect a similar reduction of $\cos w$ and $\sin w$ to a single fundamental function, without leaving the real field. This is indeed possible if we look at our figures in the light of projective geometry. In the case of the hyperbola, in fact, we could define the coordinate η, which supplied the fundamental function, as parameter in a pencil of parallels $\eta = $ constant. This means, projectively, so far as the hyperbola is concerned, that we have a pencil of lines with its vertex on the hyperbola (in particular, here, at one of the infinitely distant points). If, now, in the case of either circle or hyperbola we think of the parameter of any such pencil as a function of the area, we obtain likewise a fundamental function and one which involves only real quantities.

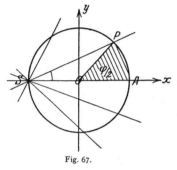

Fig. 67.

Let us think now of the circle (Fig. 67) and the pencil through the point $S(-1,0)$

$$y = \lambda(x+1),$$

where λ is the parameter. On a former occasion (p. 45), we found as the coordinates of the intersection P of the circle and the ray corresponding to λ,

$$x = \cos\varphi = \frac{1-\lambda^2}{1+\lambda^2}, \qquad y = \sin\varphi = \frac{2\lambda}{1+\lambda^2},$$

so that

$$\lambda = \lambda(\varphi) = \frac{y}{x+1}$$

is, in fact, an appropriate real fundamental function. Moreover, since $\angle PSO = \frac{1}{2} POA$, and $POA = \varphi$, it follows at once that $\lambda = \tan\varphi/2$.

The one-valued representation of $\sin \varphi$ and $\cos \varphi$ in terms of $\tan \varphi/2$ which appears in this way is often used in trigonometric calculations.

The connection between λ and the earlier fundamental function $f(\varphi)$ appears from the last formula in the form

$$\lambda = \frac{y}{x+1} = \frac{1}{i} \cdot \frac{f - f^{-1}}{f + f^{-1} + 2} = \frac{1}{i} \frac{f^2 - 1}{f^2 + 1 + 2f} = \frac{1}{i} \frac{f(\varphi) - 1}{f(\varphi) + 1},$$

or conversely,

$$f(\varphi) = x + iy = \frac{1 - \lambda^2 + 2i\lambda}{1 + \lambda^2} = \frac{1 + i\lambda}{1 - i\lambda}.$$

The introduction of λ amounts, then, simply to the determination of a linear fractional function of $f(\varphi)$ which is real along the circumference of the unit circle. In this way the formulas turn out to be real but somewhat more complicated than by the immediate use of $f(\varphi)$.

Whether one is willing to give up the advantage of reality in the face of this disadvantage, depends, of course, upon how well the person concerned knows how to deal with complex quantities. It is noteworthy, in this connection, that physicists have long since gone over to the use of complex quantities, especially in optics, for example, as soon as they have to do with equations of vibration. Engineers, in particular electrical engineers with their vector diagrams, have recently been using complex quantities advantageously. We can say then that the use of complex quantities is at last beginning to spread, even though at present the great majority still prefer the restriction to real numbers.

Passing on to a brief survey of the farther development of the theory of the goniometric functions, let us next consider certain fundamental laws.

1. The addition theorem for $\sin \varphi$ is

$$\sin(\varphi + \psi) = \sin \varphi \cos \psi + \cos \varphi \sin \psi$$

and there is a corresponding formula for $\cos(\varphi + \psi)$. These formulas appear to be more difficult than those for the exponential function, due, of course, to the fact that we are not dealing here with the true elementary function. This function, our $f(\varphi) = \cos \varphi + i \sin \varphi$, satisfies the very simple relation

$$f(\varphi + \psi) = f(\varphi) \cdot f(\psi),$$

which is precisely the formula for e^φ.

2. It is easy now to obtain expressions for the functions of multiples of an angle and of parts of an angle. Of these I shall mention only the two formulas

$$\sin \frac{\varphi}{2} = \sqrt{\frac{1 - \cos \varphi}{2}}, \qquad \cos \frac{\varphi}{2} = \sqrt{\frac{1 + \cos \varphi}{2}}$$

because they were of such importance in constructing the first trigonometric tables. An elegant expression for all these relations is given by De Moivre's formula

$$f(n \cdot \varphi) = [f(\varphi)]^n, \quad \text{where} \quad f(\varphi) = \cos\varphi + i\sin\varphi.$$

De Moivre, who was a Frenchman, but who lived in London, and was in touch with Newton, published this formula in 1730 in his book *Miscellanea analytica*.

3. From our original definition of $y = \sin\varphi$, we can of course easily derive an integral representation for the inverse $\varphi = \sin^{-1} y$. The area in Fig. 68, consisting of the sector $\varphi/2$ (AOP) of the unit circle, together with the triangle $OP'P$, is bounded by the axes, a parallel to the x axis at the distance y away, and the curve $x = \sqrt{1-y^2}$. Its area is therefore $\int_0^y \sqrt{1-y^2}\,dy$. Since the triangle has the area

$$\tfrac{1}{2} OP' \cdot P'P = \tfrac{1}{2} y \sqrt{1-y^2},$$

we have

$$\int_0^y \sqrt{1-y^2}\,dy = \tfrac{1}{2} y \sqrt{1-y^2} + \tfrac{1}{2} \varphi.$$

From this it follows by a simple transformation that

$$\varphi = \sin^{-1} y = \int_0^y \frac{dy}{\sqrt{1-y^2}}.$$

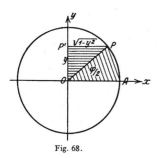

Fig. 68.

We could proceed now just as in the case of the logarithm, namely to develop the integrand by the binomial theorem, and then to integrate term by term, following Mercator. This would give us the power series for $\sin^{-1} y$, from which, by inversion, we could get the sine series itself. This is the plan that Newton himself employed, as we have seen (p. 82).

4. I prefer, however, to take the shorter way which Taylor's great discovery made possible. According to it one obtains from the above integral formula the differential quotient for the sine itself

$$\frac{d\sin\varphi}{d\varphi} = \frac{dy}{d\varphi} = \sqrt{1-y^2} = \cos\varphi,$$

from which it follows that

$$\frac{d\cos\varphi}{d\varphi} = -\sin\varphi.$$

Taylor's theorem now gives

$$\sin\varphi = \frac{\varphi}{1!} - \frac{\varphi^3}{3!} + \frac{\varphi^5}{5!} - + \cdots,$$

$$\cos\varphi = 1 - \frac{\varphi^2}{2!} + \frac{\varphi^4}{4!} - + \cdots.$$

It is easy to see that these series converge for every finite φ, including complex values, and that $\sin \varphi$ and $\cos \varphi$ are therefore defined as single-valued integral transcendental functions in the entire complex plane.

5. If we compare these series with the series for e^φ, we see that the fundamental function $f(\varphi)$ satisfies the relation

$$f(\varphi) = \cos\varphi + i \sin\varphi = e^{i\varphi}.$$

This result is unambiguous because $\sin \varphi$ and $\cos \varphi$ as well as e^Φ are single-valued integral functions.

6. It remains only to describe the nature of the complex functions $\sin w$, and $\cos w$. We notice first that each of the inverse functions $w = \sin^{-1} z$ and $w = \cos^{-1} z$ yields a Riemann surface with an infinite number of leaves and with branch points at $+1$, -1, ∞. In fact, infinitely many branch points of the first order lie over $z = +1$ and $z = -1$, while two branch points of infinitely high order lie over $z = \infty$. In order to follow better the course of the leaves in detail let us consider the division of the w plane into regions which correspond to the upper (shaded) and the lower (unshaded) z half planes. For $z = \cos w$ this division is brought about by the real axis and by the parallels to the imaginary axis through the points $w = 0$, $\pm \pi$, $\pm 2\pi$, ..., so that the resulting triangular regions (see Fig. 69), all extending to infinity, should be alternately shaded and unshaded. At the points $w = 0$,

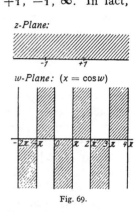

Fig. 69.

$\pm 2\pi$, $\pm 4\pi$, ... (corresponding to $z = +1$), and at the points $w = \pm \pi$, $\pm 3\pi$, ... (corresponding to $z = -1$), four of the triangles meet. These correspond to the four half leaves of the Riemann surface, which are connected at each of the corresponding branch points lying above $z = \pm 1$. If w becomes infinite within any triangle, $\cos w$ approaches the value $z = \infty$. The fact that there are two separate sets of infinitely many triangles each, all extending to infinity, corresponds to the situation that on the Riemann surface there are two separate sets of infinitely many leaves connected at $z = \infty$. For $z = \sin w$ the situation is analogous, except that the representation in the w plane is moved to the right by $\pi/2$. In these representations we find confirmation of my earlier remarks (p. 160) concerning the nature of the essential singularity at $w = \infty$ in its relation to the theorem of Picard.

2. Trigonometric Tables

After this brief survey of the theory of goniometric functions, I wish to discuss something that is of prime importance in *practical* work, namely *trigonometric tables*. At the same time I shall talk about *logarithmic tables*, which I have thus far left in the background, for the reason that from the beginning up to the present time the tabulation of logarithms has gone hand in hand with that of trigonometric values. The way in which logarithmic tables have reached their present form is of extraordinary importance and interest for the mathematician in the schools as well as in the university. I cannot describe in detail here, of course, the long history of the development of such tables, but I shall endeavor, by citing a few of the most significant works, to give you a rough historical survey. Concerning other works, some of them of equally great importance, which would round out the story, I refer you to Tropfke or, so far as logarithmic tables are concerned, to the exhaustive account in Mehmke's Encyclopedia report on *numerisches Rechnen* (Enzyklopädie, I. F.), as well as to the French revision[1] of this report by d'Ocagne.

I shall mention first the group of

A. Purely Trigonometric Tables

as they were developed before the invention of logarithms. Such tables existed in ancient times, the first of which follows.

1. *The table of chords*, by Ptolemy, which he compiled for astronomical purposes about 150 A. D. This is to be found in his work *Megale Syntaxis*, in which he developed the astronomical system bearing his name, and of which we have here a modern edition[2]. This work has come to us, by way of the Arabs, under the much used title *Almagest*, which is probably a combination of the Arabic article "al" with a mutilated form of the Greek title. The table is constructed with thirty-minute intervals. It does not give directly the sine of the angle α, but the chord of its arc (i. e. $2 \sin \alpha/2$). The values of the chords are given in three place sexagesimal fractions, that is in the form $a/60 + b/3600 + c/216000$, where a, b, c are integers between 0 and 59. The difficult thing for us, however, is that these a, b, c are written, of course, in Greek number-symbols, that is in combinations of Greek letters. The tables give also the values of the differences, which permit one to interpolate for minutes. In the calculation of his table, Ptolemy used, above all, the addition theorem for trigonometric functions, in the form of the theorem on the inscribed quadrilateral (Ptolemy's theorem). He used also the preceding formula for $\sin \alpha/2$ (i.e., the extraction of square root, in addition to the rational operations), and he employed furthermore a process of interpolation.

---

[1] Encyclopédie des Sciences Mathématiques, édition française, I, 23. See also Cajori, F., *History of Mathematics*, 1919. Macmillan; and Smith, D. E., *History of Mathematics*, 1925. Ginn.

[2] Edited by Heiberg. 1898—1903. Leipzig.

2. We advance now more than 1000 years to the time when trigonometric tables were first made in Europe. The first person who deserves mention is Regiomontanus (1436—1476), whose name was really Johannes Müller, but who changed it into the latinized form of Königsberg, his birthplace. He calculated several trigonometric tables, in which one sees distinctly the transition from the sexagesimal to the pure decimal system. At that time no one thought of the trigonometric lines as fractions corresponding to the radius one, as we do now. The values were calculated for circles with very large radii, so that they appeared as integers. To be sure, these large numbers were themselves written as decimals, but in the choice of the radius one finds a persistent suggestion of the sexagesimal system. Thus, in the first table of Regiomontanus the radius is taken as 6000000, and not until he makes the second table does he choose a pure decimal 10000000 and establish complete accord with the decimal system. By the simple insertion of a decimal point, the numbers of this table become decimals of today. These tables of Regiomontanus were first published long after his death, in the work of his teacher G. Peurbach: *Tractatus super propositiones Ptolemaei de sinubus et chordis*[1]. Notice that this work, like so many other fundamental works in mathematics*, was printed in Nürnberg in the forties of the sixteenth century. Regiomontanus himself lived mostly in Nürnberg.

3. I place before you now a work of the greatest general significance: *De revolutionibus orbium coelestium*[2] by Nic. Copernicus, the book in which the Copernican astronomical system is developed. Copernicus lived from 1473 to 1543 in Thorn, but this work appeared likewise in Nürnberg, two years after the publication of Regiomontanus' tables. Inasmuch as Copernicus never saw these tables, he was obliged to compute for himself the little table of sines which you find in his book and which was needed to work out his theory.

4. These tables by no means met the needs of the astronomers, so that we see a pupil and friend of Copernicus attempting soon a much larger work. His name was Rhäticus, which again is a latinized form of the name of his birthplace (Vorarlberg). He lived from 1514 to 1576, and was professor at Wittenberg. You must relate all these things to the general historical background of the time. Thus we are in the age of the Reformation when, as you know, Wittenberg and the free city Nürnberg were centers of intellectual life. Gradually, however, during the struggles of the Reformation, the center of gravity of the political and intellectual life moved away from the cities and toward the courts of the princes. Thus while everything heretofore had been printed in

---

[1] Norimbergae, 1541.

* I have already mentioned Cardanus and Stifel and shall soon mention others.

[2] Norimbergae, 1543.

Nürnberg, the great tables of Rhäticus now appeared under the patronage of the Elector Palatine and bore therefore his name *Opus Palatinum*[1]. They were printed shortly after the death of Rhäticus. They were much more complete than the preceding tables, containing the values of the trigonometric lines to ten plaes at intervals of ten minutes, with, to be sure, a good many errors.

5. A new edition of this table, very much improved, was published by Pitiscus of Grünberg in Silesia (1561—1613), chaplain of the Elector Palatine. This *Thesaurus Mathematicus*[2], again printed under princely subsidy, contained the trigonometric numbers to fifteen places, at intervals of ten minutes. The work was essentially freer from errors than that of Rhäticus, and was more compendious.

We must bear in mind that all these tables were constructed, in the main, with the aid solely of the half-angle formula, together with interpolation, for at that time the infinite series for $\sin x$ and $\cos x$ did not exist. We can appreciate, then, the prodigious diligence and labor which is represented in these great works.

B. Logarithmic-Trigonometric Tables

These tables were succeeded immediately by the development of the second group, the logarithmic-trigonometric tables, and it is a remarkable coincidence, the irony of history, one might say, that a year after the tables of trigonometric lines had attained, with Pitiscus, a certain completeness, the first logarithms appeared and rendered these tables superfluous, in that from then on, instead of sine and cosine, one used their logarithms. I have already mentioned the first logarithmic tables, those of Napier.

1. *Mirifici Logarithmorum Canonis Descriptio* of Napier, in 1614. Napier had in mind, primarily, the facilitating of trigonometric caculation. Consequently he did not give the logarithms of the natural numbers, but only the seven-place logarithms of the trigonometric lines, at intervals of one minute.

2. The actual construction of logarithmic tables in their present form is due mainly to the Englishman Henry Briggs (1556—1630) who was in touch with Napier. He recognized the great advantage that logarithms with base ten would have for practical calculation, since they would fit our decimal system better, and he introduced this base instead of that of Napier as early as 1617 in his *Logarithmorum Chilias Prima*, giving us the *"artificial"* or *common logarithms* which bear his name. In order to calculate these logarithms, Briggs devised a series of interesting methods which permitted the determination of each logarithm as accurately as one chose. Briggs' second considerable book bore the title

---

[1] Heidelbergae, 1596.   [2] Francofurtii, 1613.

Arithmetica logarithmica[1]. In it he tabulates the logarithms of the natural numbers themselves instead of those of the angle ratios, as Napier had done. To be sure, Briggs never finished his calculations. He gave the logarithms of the integers only from 1 to 20000 and 90000 to 100000, but to fourteen places. It is remarkable that precisely the oldest tables give the most places, whereas now we are content, for most purposes, with very few places. I shall come back to this later. Briggs also compiled the common logarithms of the trigonometric lines to ten places with ten minute intervals in his *Trigonometria Britannica*[2].

3. The gap in Briggs' table was filled by the Dutchman Adrian Vlacq, mathematician, printer, and dealer in books, who lived in Gouda near Leyden. He issued a second edition of Briggs' book[3], which contained the logarithms of all integers from 1 to 100000 but only to ten places. We may consider this as the source of all our current tables of logarithms of natural numbers.

Concerning the further development of tables, I can mention here only in a general way the points in which advances were made in later years as compared with the above mentioned early beginnings.

a) The first essential advance was in the theory. The logarithmic series furnished, namely, an extremely useful new method for the calculation of logarithms. The compilers of the first tables knew nothing about these series. As we have seen, Napier calculated his logarithms by means of the *difference equation*, that is, by successive addition of $\Delta x/x$, with the further aid of *interpolation*. The important device of *square root extraction* appeared with Briggs. He made use of the fact, which was mentioned moreover by Napier in his *Constructio* (see p. 147), that one knows $\log \sqrt{a \cdot b} = \frac{1}{2} (\log a + \log b)$ as soon as one knows the logarithms of a and b. It is probable that Vlacq also calculated in this way.

b) Essential progress was made by a more suitable arrangement in printing the tables, whereby it was made possible to combine more material, in a clearer way, in a smaller space.

c) Above all, the correctness of the tables, was considerably increased by a careful check of the older ones, thereby eliminating numerous errors, especially in the last figures.

Among the large number of tables which thus appeared, I shall mention only the most famous one.

4. This is the *Thesaurus Logarithmorum Completus* (Vollständige Sammlung grösserer logarithmisch-trigonometrischer Tafeln*), by the Austrian artillery officer Vega, which appeared in Leipzig in 1794. The original is rare, but a photostatic reprint appeared in Florence in 1896.

---

[1] Londini, 1624.   [2] Goudae, 1633.
[3] Briggs, H., *Arithmetica Logarithmica. Editio secunda aucta per* Adr. Vlacq, Goudae, 1628.
* Complete collection of larger logarithmic trigonometric tables.

The Thesaurus contains ten place logarithms of the natural numbers, and of the trigonometric lines, in an arrangement that has since become typical. Thus you find there, e.g., the small difference tables for facilitating interpolation.

If we come down now to the nineteenth century, we notice a far reaching popularization of logarithms, due partly to the fact that they were introduced into the schools in the twenties, but also to the fact that they found more and more application in physical and technical practice. At the same time we find a reduction in the number of places. For the needs of the schools, as well as those of technical practice, were better met by tables which were not too bulky, especially since three or four places were sufficient for the requisite accuracy in nearly all practical cases. To be sure, we still had, in my school days, seven-place tables, the reason assigned being that the pupils would obtain in this way an impression of the "majesty of numbers". Our minds today are in general more utilitarian, and we use throughout two, three, or at most five-place tables. I shall show you today three modern tables, selected at random. One is a handy little four place table by Schubert[1]. In it you will find all manner of devices, such as printing in two colors, repetition above and below, on every page, of guiding quantities, and the like, in order to exclude misunderstanding. The second is a modern American table by Huntington[2], which is still more cunningly arranged, where, e.g., the leaves are provided with projections and indentations to enable one to turn up at once the desired page. Finally, I am showing you a slide-rule, which, as you know, is nothing else than a three-place logarithmic table in the very convenient form of a mechanical calculator. You are all familiar, certainly, with this instrument, which every engineer nowadays has with him constantly.

We have not yet reached the end of the development, but we can see pretty clearly what its further direction will be. Of late, the calculating machine, of which I talked earlier (see p. 17 et seq.), has been coming into extensive use, and it makes logarithmic tables superfluous, since it permits a much more rapid and reliable direct multiplication. At present, however, this machine is so expensive that only large offices can afford it. When it has become considerably cheaper, a new phase of numerical calculation will be inaugurated. So far as goniometry is concerned, the old tables of Pitiscus, which became old fashioned so soon after birth, will then come into their own; for they supply directly the trigonometric ratios with which the calculating machine can operate at once, thus avoiding the use of logarithms.

---

[1] Now Schubert-Haussner, *Vierstellige Tafeln und Gegentafeln*, Sammlung Göschen, Leipzig, 1917.]

[2] Huntington, C. V., *Four-Place Tables*. Abridged edition, Cambridge, Massachusetts. 1907.

3. Applications of Goniometric Functions

It remains for me now to give you a survey of the application of goniometric functions. I shall consider three fields

A. *Trigonometry*, which, indeed, furnished the occasion for inventing the goniometric functions.

B. *Mechanics*, where, in particular, the theory of small oscillations offers a wide field for applications.

C. *Representation of periodic functions by means of trigonometric series*, which, as is well known, plays an important part in the greatest variety of problems.

Let us turn at once to the first subject.

A. Trigonometry, in particular, spherical trigonometry

We are in the presence here of a very old science, which was in full flower in ancient Egypt, where it was encouraged by the needs of two important sciences. Geodesy required the theory of the plane triangle, and astronomy needed that of the spherical triangle. For the history of astronomy we have the voluminous monograph in A. v. Braunmühl's *Vorlesungen über Geschichte der Trigonometrie*[1]. On the practical side of trigonometry the most informative book is E. Hammer's: *Lehrbuch der ebenen und sphärischen Trigonometrie*[2]; on the theoretical side, the second volume of the work I have often mentioned, the *Enzyklopädie der Elementarmathematik* of Weber-Wellstein.

Within the limits of these lectures I cannot, of course, develop systematically the whole subject of trigonometry. That would be a matter for special study. Furthermore, practical trigonometry is given full consideration here in Göttingen in the regular lectures on geodesy and spherical astronomy. I should prefer to talk to you exclusively about a very interesting chapter of theoretical trigonometry which, in spite of its great age, cannot be regarded as closed, and which, on the contrary, contains many still unsolved problems and questions, of relatively elementary character, whose study would, I think, be rewarding. I refer to spherical trigonometry. You will find this subject very fully considered in Weber-Wellstein, where importance is given to the thoughts which Study developed in his fundamental work *Sphärische Trigonometrie, orthogonale Substitutionen und elliptische Funktionen*[3]. I shall try to give you a survey of all the theories that belong here and to call your attention to the questions which are still unanswered.

The elementary notion of a spherical triangle hardly needs explana-

---

[1] Two volumes. Leipzig, 1900 and 1903.
[2] Stuttgart, 1906. [Fifth edition, 1923.]
[3] Abhandlungen der Mathematisch-physikalischen Klasse der Königlich Sächsischen Gesellschaft der Wissenschaften, vol. 20, No. 2. Leipzig, 1893.

tion. Three points on a sphere, no two of which are diametrially opposite, determine uniquely a triangle in which each angle and each side lies between 0 and π (see Fig. 70). Further investigation discloses that it is desirable to think of the sides and of the angles as unrestricted variables, which can thus be greater than π or 2π, or multiples of these values. One has to do then with sides that overlap and with angles which wind multiply around their vertices. It becomes necessary therefore to adopt conventions concerning the signs of these quantities as well as the sense in which they are measured. It is due to Möbius, the great geometer of Leipzig, that the importance of the principle of signs was consistently developed, and the way opened for the general investigation of these quantities under unrestricted variation. The part of his work which is of particular significance here is the *Entwicklung der Grundformeln der sphärischen Trigonometrie in grösstmöglicher Allgemeinheit*[1].

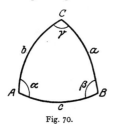
Fig. 70.

This determination of the sign begins with the assumption of a definite sense of rotation about a point A on a sphere in which the angle shall be called positive (see Fig. 71). If this sense is settled for one point, it is for every other point, since the first point can be moved continuously to that other. It is customary to select the counterclockwise rotation as positive, whereby we think of ourselves as looking at the sphere from the outside. Secondly, we must assign a sense of direction to each great circle on the sphere. We cannot be satisfied with an initial determination for one great circle and the continuous moving of it into coincidence with any second great circle, because this coincidence can be effected in two distinct ways. On this account, we shall assign a sense of direction separately to each great circle which we consider, and we shall look upon one and the same circle as, in a sense, two different configurations according as we have assigned to it the one or the other direction. With this understanding, each directed great circle a can be uniquely related to a pole P, namely to that one of its two poles, in the elementary sense, from which its sense of direction would appear positive. Conversely, every point on the sphere has a unique polar circle with a definite direction. With these considerations, the polarizing process, so important in trigonometry, is uniquely determined.

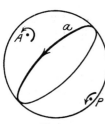
Fig. 71.

---

[1] Berichte über die Verhandlungen der Königlich Sächsischen Gesellschaft der Wissenschaften, mathematisch-physikalische Klasse, vol. 12 (1860). Reprinted in Möbius, F., Gesammelte Werke, vol. 2, p. 71. Leipzig, 1886.

If now three points A, B, C on the sphere are given, we must still make certain agreements, before a spherical triangle with these vertices is uniquely dtermined. In the first place, the direction of each great circle through A, B, C must be assigned, and we must know how many revolutions are necessary in order to bring a point from B to C, from C to A, and from A to B. The lengths a, b, c, determined in this way, which may be arbitrary real quantities, are called sides of the spherical triangle. Of course they are thought of as drawn on a sphere of radius one. The angles are then defined as follows: α is that rotation, about A in positive sense, which would bring the direction CA into coincidence with the direction AB, to which arbitrary multiples of $\pm 2\pi$ may be added. The other angles are defined analogously. If we now examine an ordinary elementary triangle, as shown in Fig. 72, and determine the directions of the sides so that a, b, c are less than π, we find that the angles α, β, γ are, according to our new definiton, the *exterior angles* instead of the *interior* angles as in the usual consideration of the elementary triangle.

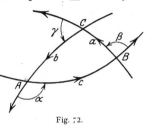

Fig. 72.

It has been known for a long while that by replacing the customary angles of a spherical triangle by their supplements, in this manner, the formulas of spherical trigonometry turn out to be more symmetrical and perspicuous. The deeper reason for this appears from the following consideration. The polarizing process described above, by virtue of the conventions of Möbius, furnishes uniquely, for every given triangle, another triangle called the *polar triangle* of the first; and it is easy to see that, in view of our new definition, this polar triangle has for its sides and angles the angles and sides, respectively, of the original triangle. According to our agreements, then, every formula of spherical trigonometry must still hold if we interchange in it a, b, and c with α, β, and ν, respectively, so that there must always be this simple symmetry. If, on the other hand, the sides and angles are measured in the usual way, this symmetry is lost; for the relation between triangle and polar triangle depends upon how one chooses the sides and angles in a given case, and upon how one resolves the ambiguity of the pole in the case of a non directed given circle.

It is clear now that, of the six parts of a spherical triangle defined in this way, only three can be independent continuous variables, e.g. two sides and the included angle. The formulas of spherical trigonometry represent a number of relations between these parts or, to be more exact, of algebraic relations between their twelve sines and cosines, in which only three of these twelve magnitudes can be allowed to vary arbitrarily, while the other nine depend algebraically upon them. If we go over to the sine and cosine, we can ignore the additive arbitrary multiples of 2π. Let us now think of trigonometry as the aggregate of all possible such algebraic relations of this kind. Then we can state its problem, according to the modern manner of thinking, as follows. If we interpret the quantities

$x_1 = \cos a$, $x_2 = \cos b$, $x_3 = \cos c$, $x_4 = \cos \alpha$, $x_5 = \cos \beta$, $x_6 = \cos \gamma$,
$y_1 = \sin a$, $y_2 = \sin b$, $y_3 = \sin c$, $y_4 = \sin \alpha$, $y_5 = \sin \beta$, $y_6 = \sin \gamma$

as coordinates in a twelve dimensional space R_{12}, then the totality of those of its points which correspond to actually possible spherical triangles $a, \ldots, \gamma$ represents a three-dimensional algebraic configuration M_3 of this R_{12}, and the problem is to study this M_3 in the R_{12}. In this manner spherical trigonometry is coordinated with general analytic geometry of hyperspace.

Now this M_3 must have various simple symmetries. Thus the polarizing process showed that the interchange of a, b, c with α, β, γ, always yielded a spherical triangle. Translated into our new language, this states that when one interchanges $x_1, x_2, x_3, y_1, y_2, y_3$ with $x_4, x_5, x_6, y_4, y_5, y_6$, respectively, any point of M_3 goes over into another point belonging to it. Further, corresponding to the division of space into eight octants by the planes of the three great circles, there exists for any triangle seven auxiliary triangles whose parts arise from those of the initial triangle through change of sign and the addition of π. This yields for every point of M_3 seven further points whose coordinates $x_1, \ldots, x_6$ arise as a result of sign change. The totality of these symmetries leads to a certain group of substitutions and sign changes of the coordinates of R_{12}, which transforms M_3 into itself.

The most important question now is that concerning the algebraic equations which are satisfied by the coordinates of M_3, and which constitute the totality of trigonometric formulas. Since $\cos^2 \alpha + \sin^2 \alpha = 1$, we have, to start with, the six quadratic relations

$$\text{(1)} \qquad x_i^2 + y_i^2 = 1, \qquad (i = 1, 2, \ldots, 6),$$

or, speaking geometrically, six cylindrical surfaces $F^{(2)}$ of order two passing through M_3.

Six further formulas are supplied by the cosine theorem of spherical trigonometry, which in our notation, is

$$\cos a = \cos b \cos c - \sin b \sin c \cos \alpha,$$

from which one gets by polarization

$$\cos \alpha = \cos \beta \cos \gamma - \sin \beta \sin \gamma \cos a.$$

These equations, together with the four others which arise through cyclic permutation of a, b, c and α, β, γ determine, all told, six cubic surfaces $F^{(3)}$ passing through M_3:

(2) $\quad x_1 = x_2 x_3 - y_2 y_3 x_4, \quad x_2 = x_3 x_1 - y_3 y_1 x_5, \quad x_3 = x_1 x_2 - y_1 y_2 x_6,$

(3) $\quad x_4 = x_5 x_6 - y_5 y_6 x_1, \quad x_5 = x_6 x_4 - y_6 y_4 x_2, \quad x_6 = x_4 x_5 - y_4 y_5 x_3.$

Finally, we can make use of the sine theorem, which can be expressed by the vanishing of the minors of the following matrix

$$\begin{vmatrix} \sin a, & \sin b, & \sin c \\ \sin \alpha, & \sin \beta, & \sin \gamma \end{vmatrix} = \begin{vmatrix} y_1, & y_2, & y_3 \\ y_4, & y_5, & y_6 \end{vmatrix}$$

or, written at length,

(4) $\qquad y_2 y_6 - y_3 y_5 = y_3 y_4 - y_1 y_6 = y_1 y_5 - y_2 y_4 = 0$.

These expressions represent three quadratic surface $F^{(2)}$, of which only two, to be sure, are independent. Thus we have set up altogether fifteen equations for our M_3 in R_{12}.

Now, in general, $12 - 3 = 9$ equations do not, by any means, suffice to determine a three dimensional algebraic configuration in R_{12}. Even in the ordinary geometry of R_3, not every space curve can be represented as the complete intersection of two algebraic surfaces. The simplest example here is the space curve of order three which requires for its determination at least three equations. It is easy to see that, in our case also, the nine equations (1) and (2) do not determine M_3. It is well known, namely, that the sine theorem can be derived from the cosine theorem only to within the sign, which one then determines, ordinarily, by geometric considerations. We should like to know then how many, and which, of the trigonometric equations really determine our M_3 completely. In this connection I should like to formulate four definite questions to which the literature thus far appears to give no precise answer. It could be a worth-while task to investigate them thoroughly. That would probably not be especially difficult, after one had acquired a certain skill in handling the formulas of spherical trigonometry. My questions are:

1. What is the order of M_3?
2. What are the equations of lowest degree by means of which M_3 can be completely represented?
3. What is the complete system of linearly independent equations which represent M_3, i.e., of equations $f_1 = 0, \ldots, f_n = 0$ such that the equation of every other surface passing through M_3 could be written in the form $m_1 f_1 + \ldots + m_n f_n = 0$, where $m_1, \ldots, m_n$ are integers? It is possible that more equations may be needed here than in 2.
4. What algebraic identities (so called syzygies) exist between these n formulas $f_1, \ldots, f_n$?

One could gain familiarity with these things by consulting investigations which have been made in exactly the same direction but in which the questions have been put somewhat differently. These appear in the Göttingen dissertation[1], 1894, of Miss Chisholm (now

---

[1] *Algebraisch-gruppentheoretische Untersuchungen zur sphärischen Trigonometrie*, Göttingen, 1895.

Mrs. Young), who, by the way, was the first woman to pass the normal examination in Prussia for the doctor's degree. The most noteworthy of Miss Chisholm's various preliminary assumptions is her selection of the cotangents of the half angles and sides as independent coordinates. Since tan $(\alpha/2)$ and likewise, of course, ctn $(\alpha/2)$, is a fundamental function, in terms of which sin α and cos α can be uniquely expressed, it is possible to write all the trigonometric equations as algebraic relations between ctn $(a/2)$, ..., ctn $(\gamma/2)$. The spherical triangles constitute now a three-dimensional configuration M_3 in a six dimensional space R_6 which has ctn $(a/2)$, ..., ctn $(c/2)$, ctn $(\alpha/2)$, ..., ctn $(\gamma/2)$, as coordinates. Miss Chisholm shows that this M_3 is of order eight and that it can be fully represented as the complete intersection of three surfaces of degree two (quadratic equations) of R_6; and she investigates also the questions which arise here, which are analogous to those stated above.

In my lectures on the hypergeometric function[1], I called the group of formulas of spherical trigonometry which I have discussed above, and which connect the sines and the cosines of the sides and angles, *formulas of the first kind*, in distinction from an essentially different group of formulas which I called *formulas of the second kind*. The latter are algebraic equations between the trigonometric functions of the half angles and sides. In studying them it will be best to select the twelve quantities

$$\cos\frac{a}{2}, \quad \sin\frac{a}{2}, \ldots; \quad \cos\frac{\alpha}{2}, \quad \sin\frac{\alpha}{2}, \ldots$$

as coordinates in a new twelve space R'_{12}, in which the spherical triangles again constitute a three-dimensional configuration M'_3. It is here that those elegant formulas appear which, at the beginning of the last century, were published independently and almost simultaneously by Delambre (1807), Mollweide (1808) and finally Gauss 1809 [in the *Theoria motus corporum coelestium*, No. 54[2]]. These are twelve formulas which arise by cyclic permutation in:

$$\frac{\sin\frac{\beta+\gamma}{2}}{\sin\frac{\alpha}{2}} = \pm\frac{\cos\frac{b-c}{2}}{\cos\frac{a}{2}}, \quad \frac{\sin\frac{\beta-\gamma}{2}}{\sin\frac{\alpha}{2}} = \mp\frac{\sin\frac{b-c}{2}}{\sin\frac{a}{2}},$$

$$\frac{\cos\frac{\beta+\gamma}{2}}{\cos\frac{\alpha}{2}} = \mp\frac{\cos\frac{b+c}{2}}{\cos\frac{a}{2}}, \quad \frac{\cos\frac{\beta-\gamma}{2}}{\cos\frac{\alpha}{2}} = \pm\frac{\sin\frac{b+c}{2}}{\sin\frac{a}{2}}.$$

---

[1] Winter semester 1893—1894. Elaborated by E. Ritter.—Reprinted Leipzig, 1906.

[2] Reprinted in Werke, Leipzig, 1906, vol. 7, p. 67.

That which is essential and new in them, as opposed to the formulas of the first kind, is the double sign, with respect to which the following is true. For one and the same triangle, the same sign, either the upper or the lower, holds for all twelve formulas, and there are triangles of both sorts. The M'_3 of spherical triangles in the above defined R'_{12} satisfies, in other words, two entirely different systems of twelve cubic equations each, and divides therefore into two separate algebraic configurations $\overline{M_3}$, for which the one sign holds, and $\overline{\overline{M_3}}$, for which the other holds. By virtue of this remarkable fact these formulas take on the greatest significance for the theory of spherical triangles. They are much more than mere transformations of the old equations which might at most serve to facilitate trigonometric calculation. To be sure, Delambre and Mollweide did consider these formulas only from this practical standpoint. It was Gauss who had the deeper insight, for he draws attention to the possibility of a change of sign "if one grasps in its greatest generality the idea of spherical triangle". It seems to me proper, therefore, that the formulas should bear Gauss's name, even if he did not have priority of publication.

It was Study who first recognized the full range of this phenomenon, and who developed it in his memoir of 1893, which I mentioned on p. 175. His chief result can be stated most conveniently if we consider the six space R_6 which has for coordinates the quantities $a, b, c, \alpha, \beta, \gamma$ themselves, thought of as unrestricted variables. I call them *transcendental parts* of the triangle in destinction from the *algebraic* parts $\cos a, \ldots$, or $\cos(a/2), \ldots$, because the former are transcendental functions, while the latter are algebraic functions of the ordinary space coordinates of the vertices of the triangle. In this R_6, the aggregate of all spherical triangles appears as the *transcendental configuration* $M_3^{(t)}$ whose image in R'_{12} is the algebraic M'_3 considered above. Since however the latter split into two parts and the mapping functions $\cos(a/2), \ldots$ are single valued continuous functions of the transcendental coordinates, the transcendental $M_3^{(t)}$ must split into at least two separated parts. Study's theorem is as follows: *The transcendental configuration $M_3^{(t)}$ of the quantities $a, b, c, \alpha, \beta, \gamma$, belonging to a spherical triangle of the most general sort, divides into two separate parts corresponding to the double sign in the Gaussian formulas, and each of these parts is a connected continuum.* The essential thing here is the exclusion of any farther division. It would not be possible, by farther manipulation of the trigonometric formulas, to bring about similar and equally significant groupings of spherical triangles. The triangles of the first of these parts, that corresponding to the upper sign in the Gaussian formulas, are called *proper* triangles, those of the other, *improper*, and we may state Study's theorem briefly as follows: *The totality of all spherical triangles resolves itself into a continuum of proper and one of improper triangles.* You will find further details, and a proof of this theorem, in Weber-Wellstein[1]. I am attempting here only to state the results clearly.

I must now say something further concerning the difference between the two sorts of triangles. If a spherical triangle is given, i.e., an admissible set of values of $a, b, c, \alpha, \beta, \gamma$, whose cosines and sines satisfy the formulas of the first sort, and which therefore represents a point of $M_3^{(t)}$, how can we decide whether the triangle is proper or improper? In order to answer this question we first find the smallest positive residues $a_0, b_0, c_0, \alpha_0, \beta_0, \gamma_0$ of the given numbers, with respect to the modulus 2π:

$$a_0 \equiv a \,(\text{mod}\, 2\pi), \ldots, \qquad \alpha_0 \equiv \alpha \,(\text{mod}\, 2\pi), \ldots$$
$$0 \leq a_0 < 2\pi, \ldots, \qquad 0 \leq \alpha_0 < 2\pi, \ldots$$

Their sines and cosines coincide with those of $a, \ldots, \alpha, \ldots$ so that they also represent a triangle which we shall call the reduced, or the Moebius, triangle corresponding to the given one, since Moebius himself did not consider the parts as varying beyond 2π. Then we can determine, by means of a table, whether the Moebius triangle is proper or improper. You will find this, in a form somewhat less clear, in Weber-Wellstein (p. 352, 379, 380), as well as figures (p. 348, 349) of the types of proper and improper triangles. As is usual, I shall call an angle *reentrant* when it lies between π and 2π and I shall, for the sake of brevity, apply this term also to the sides of the spherical triangle. Then there are, altogether, four typical cases of each sort.

I. *Proper Moebius triangles*:
1. 0 sides reentrant; 0 angles reentrant.
2. 1 side reentrant; 2 adjacent angles reentrant.
3. 2 sides reentrant; 1 included angle reentrant.
4. 3 sides reentrant; 3 angles reentrant.

II. *Improper Moebius triangles*:
1. 0 sides reentrant; 3 angles reentrant.
2. 1 side reentrant; 1 opposite angle reentrant.
3. 2 sides reentrant; 2 opposite angles reentrant.
4. 3 sides reentrant; 0 angles reentrant.

There are no cases other than these, so that this table enables us actually to determine the character of a Moebius triangle.

The transition to the general triangle $a, \ldots, \alpha, \ldots$ from the corresponding reduced triangle is made, after what was said above, by means of the formulas:

$$a = a_0 + n_1 \cdot 2\pi, \quad b = b_0 + n_2 \cdot 2\pi, \quad c = c_0 + n_3 \cdot 2\pi,$$
$$\alpha = \alpha_0 + \nu_1 \cdot 2\pi, \quad \beta = \beta_0 + \nu_2 \cdot 2\pi, \quad \gamma = \gamma_0 + \nu_3 \cdot 2\pi.$$

---

[1] Vol. 2, second edition (1907), p. 385 (§ 47).

We may then make use of the following theorem *The character of the general triangle is the same as or the reverse of that of the reduced triangle according as the sum of the six integers* $n_1 + n_2 + n_3 + \nu_1 + \nu_2 + \nu_3$ *is even or odd.* Thus the character of every triangle as proper or improper can be determined.

I shall conclude this chapter with a few remarks about the area of spherical triangles. Nothing is said about this in Study or in Weber-Wellstein. It does come up for consideration in my *Älteren funktionentheoretischen Untersuchungen über Kreisbogendreiecke*[*]. Up to this point we have considered the triangle merely as an aggregate of three angles and three sides which satisfy the sine and consine laws. In my investigations I was concerned with a definite area bounded by these sides,—in a certain sense with a membrane stretched between these sides and involving appropriate angles.

Of course we can now no longer think of α, β, γ as the exterior angles of the triangle, as we did before for reasons of symmetry. We shall talk, rather, of those angles which the membrane itself forms at the vertices, and I shall call them *interior angles* of the triangle. I shall denote them, as is my habit by $\lambda\pi, \mu\pi, \nu\pi$ (see Fig. 73). These angles can also be thought of as unrestricted positive variables, since the membrane might wind about the vertices. In accordance with this, I shall denote the absolute lengths of the sides by $l\pi, m\pi, n\pi$, which are also unrestricted positive variables. But it will be no longer possible for the sides and the angles to "overlap" independently of one another, i. e., to contain arbitrary multiples of 2π, as they could before, for the fact that a singly-connected membrane should exist with these sides

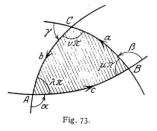

Fig. 73.

and angles finds its expression in certain relations between the numbers of these overlappings. In my memoir *Über die Nullstellen der hypergeometrischen Reihe*[1] I called these *supplementary relations* of spherical trigonometry. If we denote by $E(x)$ the largest positive integer which x exceeds, $[E(x) < x]$, these relations are

$$E\left(\frac{l}{2}\right) = E\left(\frac{\lambda - \mu - \nu + 1}{2}\right),$$
$$E\left(\frac{m}{2}\right) = E\left(\frac{-\lambda + \mu - \nu + 1}{2}\right),$$
$$E\left(\frac{n}{2}\right) = E\left(\frac{-\lambda - \mu + \nu + 1}{2}\right),$$

---
[*] Earlier function-theoretic investigations of spherical triangles.
[1] Mathematische Annalen, vol. 37 (1888). [Reprinted in Klein, F., Gesammelte Mathematische Abhandlungen, vol. 2 (1921), p. 550.]

and since $E\,(l/2)$, for example, gives the multiple of $2\,\pi$ which is contained in the side $l\,\pi$, these relations determine precisely the desired "overlap" numbers of the sides $l\pi$, $m\pi$, $n\pi$ when one knows the angles $\lambda\pi$, $\mu\pi$, $\nu\pi$ together with their overlap numbers. It is easy to see, in particular, that of the three numbers $\lambda - \mu - \nu$, $-\lambda + \mu - \nu$, $-\lambda - \mu + \nu$, one at most can be positive. Consequently only one of the three arguments on the right sides can exceed unity, and since $E\,(x) = 0$ for $x \leqq 1$, it is possible for only one of the overlap numbers to be different from zero. In other words only one side, at most, of a triangular membrane can overlap (be greater than 2) and that side must be opposite the largest angle.

For the proof of these supplementary relations I refer you to my mimeographed lectures *Über die hypergeometrische Funktion*[1] (p. 384), although the edition is long since exhausted. There, as well as in my memoir in volume 37 of the Mathematische Annalen, the initial assumptions were somewhat broader than the present ones, in that spherical triangles were considered which are bounded by arbitrary circles on the sphere, not necessarily by great circles. I shall sketch briefly the train of thought of the proof. We start with an elementary triangle, in which a membrane can certainly be stretched, and obtain from it step by step the most general admissible triangular membrane by repeatedly attaching circular membranes, either at the sides, or, with branchpoints, at the vertices. Fig. 74 shows, as an example, (in stereographic projection) a triangle ABC which arises from an elementary triangle by attaching the hemisphere which is bounded by the great circle AB, whereby the side AB overlaps as well as the angle C. It is clear that the supplementary relations continue to hold here, and one sees in the same way that they retain their validity for the most general triangular membrane which can be built up by this process.

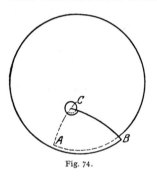

Fig. 74.

We must now inquire how these triangles, which satisfy the supplementary relations, fit the general theory which we have discussed already. They are obviously only special cases, (because the overlap numbers of the sides and angles are, in general, entirely arbitrary) — special cases which are characterized by the possibility of framing a stretched membrane in a triangle. At first one can really be puzzled here, for we have seen that the totality of all proper triangles (some of which do not need to satisfy the supplementary relations) constitutes

---

[1] These lectures were referred to on p. 180.

a continuum, and that any one of them could be derived, therefore, from an elementary triangle by a continuous deformation. One would think, naturally, that it would be impossible, during this deformation to lose the membrane which was stretched in the initial elementary triangle. The explanation of this difficulty appears if we extend Moebius' principle of sign-change to areas, by agreeing that an area is to be called positive or negative according as its boundary is traversed in the positive (counter clockwise) or negative sense. Accordingly, when a curve which crosses itself bounds several partial areas, the entire area is the algebraic sum of the several parts, each of these determined, as to sign, by the sense in which its boundary is traversed. In Fig. 75 this would be the difference, in Fig. 76, the sum of the parts which are distinguished by different shading. These agreements are, of course, merely the geometric expression of that which the analytic definition itself supplies.

If we apply this, in particular, to triangles formed by circular arcs, it turns out, in fact, that with every proper triangle we can associate an area on

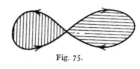
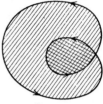

Fig. 75.   Fig. 76.

the sphere such that, when one circuit of the triangle is made, different parts of this area are combined with different signs because the boundaries of these parts are traversed in different senses. Those triangles for which the supplementary relations hold are special, then, in that their areas consist of a single piece of membrane bounded by a positive circuit. It is this property which gives them their great significance for the function-theoretic purposes to which I put them in my earlier studies.

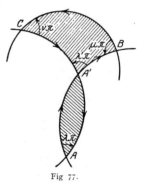

Fig 77.

I will now illustrate this situation by means of an example. Let us consider the triangle ABC in stereographic projection (Fig. 77) where, of the points of intersection A, A' of the great circles BA, CA, A is the one more remote from the arc BC. If we now transfer the general definition of the exterior angles (p. 177) to their supplements, the interior angles, we find that $\mu\pi$ and $\nu\pi$ measure the rotation of BC into BA and of CA into CB, respectively, and are, therefore, positive in our case. Similarly $\lambda\pi$ measures the rotation of AB into AC and is therefore negative. Put $\lambda = -\lambda'$, $\lambda' > 0$. Then the triangle $A'BC$ is obviously an elementary triangle with angles $\lambda'\pi$, $\mu\pi$, $\nu\pi$, all of which are positive. If we now make a circuit about the triangle ABC, the boundary of the elementary triangle $A'BC$ will be traversed in the positive sense but that of the spherical sector AA' in the negative, and the area of the triangle ABC, in the Moebius sense, will be the difference of these two areas. This breaking up of the triangular membrane into a positive and a negative part can be visualized, perhaps, by supposing the membrane twisted at A' so that the rear or negative side of the sector is brought to the front. It is not hard to construct more difficult examples after this pattern.

I shall now show, by means of this same example, that with this general definition of area, the formulas for the area of elementary triangles still remain valid. As you know, the area of a spherical triangle with angles $\lambda\pi, \mu\pi, \nu\pi$, on a sphere with radius one, is given by the so-called spherical excess $(\lambda + \mu + \nu - 1)\pi$ where $\lambda, \mu, \nu > 0$. Let us now see that this formula holds also for the above triangle ABC. It is clear that the area of the elementary triangle $A'BC$ is $(\lambda' + \mu + \nu - 1)\pi$. From this we must subtract the area of the sector AA' whose angle is $\lambda'\pi$. But this is $2\lambda'\pi$, because the area of a sector is proportional to its angle; and it becomes 4π when the angle is 2π (the entire sphere). We get then, as the area of ABC,

$$(\lambda' + \mu + \nu - 1)\pi - 2\lambda'\pi = (-\lambda' + \mu + \nu - 1)\pi = (\lambda + \mu + \nu - 1)\pi.$$

It is probable, if we had a general proper triangle with arbitrary sides and angles, and if we should try to fit into it a multi-parted membrane and determine its area (which, according to the sign rule, would be the algebraic sum of the parts), that the result would show the general validity of the formula $(\lambda + \mu + \nu - 1)\pi$, where, of course, $\lambda\pi, \ldots$ are the real angles of the membrane, and not, as before, the exterior angles. The investigation suggested here has not been carried out, however. It would certainly not offer great difficulties, and I should be glad if it were undertaken. At the same time, it would be important to determine, from the present standpoint, the rôle of the improper triangles.

With this I shall leave the subject of trigonometry and go over to the second important application of goniometric functions, one which also falls within the field of the schools.

B. Theory of small oscillations, especially those of the pendulum

I shall recall briefly the deduction of the law of the pendulum as we are in the habit of giving it at the university, by means of infinitesimal calculus. A pendulum (see Fig. 78) of mass m hangs by a thread of length l, its angle of deflection from the normal being φ. Since the force of gravity acts vertically downwards, it follows from the fundamental laws of mechanics that the motion of the pendulum is determined by the equation

(1) $$\frac{d^2\varphi}{dt^2} = -\frac{g}{l}\sin\varphi.$$

For small amplitudes we may replace $\sin\varphi$ by φ without serious error. This gives for so called infinitely small oscillation of the pendulum

(2) $$\frac{d^2\varphi}{dt^2} = -\frac{g}{l}\varphi.$$

The general integral of this differential equation is given, as you know, by goniometric functions, which are important here, as I said before. precisely by reason of their differential properti,es The general integral is

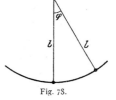

Fig. 78.

$$\varphi = A\sin\sqrt{\frac{g}{l}}\,t + B\cos\sqrt{\frac{g}{l}}\,t,$$

where A, B are arbitrary constants. If we introduce appropriate new constants C, t_0, we find

(3) $$\varphi = C\cdot\cos\sqrt{\frac{g}{l}}\,(t - t_0),$$

where C is called the *amplitude* and t_0 the *phase* of the oscillation. From this we get, for the duration of a complete oscillation, $T = 2\pi\sqrt{l/g}$.

Now these are very simple and clear considerations, and if we went more fully into the subject they could of course be given graphical form. But how different they appear from the so called *elementary* treatment of the pendulum law which is widely used in school instruction. In this, one endeavors, at all costs, to avoid a consistent use of infinitesimal calculus, although it is precisely here that the essential nature of the problem demands emphatically the application of infinitesimal methods. Thus one uses *methods contrived ad hoc*, which involve infinitesimal notions without calling them by their right name. Such a plan is, of course, extremely complicated, if it is to be at all exact. Consequently it is often presented in a manner so incomplete that it cannot be thought of, for a moment, as a proof of the pendulum law. Then we have the curious phenomenon that one and the same teacher, during one hour, the one devoted to mathematics, makes the very highest demands as to the logical exactness of all conclusions. In his judgment, still steeped in the traditions of the eighteenth century, his demands are not satisfied by the infinitesimal calculus. In the next hour, however, that devoted to physics, he accepts the most questionable conclusions and makes the most daring application of infinitesimals.

To make this clearer, let me give, briefly, the train of thought of such an *elementary deduction of the pendulum law*, one which is actually found in text books and used in instruction. One begins with a *canonical* pendulum, i.e. a pendulum in space whose point moves with uniform velocity v in a circle about the vertical, as axis, so that the suspending thread describes a circular cone (see Fig. 79). This is the motion which is called in mechanics *regular precession*. The possibility of such motion is, of course, assumed in the schools as a datum of experience and the question is asked merely concerning *the relation which obtains between the velocity v and the constant deflection of the pendulum*, $\varphi = \alpha$ (angular opening of the cone which is described by the thread).

One notices, first, that the point of the pendulum describes a circle of radius $r = l \sin \alpha$, for which one may write $r = l \cdot \alpha$ when α is sufficiently small. Then one talks of *centrifugal force* and reasons that the point, with mass m, revolving with velocity v, must exert the centrifugal force

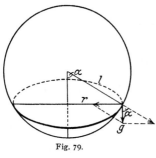

Fig. 79.

$$m \frac{v^2}{r} = m \frac{v^2}{l \cdot \alpha}$$

In order to maintain the motion there must be an equal *centripetal* force directed toward the center of the circular path. This is found by resolving the force of gravity into two components, one directed along the thread of the pendulum, the other, the desired force, acting in the plane of the circular path and directed toward its center, having the magnitude $m \cdot g \cdot \tan \alpha$ (see Fig. 79). This can be replaced by $mg \cdot \alpha$ when α is sufficiently small. We obtain, then, the desired relation in the form

$$m \cdot \frac{v^2}{l \alpha} = mg \cdot \alpha, \quad \text{or} \quad v = \alpha \sqrt{g \cdot l}.$$

The time of oscillation T of the pendulum, that is, the time in which the entire circumference of the circle $2 \pi r = 2 \pi l \alpha$ is traversed, is then

$$T = \frac{2 \pi l \alpha}{v} = 2 \pi \sqrt{\frac{l}{g}}.$$

In other words, when the angle of oscillation α is sufficiently small, the canonical pendulum performs a regular precession in this time, which is independent of α.

To criticize briefly this part of the deduction, we might admit the validity of replacing $\sin \alpha$ and $\tan \alpha$ by α itself, which we did ourselves in our exact deduction (p. 187); for this permits the transition from "finite" to "infinitely small" oscillations. On the other hand, we must call attention to the fact that the formula used above for centrifugal force can be deduced in "elementary" fashion only by neglecting all sorts of small quantities; and the exact justification for this is founded precisely on differential calculus. The very definition of centrifugal

Small Oscillations.

force, for example, requires in fact the notion of the second differential coefficient, so that the elementary deduction must also smuggle this in. And since in doing this, one is unable to say clearly and precisely what one is talking about, there arise the greatest obstacles to understanding, which are not present at all when the differential calculus is used. I do not need to go into detail here because I can refer you to some very readable articles on school programs, by the deceased realgymnasium director H. Seeger[1], in Güstrow and to a very interesting study by H. E. Timerding: *Die Mathematik in den physikalischen Lehrbüchern*[2]. In Seeger you will find, among other things, an exhaustive criticism of the deductions of the formula for centrifugal force, in a manner corresponding to our standpoint. In Timerding there are extensive studies of the mathematical methods which are traditionally used in the teaching of physics*. Let me now continue with the discussion of pendulum oscillations.

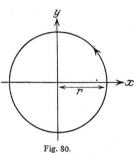

Fig. 80.

The considerations set forth above show the possibility of uniform motion in a circle. If we now set up an xy coordinate system (see Fig. 80) in the plane of this circle (i.e., in view of our approximation, the tangent plane to the sphere), this motion will, in the language of analytic mechanics, be given by the equations

(4) $$\begin{cases} x = l \cdot \alpha \cdot \cos \sqrt{\frac{g}{l}}\,(t - t_0) \\ y = l \cdot \alpha \cdot \sin \sqrt{\frac{g}{l}}\,(t - t_0)\,. \end{cases}$$

But we wish the plane oscillations of the pendulum; that is, the point of the pendulum in our xy plane is to move on a straight line, the x axis. The equations of its motion must be

(5) $$x = l \cdot C \cos \sqrt{\frac{g}{l}}\,(t - t_0)\,, \qquad y = 0,$$

---

[1] *Über die Stellung des hiesigen Realgymnasiums zu einem Beschlusse der letzten Berliner Schulkonferenz* (Güstrow, 1891, Schulprogramm No. 649). *Über die Stellung des hiesigen Realgymnasiums zu dem Erlass des preussischen Unterrichtsministeriums von 1892* (1893, No. 653). *Bemerkungen über Abgrenzung und Verwertung des Unterrichts in den Elementen der Infinitesimalrechnung* (1894, No. 658).

[2] Bd. III, Heft 2 der "Abhandlungen des deutschen Unterausschusses der Internationalen mathematischen Unterrichtskommission". Leipzig u. Berlin 1910.

* See also Report on the Correlation of Mathematics and Science Teaching by a joint committee of the British Mathematical Association and the Science Masters Association 1908. Reprinted 1917. Bell and Sons, London.

in order that the correct equation (3) shall result when $\varphi = x/l$. Thus we must pass from equations (4) to (5) without, however, making use of the dynamical differential equations. This is made possible by setting up the principle of superposition of small oscillations, according to which the motion $x + x_1$, $y + y_1$ is possible when the motions x, y and x_1, y_1 are given. We may combine, namely, the counterclockwise pendulum motion (4) with the clockwise motion

$$x_1 = l \cdot \alpha \cdot \cos\sqrt{\frac{g}{l}}(t - t_0), \quad y_1 = -l \cdot \alpha \cdot \sin\sqrt{\frac{g}{l}}(t - t_0).$$

Then, if we put $\alpha = C/2$, the motion $x + x_1$, $y + y_1$ is precisely the oscillating motion (5) which was desired.

In criticizing what precedes, we inquire, above all, how the principle of superposition is to be established, or at least made plausible, without the differential calculus. With these elementary presentations there remains always the doubt as to whether or not our neglecting of successive small quantities may not finally accumulate to a noticeable error, even if each is permissible singly. But I do not need to carry this out in detail, for these questions are so thoroughly elementary that each of you can think them through when you feel so inclined. Let me, in conclusion, state with emphasis that we are concerned in this whole discussion with a central point in the problem of instruction. First, the need for considering the infinitesimal calculus is evident. Moreover, it is clear that we need also a general introduction of the goniometric functions, independently of the geometry of the triangle, as a preparation for such general applications.

I come now to the last of the applications of the goniometric functions which I shall mention.

C. Representation of periodic functions by means of series of goniometric functions (trigonometric series)

As you know, there is frequent occasion in astronomy, in mathematical physics, etc., to consider periodic functions, and employ them in calculation. The method indicated in the title is the most important and the one most frequently used. For convenience we shall suppose the unit so chosen that the given periodic function $y = f(x)$ has the period 2π (see Fig. 81). The question then arises as to whether or not we can approximate to this function by means of a sum of cosines and sines of integral multiples of x, from the first, to the second, ..., in general to the n-th, each, with a

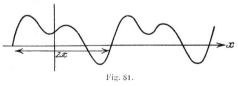
Fig. 81.

properly chosen constant factor. In other words, can one replace $f(x)$, to within a sufficiently small error, by an expression of the form

(1) $$\begin{cases} S_n(x) = \frac{a_0}{2} + a_1 \cos x + a_2 \cos 2x + \cdots + a_n \cos nx \\ \qquad\quad + b_1 \sin x + b_2 \sin 2x + \cdots + b_n \sin nx. \end{cases}$$

The factor $\frac{1}{2}$ is added to the constant term to enable us to give a general expression for the coefficients.

First I must again complain about the presentation in the text books, this time the texts in differential and integral calculus. Instead of putting into the foreground the elementary problem which I have outlined above, they often seem to think that the only problem which is of any interest at all is the theoretical question, connected with the one we have raised, whether $f(x)$ can be *exactly* represented by an *infinite* series. A notable exception to this is Runge in his *Theorie und Praxis der Reihen*[1]. As a matter of fact, that theoretical question is, in itself, thoroughly uninteresting for practical purposes, since we are concerned in practice with a finite number of terms, and not too many at that. Moreover it does not even permit a conclusion *a posteriori* as to the practical usableness of the series. One may by no means conclude from the convergence of a series that its first few terms afford even a fair approximation to the sum. Conversely, the first few terms of a divergent series may be useful, under certain conditions, in representing a function. I am emphasizing these things because a person who knows only the usual presentation and who wishes then to use finite trigonometric series in, say, the physical laboratory, is apt to be deceived and to reach conclusions that are unsatisfactory.

The customary neglect of finite trigonometric sums seems still more remarkable when one recalls that they have long been completely treated. The astronomer Bessel gave the authoritative treatment in 1815. You will find details concerning the history and literature of these questions in the encyclopedia reference by Burkhardt on *trigonometrische Interpolation* (Enzyklopädie II A 9, p. 642 et seq.). Moreover, the formulas that concern us here are essentially the same as those that arise in the usual convergence proofs. It is only that the thoughts which we shall attach to them have another shade of meaning and are designed to adapt the material more for practical use.

I turn now to a detailed consideration of our problem, and I shall inquire first as to the most appropriate determination of the coefficients $a, b, \ldots$ for a given number n of terms. Bessel developed an idea here which involves the method of least squares. The error that is made when, for a particular x, we replace $f(x)$ by the sum $S_n(x)$ of the first

---

[1] Sammlung Schubert No. 32, Leipzig, 1904. — See also Byerly, W. E., *Fourier's Series and Spherical Harmonics*.

$2n+1$ terms of the trigonometric series, is $f(x) - S_n(x)$, and a measure of the closeness of representation throughout the interval $0 \leq x \leq 2\pi$ (the period of $f(x)$) will be the sum of the squares of all the errors, that is, the integral

$$J = \int_0^{2\pi} [f(x) - S_n(x)]^2 \, dx.$$

The most appropriate approximation to $f(x)$ will therefore be supplied by that sum $S_n(x)$ for which this integral J has a minimum. It was from this condition that Bessel determined the $2n+1$ coefficients a_0, $a_1, \ldots, a_n, b_1, \ldots, b_n$. Since we are to consider J as a function of the $2n+1$ quantities $a_0, \ldots, b_n$, we have, as *necessary conditions* for a minimum:

(2) $$\begin{cases} \dfrac{\partial J}{\partial a_0} = 0, \quad \dfrac{\partial J}{\partial a_1} = 0, \ldots \dfrac{\partial J}{\partial a_n} = 0, \\ \dfrac{\partial J}{\partial b_1} = 0, \ldots \dfrac{\partial J}{\partial b_n} = 0. \end{cases}$$

Since J is an essentially positive quadratic function of $a_0, \ldots, b_n$, it is easy to see that the values of the variables determined by these $2n+1$ equations really yield a minimum.

If we differentiate under the sign of integration, the equations (2) take the form

(2′) $$\begin{cases} \int_0^{2\pi} [f(x) - S_n(x)] \, dx = 0, \\ \int_0^{2\pi} [f(x) - S_n(x)] \cos x \, dx = 0, \ldots, \int_0^{2\pi} [f(x) - S_n(x)] \cos nx \, dx = 0, \\ \int_0^{2\pi} [f(x) - S_n(x)] \sin x \, dx = 0, \ldots, \int_0^{2\pi} [f(x) - S_n(x)] \sin nx \, dx = 0. \end{cases}$$

Now the integrals of the products of $S_n(x)$ by a cosine or a sine can be much simplified. We have, namely, for $\nu = 0, 1, \ldots, n$,

$$\int_0^{2\pi} S_n(x) \cos \nu x \, dx = \frac{a_0}{2} \int_0^{2\pi} \cos \nu x \, dx + a_1 \int_0^{2\pi} \cos x \cos \nu x \, dx + \cdots + a_n \int_0^{2\pi} \cos nx \cos \nu x \, dx$$
$$+ b_1 \int_0^{2\pi} \sin x \cos \nu x \, dx + \cdots + b_n \int_0^{2\pi} \sin nx \cos \nu x \, dx.$$

According to known elementary integral properties of the goniometric functions, all the terms on the right vanish, with the exception of the cosine term with index ν, which takes the value $a_\nu \cdot \pi$, so that

$$\int_0^{2\pi} S_n(x) \cos \nu x \, dx = a_\nu \cdot \pi, \qquad (\nu = 0, 1, \ldots, n).$$

This result holds also for $\nu = 0$, by virtue of our having given to a_0 the factor $\frac{1}{2}$. Similarly, we have also

$$\int_0^{2\pi} S_n(x) \sin \nu x \, dx = b_\nu \cdot \pi, \qquad (\nu = 1, \ldots, n).$$

From these simple relations, it follows that each of the equations (2′) contains only one of the $2n+1$ unknowns. We can therefore write down their solutions immediately in the form

$$(3) \quad \begin{cases} a_\nu = \dfrac{1}{\pi} \displaystyle\int_0^{2\pi} f(x) \cos \nu x \, dx, & (\nu = 0, 1, \ldots, n), \\[1ex] b_\nu = \dfrac{1}{\pi} \displaystyle\int_0^{2\pi} f(x) \sin \nu x \, dx, & (\nu = 1, \ldots, n). \end{cases}$$

If we make use of these values of the coefficients in $S_n(x)$, as we shall from now on, J actually becomes a minimum, and its value is found to be

$$\int_0^{2\pi} f(x)^2 \, dx - \pi \left\{ \frac{a_0^2}{2} + \sum_{\nu=1}^n (a_\nu^2 + b_\nu^2) \right\}.$$

It is important to notice that the values of the coefficients a, b which result from our initially assumed form of $S_n(x)$ are independent of the special number n, and that, furthermore, the coefficient belonging to a term $\cos \nu x$ or $\sin \nu x$ has precisely the same value, whether one uses this term alone or together with any of the others, in approximating to $f(x)$ according to the same principle. If we attempt, namely, to make the best possible approximation to $f(x)$ means of a single cosine term $a_\nu \cos \nu x$, that is, so that

$$\int_0^{2\pi} [f(x) - a_\nu \cos \nu x]^2 \, dx = \text{Minimum}$$

we find for a_ν the same value that was deduced above. This fact makes this method of approximation especially convenient in practice. If, for example, one has been led to represent a function by a single multiple of $\sin x$, because its behaviour resembled the sine, and finds that the approximation is not close enough, one can add on more terms, always according to the principle of least squares, without having to alter the term already found.

I must now show how the sums $S_n(x)$, determined in this way, actually tend toward the function $f(x)$. For such an inquiry it seems to me desirable to proceed, in a sense, experimentally, after the method of natural scientists, namely by first drawing for a few concrete cases the approximating curves $S_n(x)$. This gives a vivid picture of what happens, and, even for persons without special mathematical gift, it will awaken interest, and will show the need of mathematical explanation.

In a former course of lectures (Winter semester 1903-1904) when I discussed these things in detail, my assistant, Schimmack, made such drawings, some of which I shall show you in the original and on the screen.

1. We get simple and instructive examples of the desired kind if we take curves made up of straight line segments. For example, consider the curve $y = f(x)$ as coinciding with $y = x$, from $x = 0$ to $x = \pi/2$; with $y = \pi - x$, from $x = \pi/2$ to $x = 3\pi/2$; with $y = x - 2\pi$ from $x = 3\pi/2$ to $x = 2\pi$; and as periodically repeating itself beyond the interval considered $(0, 2\pi)$. If we calculate the coefficients, we find all the coefficients a_ν are zero, since $f(x)$ is an odd function, and there remain only the sine terms. The desired series has the form

$$S(x) = \frac{4}{\pi}\left(\frac{\sin x}{1^2} - \frac{\sin 3x}{3^2} + \frac{\sin 5x}{5^2} - + \cdots\right).$$

In Fig. 82 the course of the first and second partial sums is sketched. The partial sums approach the given curve $y = f(x)$ more and more

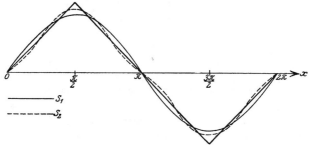

Fig. 82.

closely in that the number of their intersections with it increase continually It should be noticed especially that the approximating curves crowd more and more into the corners of the curve at $\pi/2, 3\pi/2, \ldots$, although they themselves, as analytic functions, can have no corners.

2. Let $f(x)$ be defined as x from $x = 0$ to $x = \pi$, and as $x - 2\pi$ from $x = \pi$ to $x = 2\pi$, with a gap at $x = \pi$. The curve consists, then, of parallel straight line segments through the points $x = 0, 2\pi, 4\pi, \ldots$ of the x axis. If at the points of discontinuity we insert vertical lines joining the ends of the discontinuous segments, the function will be represented by an unbroken line (see Fig. 83). It looks like the m strokes which you all practiced when you were larning to write. Again the function is odd, so that the cosine terms drop out, and the series becomes

$$S(x) = 2\left(\frac{\sin x}{1} - \frac{\sin 2x}{2} + \frac{\sin 3x}{3} - \frac{\sin 4x}{4} + - \cdots\right).$$

Fig. 83 represents the sums of the first two, three, and four terms. It is especially interesting here, also, to notice how they try to imitate the discontinuities of $f(x)$, e.g., by going through zero at $x = \pi$ with ever increasing steepness.

3. As a last example (see Fig. 84) I shall take a curve which is equal to $\pi/2$ between 0 and $\pi/2$, equal to 0 between $\pi/2$ and $3\pi/2$, and finally equal to $-\pi/2$ between $3\pi/2$ and 2π, and which continues periodically beyond that. If we again insert vertical segments at the places of discontinuity we get a hookshaped curve. Here also only the sine coefficients are different from zero, since we have an odd function, and the series becomes

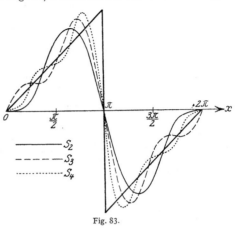

Fig. 83.

$$S(x) = \sin x + 2 \cdot \frac{\sin 2x}{2} + \frac{\sin 3x}{3} + 0 + \frac{\sin 5x}{5} + 2 \cdot \frac{\sin 6x}{6}$$
$$+ \frac{\sin 7x}{7} + 0 + \frac{\sin 9x}{9} + \cdots .$$

The law of the coefficients is not so simple here as it was before and hence the successive approximating curves (Fig. 84 shows the third, fifth and sixth) are not so comparable graphically as they were in the preceding cases.

We turn now to the question as to how large the error is, in general, when we replace

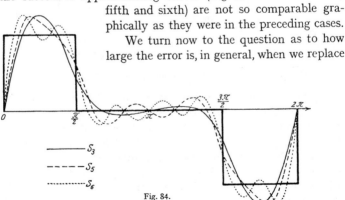

Fig. 84.

$f(x)$, at a definite place, by the sum $S_n(x)$. Up to this point we have been concerned only with the integral of this error, taken for the entire interval. Let us consider the integrals (3) (p. 193) for the coefficients a_ν, b_ν and replace the variable of integration by ξ, to distinguish it from x, which we use to denote a definite point. Then we can write our finite sum (1) as

$$S_n(x) = \frac{1}{\pi}\int_0^{2\pi} d\xi \cdot f(\xi) \cdot [\tfrac{1}{2} + \cos x \cos \xi + \cos 2x \cos 2\xi + \cdots + \cos nx \cos n\xi \\ + \sin x \sin \xi + \sin 2x \sin 2\xi + \cdots + \sin nx \sin n\xi],$$

or, if we combine summands which are in the same column, we have

$$S_n(x) = \frac{1}{\pi}\int_0^{2\pi} d\xi \cdot f(\xi)[\tfrac{1}{2} + \cos(x-\xi) + \cos 2(x-\xi) + \cdots + \cos n(x-\xi)].$$

The series in the parenthesis can be summed easily, perhaps most conveniently by using the complex exponential function. I cannot go into the details here, but we get, if we also use the fact that the periodicity of the integrand enables us to integrate from $-\pi$ to $+\pi$:

$$S_n(x) = \frac{1}{2\pi}\int_{-\pi}^{+\pi} d\xi \cdot f(\xi) \frac{\sin\frac{2n+1}{2}(\xi-x)}{\sin\tfrac{1}{2}(\xi-x)}.$$

To enable us to judge as to the value of this integral, let us first draw the curves

$$\zeta = \pm \frac{1}{2\pi} \frac{1}{\sin\tfrac{1}{2}(\xi-x)}$$

for the interval $x - \pi \leq \xi \leq x + \pi$ of the ξ axis. They obviously have branches resembling a hyperbola (see Fig. 85), and between these branches the curve

$$\eta = \frac{1}{2\pi}\frac{\sin\frac{2n+1}{2}(\xi-x)}{\sin\tfrac{1}{2}(\xi-x)} = \zeta \cdot \sin\frac{2n+1}{2}(\xi-x)$$

oscillates back and forth with increasing frequency as n gets larger. For $\xi = x$ it has the value $\eta = (2n+1)/(2\pi)$ which increases with n. If we now put $f(\xi) = 1$, for the sake of simplicity, then $S_n(x) = \int_{-\pi}^{+\pi} \eta \cdot d\xi$ will represent simply the area lying between the η curve and the ξ axis (shaded in the figure). Now anyone who has moderate feeling for continuity will see at once that if n increases sufficiently the oscillation areas to the right, as well as those to the left, being alternately positive and negative, will compensate each other and that only the area of the long narrow central arch will remain. But it is easy to see that with increasing n this approaches the value $f(x) = 1$, as it should. And, in general, things turn out in this same way, provided $f(x)$ does not oscillate too strongly at $x = \xi$.

It is just such considerations, developed for more precise use, which form the basis for Dirichlet's proof of convergence of the infinite trigonometric series.

This proof was published[1] for the first time by Dirichlet in 1829 in volume 4 of Crelle's Journal. Later (1837) he gave a more popular presentation[2] in the *Repertorium der Physik* by Dove and Moser. The proof is given nowadays in most textbooks[*], and I do not need to dwell upon it here. But I must mention certain sufficient conditions which the function $f(x)$ must satisfy if it is to be represented by an infinite trigonometric series. Again think of $f(x)$ as given in the interval

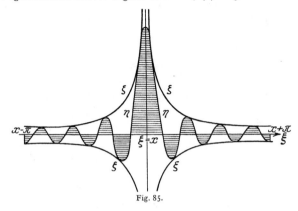

Fig. 85.

$0 \leq x \leq 2\pi$ and as periodically continued beyond. Dirichlet makes, then, the following two assumptions which are called today simply Dirichlet's conditions:

a) The given function $f(x)$ is *segmentally continuous*, i.e., it has in the interval $(0, 2\pi)$ only a finite number of discontinuities, and is otherwise continuous up to the points where it jumps.

b) The given function $f(x)$ is segmentally monotone, i.e., one can divide the interval $(0, 2\pi)$ into a finite number of sub-intervals, in every one of which $f(x)$ either does not increase or does not decrease. In other words, $f(x)$ has only a finite number of maxima and minima. (This would exclude, for example, such a function as $\sin 1/x$, for which $x = 0$ is a limit point of extrema.)

Dirichlet shows that, under these conditions, the infinite series represents the function $f(x)$ exactly for all values of x for which $f(x)$ is continuous. That is
$$\lim_{n=\infty} S_n(x) = f(x).$$
Moreover Dirichlet proves that, at a point of discontinuity, the series converges also, but to a value which is the arithmetic mean of the two

---

[1] Reprinted in Dirichlet, Werke, vol. 1, p. 117, Berlin, 1889.

[2] *Über die Darstellung ganz willkürlicher Funktionen durch Sinus- und Kosinusreihen.* Reprinted, Werke, vol. 1, p. 133–160, and Ostwalds Klassiker No. 116, Leipzig, 1900.

[*] See Byerly, *Fourier's Series and Spherical Harmonics.*

values which $f(x)$ approaches when x approaches the discontinuity from the one side or the other. This fact is usually expressed in the form

$$\lim_{n=\infty} S_n(x) = \frac{f(x+0) + f(x-0)}{2}.$$

Fig. 86 exhibits such discontinuities and the corresponding mean values.

These conditions of Dirichlet are sufficient, but by no means necessary, in order that $f(x)$ may be represented by the series $S(x)$. On the other hand, mere continuity of $f(x)$ is not sufficient. In fact it is possible to give examples of continuous functions where oscillations cluster so strongly that the series $S(x)$ diverges.

After these theoretical matters I shall now return to the practical side of trigonometric series. For a detailed treatment of the questions that arise here I refer you to the book by Runge which I mentioned before (see p. 191). You will find there a full treatment of the question as to the numerical calculation of the coefficients in the series, i.e., the question as to how, when a function is given, one can rapidly evaluate the integrals for a_ν, b_ν in the most suitable way.

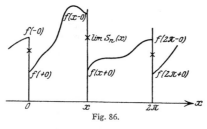
Fig. 86.

Special mechanical devices called harmonic analyzers have been constructed for calculating these coefficients. This name has reference to the relation which the development of a function $f(x)$ into a trigonometric series has to acoustics. Such a development corresponds to the separation of a given tone $y = f(x)$ (where x is the time and y the amplitude of the tone vibration) into "pure tones", that is, into pure cosine and sine vibrations. In our collection we have an analyzer by Coradi in Zürich, by means of which one can determine the coefficients of six cosine and sine terms ($\nu = 1, 2, \ldots, 6$), i.e. twelve coefficients in all. The coefficient $a_0/2$ must be separately determined by a planimeter. Michelson and Stratton have made an apparatus with which 160 coefficients ($\nu = 1, 2, \ldots, 80$) can be determined. It is described in Runge's book. Conversely, this apparatus can also sum a given trigonometric series of 160 terms, i.e. calculate the function from the given coefficients a_ν, b_ν. This problem also, of course, is of the greatest practical importance.

The apparatus of Michelson and Stratton called attention anew to a very interesting phenomenon, one which had been noticed earlier[1] but

---

[1] According to Enzyklopädie vol. 2, 12 *(Trigonometrische Reihen und Integrale)*, p. 1048, H. Wilbraham was already familiar with the phenomenon under discussion here and had treated it with a view to calculation.

which, with the passage of decades, had, curiously enough, been forgotten. In 1899 Gibbs again discussed it in *Nature*[1], whence it is called Gibb's phenomenon. Let me say a few words about it. The theorem of Dirichlet gives as the value of the infinite series, for a fixed value x, the expression $[f(x+0) + f(x-0)]$. In the second example discussed above (to have a concrete case in mind) the series gives the values at the isolated points $\pi, 3\pi, \ldots$ of the function pictured in Fig. 87.

Now the way in which we explained the matter of trigonometric approximation was different from the Dirichlet procedure, where x is kept fixed while n becomes infinite. We thought of n as fixed, considered $S_n(x)$ with variable x, and drew the successive approximating curves $S_1(x), S_2(x), S_3(x), \ldots$ We may now inquire, what happens to these curves when n becomes infinite; or, to put it arithmetically, what is the limit of $S_n(x)$ when n becomes infinite, x being variable?

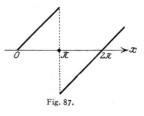

Fig. 87.

It is clear, intuitively, that the limit function cannot exhibit isolated points, as before, but must be represented by a connected curve. It would appear probable that this limit curve must consist of the continuous branches of $y = f(x)$, together with the vertical segments which join $f(x+0)$ and $f(x-0)$ at the points of discontinuity, that is, in our example, the curve would be shaped like a German m, as is shown in Fig. 83. The fact is, however, that the vertical part of the limit curve projects beyond $f(x+0)$ and $f(x-0)$, by a finite amount, so that the limit curve has the remarkable form sketched in Fig. 88.

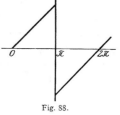

Fig. 88.

This little superimposed tower was noticed in the curves which the Michelson machine drew; in other words it was disclosed experimentally. At first it was ascribed to imperfections in the apparatus, but finally Gibbs recognized it as necessary. If $D = |f(x+0) - f(x-0)|$ is, in general, the magnitude of the jump, then the length of the extension is, according to Gibbs:

$$-\frac{D}{\pi}\int_{\pi}^{\infty}\frac{\sin\xi}{\xi}d\xi \approx \frac{1}{\pi} 0.28 D \approx 0.09 D.$$

As to the proof of this statement, it is sufficient to give it for a single discontinuous function, e.g., the one in our example, since all other functions with the same spring can be obtained from it by the addition of continuous functions. This proof is not very difficult. It results

---

[1] Vol. 59 (1898—99), p. 200. Scientific papers II, p. 158. New York 1906.

immediately from consideration of the integral formula for $S_n(x)$ (see p. 196). Furthermore, if one draws a sufficient number of the approximating curves one sees quite clearly how the Gibbs point arises.

It would lead me too far afield if I were to consider further the many interesting niceties in the behaviour of the approximating curves. I am glad to refer you to the full and very readable article by Fejér in Vol. 64 (1907) of the Mathematische Annalen.

With this I shall conclude the special discussion of trigonometric series in order to wander in a field which as to its content and its history is closely related to them.

Excursus Concerning the General Notion of Function

We must be all the more willing, in these lectures, to discuss the notion of function, since our school reform movement advocates giving this important concept a prominent place in instruction.

If we follow again the historical development, we notice first that the older authors, like Leibniz and the Bernoullis, use the function concept only in isolated examples, such as powers, trigonometric functions, and the like. A general formulation is met first in the eighteenth century.

1. With Euler, about 1750 (to use only round numbers), we find two different explanations of the word function.

a) In his *Introductio* he defines, as a function y of x, every analytic expression in x, i. e., every expression which is made up of powers, logarithms, trigonometric functions, and the like; but he does not indicate precisely what combinations are to be admitted. Moreover, he had, already, the familiar division into algebraic and transcendental functions.

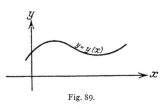

Fig. 89.

b) At the same time, a function $y(x)$ (see Fig. 89) was defined for him whenever a curve was arbitrarily drawn (libero manus ductu) in an x, y coordinate system.

2. Lagrange, about 1800, in his *Théorie des fonctions analytiques* restricts the notion function, in comparison with Euler's second definition, by confining it to so called *analytic functions*, which are defined by a power series in x. Modern usage has retained the words analytic functions with this same meaning, where, of course, one must recognize that this includes only a special class of the functions that really occur in analysis. Now a power series

$$y = P(x) = a_0 + a_1 x + a_2 x^2 + \cdots$$

defines a function primarily only within the region of its convergence, i.e., in a certain region around $x = 0$. A method was soon found, however, for extending beyond this the region of definition for the function. If, say, x_1 (see Fig. 90) is within the region of convergence of $P(x)$, and if $P(x)$ is resolved into a new series

$$y = P_1(x - x_1)$$

which proceeds according to powers of $(x - x_1)$, it is possible that this may converge in a region extending beyond the first one, and so may define y in a larger field. A repetition of this process may extend the field still farther. This method of *analytic continuation* is well known to any one who is familiar with complex function theory.

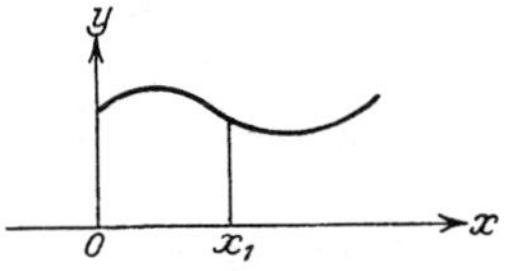

Notice, in particular, that every coefficient in the power series $P(x)$, and therefore the entire function y is determined when the behavior of the function y along an arbitrarily small segment of the x axis is known, say in the neighborhood of $x = 0$. For then the values of all the derivatives of y are known for $x = 0$, and we know that

$$y(0) = a_0, \quad y'(0) = a_1, \quad y''(0) = 2a_2, \ldots$$

Fig. 90.

Thus an analytic function, in the Lagrange sense, is determined throughout its entire course by the shape of an arbitrarily small segment. This property is completely opposed to the behavior of a function in the sense of Euler's second definition. There, any part of a curve can be continued at will.

3. The further development of the function concept is due to J. J. Fourier, one of the numerous important mathematicians who worked in Paris at the beginning of the nineteenth century. His chief work is the *Théorie analytique de la chaleur*[1] which appeared in 1822. Fourier made the first communication, however, concerning his theories, to the Paris Academy in 1807. This work is the source of that far reaching method, so much used in mathematical physics today, which can be characterized as the resolution of all problems to the integration of partial differential equations with initial conditions, to a so called *boundary-value problem*.

Fourier treated, in particular, the problem of heat conduction which, for a simple case, may be stated as follows. The boundary of a circular plate is kept at a constant temperature, e.g., one part at the freezing, the other at the boiling point (see Fig. 91). What stationary temperature is ultimately brought about by the resulting flow of heat? Boundary values are introduced here which can be assigned independently of each other at different parts of the boundary. Thus Euler's second definition

---

[1] Reprinted in Fourier, Œuvres, vol. I. Paris 1888. Translated into German by Weinstein. Berlin 1884.

of function comes appropriately into the foreground, as opposed to that of Lagrange.

This definition is retained essentially by Dirichlet in the works which we mentioned (p. 197), except that it is translated into the language of analysis or, to use a modern term, it is arithmetized. This is in fact, necessary. For no matter how fine a curve be drawn, it can never define exactly the correspondence between the values of x and y. The stroke of the pen will always have a certain width, from which it follows that the lengths x and y which correspond to one another can be measured exactly only to a limited number of decimal places.

Dirichlet formulated the arithmetic content of Euler's definition in the following way. If in any way a definite value of y is determined, corresponding to each value of x in a given interval, then y is called a function of x. Although he announced this very general notion of a function, nevertheless he always thought primarily of continuous functions, or of such as were not all too discontinuous, as was done then quite generally. People considered complicated clusterings of discontinuities as thinkable, but they hardly believed that they deserved much attention.

Fig. 91.

This standpoint finds expression when Dirichlet speaks of the development into series of "entirely arbitrary functions" (just as Fourier had said "fonctions entièrement arbitraires) even when he formulated very precisely his Dirichlet conditions, which must be satisfied by all the functions he considered.

5. We must now take account of the fact that at about this time, say around 1830, the independent development of the theory of functions of a complex argument began; and that in the next three decades it became the common property of mathematicians. This development was connected, above all, with the names Cauchy, Riemann, and Weierstrass. The first two start, as you know, from the partial differential equations which bear their names, and which must be satisfied by the real and imaginary parts u, v of the complex function

$$f(x + iy) = u + iv,$$

while Weierstrass defines the function by means of a power series and the aggregate of its analytic continuations, so that he, in a sense, follows Lagrange.

Now it is remarkable that this passage into the complex domain brings about an agreement and connection between the two function concepts considered above. I shall give a brief sketch of this.

Let us put $z = x + iy$, and consider the power series

(1) $$f(z) = u + iv = c_0 + c_1 z + c_2 z^2 + \cdots,$$

as converging for small $|z|$ so that, in the terminology of Weierstrass, it defines an element of an analytic function. We consider its values on a sufficiently small circle of radius r, about $z = 0$, which lies entirely within the region of convergence (see Fig. 92), i. e., we put $z = x + iy = r(\cos\varphi + i \sin\varphi)$ in the power series, and we get

$$f(z) = c_0 + c_1 r (\cos\varphi + i\sin\varphi) + c_2 r^2 (\cos 2\varphi + i \sin 2\varphi) \cdots .$$

If we separate the coefficients into real and imaginary parts:

$$c_0 = \frac{\alpha_0 - i\beta_0}{2}, \quad c_1 = \alpha_1 - i\beta_1, \quad c_2 = \alpha_2 - i\beta_2, \ldots ,$$

we get as the real part of $f(z)$

(2) $$\begin{cases} u = u(\varphi) = \frac{\alpha_0}{2} + \alpha_1 r \cos\varphi + \alpha_2 r^2 \cos 2\varphi + \cdots \\ \qquad\qquad + \beta_1 r \sin\varphi + \beta_2 r^2 \sin 2\varphi + \cdots . \end{cases}$$

The sign of the imaginary part in the c was taken negative in order that all the signs should be positive. Thus the power series for $f(z)$ yields for the values, on our circle, of the real part u, thought of as a function of the angle φ, a trigonometric series of exactly the former sort, whose coefficients are α_0, $r^\nu \alpha_\nu$, $r^\nu \beta_\nu$.

Of course, these values u will be analytic functions of φ, in the sense of Lagrange, as long as the circle (r) lies entirely within the region of convergence of the power series (1). But if we allow it to coincide with the circle of convergence of the series (1) which bounds its region

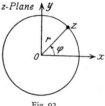

Fig. 92.

of convergence, then the series (1) and consequently also the series (2) will not necessarily converge any longer. Meantime it can happen that the series (2) continues to converge, in which case the boundary values $u(\varphi)$ cannot be analytic functions in the sense of Dirichlet.

If we proceed conversely and assign to circle (r) an arbitrary distribution of values $u(\varphi)$ which satisfy only the conditions of Dirichlet, then they can be developed into a trigonometric series of the form (2), so that the quantities $\alpha_0, \alpha_1, \ldots, \beta_1, \beta_2, \ldots$ and hence the coefficients of the power series (1) (to within an arbitrary additive constant $-(i\beta_0)/2$) will be determined. It can be shown that this power series actually converges within the circle (r) and that the real part of the analytic function which it determines has the values $u(\varphi)$ as boundary values on the circle (r), or, to be more exact, that it approaches the value $u(\bar\varphi)$ whenever a position $\bar\varphi$ is approached for which $u(\bar\varphi)$ is continuous.

The proofs of these facts are all contained in the investigations concerning the behavior of power series on the circle of convergence. I cannot, of course, give them here. But these remarks may serve to show how, in this way, the Fourier-Dirichlet function concept and that of Lagrange merge into each other in that the arbitrariness in the behaviour of the trigonometric series $u(\varphi)$ on the boundary of the circle is concentrated, for the power series, into the immediate neighborhood of the center.

6. Modern science has not stopped with the formulation of these concepts. Science never rests, even though the individual investigator may become weary. During the last three decades mathematicians, taking a standpoint quite different from that of Dirichlet, have siezed upon functions having the greatest possible discontinuity, which, in particular, do not satisfy the Dirichlet conditions. The most remarkable types of function have been found, which contain the most disagreeable singularities "balled into horrid lumps". It becomes a problem then to determine how far the theorems which hold for "reasonable" functions still have validity for such abnormities.

7. In connection with this, there has arisen, finally, a still more far reaching generalization of the notion of function. Up to this time, a function was thought of as always defined at every position in the continuum made up of all the real or complex values of x, or at least at every position in an entire interval or region. But recently the theory of point sets, invented by G. Cantor, has made its way more and more to the foreground, in which the continuum of all x is only an obvious example of a set of points. From this new standpoint functions are being considered which are defined only for the positions x of some arbitrary set, so that in general y is called a function of x when to every element of a set x of things (numbers or points) there corresponds an element of a set y.

Let me point out a difference between this newest development and the older one. The notions considered under headings 1. to 5. have arisen and have been developed with reference primarily to applications in nature. We need only think of the title of Fourier's work. But the newer investigations mentioned in 6. and 7. are the result purely of the love of mathematical research, which has taken no account whatever of the needs of natural phenomena, and the results have indeed found as yet no direct application. The optimist will think, of course, that the time for such application is bound to come.

We shall now put our customary question as to how much of all this should be taken up by the schools. What should the teacher and what should the pupils know?

In this connection I should like to say that it is not only excusable but even desirable that the schools should always lag behind the most recent advances of our science by a considerable space of time, certainly several decades; that, so to speak, a certain hysteresis should take place. But the hysteresis which actually exists at the present time is in some respects unfortunately much greater. It embraces more than a century, in so far as the schools, for the most part, ignore the entire development since the time of Euler. There remains, therefore, a sufficiently large field for the work of reform. And what we demand in the way of reform is really quite modest, if you compare it with the present state of the science. We desire merely that the general notion of function, according to the one or the other of Euler's interpretations, should permeate as a ferment the entire mathematical instruction in the higher schools. It should not, of course, be introduced by means of abstract definitions, but should be transmitted to the student as a living possession, by means of elementary examples, such as one finds in large number in Euler. For the teacher of mathematics, however, something more than this seems desirable, at least a knowledge of the elements of complex function theory; and although I should not make the same demand regarding the most recent concepts in the theory of point sets, still it seems very desirable that among the many teachers there should always be a small number who devote themselves to these things with the thought of independent work.

I should like to add to these last remarks a few words concerning the important rôle that has been played in this entire development by the theory of trigonometric series. You will find extensive references to the literature of the subject in Burkhardt's *Entwickelungen nach oszillierenden Funktionen* (especially in chapters 2, 3, 7), that "giant report", as his friends call it, which since 1901 has been appearing serially in volume 10 of the *Jahresbericht der deutschen Mathematikervereinigung*[1]. It combines, in more than 9000 references, an amount of pertinent literature such as you will hardly find elsewhere.

The first to come upon the representation of general functions by means of trigonometric series was Daniel Bernoulli, the son of John Bernouilli. He noticed, about 1750, in his study of the acoustic problem of vibrating strings, that the general vibration of a string could be represented by the superposition of those sine vibrations which corresponded to the fundamental tone and the overtones. That involves precisely the development into a trigonometric series of the function which represents the form of the string.

Although advances were soon made in knowledge of these series, still no one really believed that arbitrary functions graphically given, could be represented by them. At bottom, here, there was an undefined presentiment of considerations which have become quite clear to us now through the theory of point sets. Perhaps one assumed, without,

---

[1] Completed in two half volumes as Heft 2 of this volume. Leipzig 1908. [A short summary appears in the Enzyklopädie der mathematischen Wissenschaften, vol. 2. Burkhardt's report goes to 1850. The development from 1850 on is sketched by Hilb and Riesz in their article in the Enzyklopädie, vol. 2, C 10.]

of course, being able to give precise expression to the feeling, that the "set" of all arbitrary functions, even if discontinuities are excluded, was greater than the "set" of all possible systems of numbers $a_0, a_1, a_2, \ldots, b_1, b_2, \ldots$, which represents the totality of trigonometric series.

It is only the precise concepts of the modern theory of point sets that have cleared this up, and have shown that that judgment was false. Let me, at this place, elaborate somewhat this important point. It is easy to see that the entire course of a continuous function arbitrarily defined in a given interval, say from 0 to 2π, is completely known if one knows its values at all the rational positions of that interval (see Fig. 93). For, since the set of these rational points is dense, we can effect an arbitrarily close approximation for any irrational position, in terms of function values at rational ones, so that, by virtue of the continuity of the function, the value of $f(x)$ is known as the limit of the function values at the approximating points. Furthermore, we know that the set of all rational numbers is *denumerable* (see appendix II, p. 252), i. e., that they can be arranged in a series in which a definite first element is followed by a definite second, this by a definite third, and so on. From this it follows, however, that the assignment of the arbitrary continuous function means nothing more than the assignment of an appropriate denumerable set of constants—the function values at the ordered rational points. But in the same way, by means, namely, of the denumerable series of constants $a_0, a_1, b_1, a_2, b_2, \ldots$, we can assign a definite trigonometric series, so that the doubt as to whether the totality of continuous functions was, in the nature of things, essentially greater than that of the series, is groundless. Similar considerations hold for functions which are discontinuous but which satisfy the Conditions of Dirichlet. We shall have occasion later to give detailed consideration to these matters.

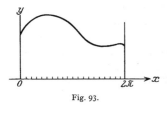

Fig. 93.

The man who abruptly brushed aside all these misgivings was Fourier and it was just this which made him so significant in the history of trigonometric series. Of course, he did not base his conclusions on the theory of point sets, but he was the first one who had the courage to *believe* in the general power of series for purposes of representation. Fortified by this belief he set up a number of series by actual calculation, using characteristic examples of discontinuous functions, as we did a short time back. The proofs of convergence, as we have noted, were first given later, by Dirichlet, who, moreover, was a pupil of Fourier. This stand of Fourier's had a revolutionary effect. That it should be possible to represent by series of analytic functions such arbitrary functions as these, which obeyed in different intervals such entirely different laws, this was something quite new and unexpected to the mathematicians of that time. In recognition of the disclosure of this possibility, the name of Fourier was given to the trigonometric series which he employed, a name which has persisted to this day. To be sure every such personal designation implies a marked one-sidedness, even when it is not outright injustice.

In conclusion, I must mention briefly a second accomplishment of Fourier. He considered, namely, the limiting case of the trigonometric series when the period of the function to be represented is allowed to become infinite. Since a function with an infinite period is simply a non periodic function, arbitrary along the entire x axis, this limiting case supplies a means of representing non periodic functions. The transition is brought about by introducing a linear transformation of the argument of the series, which effects a representation of functions with a period l instead of 2π, and then letting l become infinite. The series then goes over into the so called Fourier integral

$$f(x) = \int_0^\infty [\varphi(\nu)\cos\nu x + \psi(\nu)\sin\nu x]\,d\nu,$$

when $\varphi(\nu)$, $\psi(\nu)$ are expressed in definite manner as integrals of the function $f(x)$ from $-\infty$ to $+\infty$. The new thing here is that the index ν takes *continuously* all values from 0 to ∞, not merely the values 0, 1, 2, ...; and that, correspondingly, $\varphi(\nu)d\nu$ and $\psi(\nu)d\nu$ take the place of a_ν, b_ν.

We shall now leave the elementary transcendental functions, which have hitherto been our chief concern in our remarks on analysis, and go over to a new concluding chapter.

III. Concerning Infinitesimal Calculus Proper

Of course I shall assume that you all know how to differentiate and integrate, and that you have frequently used both processes. We shall be concerned here solely with more general questions, such as the logical and psychological foundations, instruction, and the like.

1. General Considerations in Infinitesimal Calculus

I should like to make a general preliminary remark concerning the range of mathematics. You can hear often from non mathematicians, especially from philosophers, that mathematics consists exclusively in drawing conclusions from clearly stated premises; and that, in this process, it makes no difference what these premises signify, whether they are true or false, provided only that they do not contradict one another. But a person who has done productive mathematical work will talk quite differently. In fact those persons are thinking only of the crystal lized form into which finished mathematical theories are finally cast.

The investigator himself, however, in mathematics, as in every other science, does not work in this rigorous deductive fashion. On the contrary, he makes essential use of his phantasy and proceeds inductively, aided by heuristic expedients. One can give numerous examples of mathematicians who have discovered theorems of the greatest importance, which they were unable to prove. Should one, then, refuse to recognize this as a great accomplishment and, in deference to the above definition, insist that this is not mathematics, and that only the successors who supply polished proofs are doing real mathematics? After all, it is an arbitrary thing how the word is to be used, but no judgment of value can deny that the inductive work of the person who first announces the theorem is at least as valuable as the deductive work of the one who first proves it. For both are equally necessary, and the discovery is the presupposition of the later conclusion.

It is precisely in the discovery and in the development of the infinitesimal calculus that this inductive process, built up without compelling logical steps, played such a great rôle; and the effective heuristic aid was very often sense perception. And I mean here immediate sense perception, with all its inexactness, for which a curve is a stroke of definite width, rather than an abstract perception which postulates a completed passage to the limit, yielding a one dimensional line. I should like to corroborate this statement by outlining to you how the ideas of the infinitesimal calculus were developed historically.

If we take up first the notion of an integral, we notice that it begins historically with the problem of measuring areas and volumes (quadrature and cubature). The abstract logical definition determines the integral $\int_a^b f(x)\, dx$, i.e., the area bounded by the curve $y = f(x)$, the x axis, and the ordinates $x = a$, $x = b$, as the limit of the sum of narrow rectangles inscribed in this area when their number increases and their width decreases without bound. Sense perception, however, makes it natural to define this area, not as this exact limit, but simply as the sum of a large number of quite narrow rectangles. In fact, the necessary inexactness of the drawing would inevitably set bounds to the further narrowing of the rectangles (see Fig. 94).

This naïve method characterizes, in fact, the thinking of the greatest investigators in the early period of infinitesimal calculus. Let me mention, first of all, Kepler who in his *Nova stereometria doliorum vinariorum*[1] was concerned with the volumes of bodies. His chief interest here was in the measuring of casks, and in determining their most suitable shape. He took precisely the naïve standpoint indicated above.

---

[1] Linz on the Danube, 1615. German in Ostwalds Klassikern, No. 165. Leipzig, 1908.

He thought of the volume of the cask, as of every other body (see Fig. 95), as made up of numerous thin sheets suitably ranged in layers, and considered it as the sum of the volumes of these leaves, each of which was a cylinder. In a similar way he calculated the simple geometric bodies, e. g., the sphere. He thought of this as made up of a great many small pyramids with common vertex at the center (see Fig. 96). Then its volume, according to the well known formula for the pyramid, would be $r/3$ times the sum of the bases of all the small pyramids. By writing for the sum of these little facets simply the surface of the sphere, or $4\pi r^2$, he obtained $4\pi r^3/3$, the correct formula for the volume.

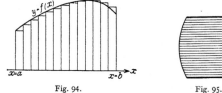

Fig. 94.

Fig. 95.   Fig. 96.

Moreover, Kepler emphasizes explicitly the practical heuristic value of such considerations, and refers, so far as rigorous mathematical proofs are concerned, to the so called *method of exhaustion*. This method, which had been used by Archimedes, determines, for example, the area of the circle by following carefully the approximations to the area by means of inscribed and circumscribed polygons with an increasing number of sides. The essential difference between it and the modern method lies in the fact that it tacitly assumes, as self evident, the existence of a number which measures the area of the circle, whereas the modern infinitesimal calculus declines to accept this intuitive evidence, but has recourse to the abstract notion of limit and defines this number as the limit of the numbers that measure the areas of the inscribed polygons. Granted, however, the existence of this number, the method of exhaustion is an exact process for approximating to areas by means of the known areas of rectilinear figures, one which satisfies rigorous modern demands. The method is, however, very tedious in many cases, and ill suited to the *discovery* of areas and volumes. One of Archimedes writings[1], discovered by H. Heiberg in 1906, shows, in fact, that he did not use the method of exhaustion at all in his investigations. After he had first obtained his results by some other method, he developed the proof by exhaustion in order to meet the demands of that time as to rigor. For the discovery of his theorems he used a method which included considerations of the center of gravity and the law of the lever, and also of intuition, such as, for example, that triangles and parabolic

---

[1] Already referred to on p. 80.

segments consist of series of parallel chords, or that cylinders, spheres, and cones are made up of series of parallel circular discs.

Returning now to the seventeenth century, we find considerations analogous to those of Kepler in the book of the Jesuit Bonaventura Cavalieri: *Geometria indivisibilibus continuorum nova quadam ratione promota*[1] where he sets up the principle called today by his name: *Two bodies have equal volumes if plane sections equidistant from their bases have equal areas.* This principle of Cavalieri is, as you know, much used in our schools. It is believed there that integral calculus can be avoided in this way, whereas this principle belongs, in fact, entirely to the calculus. Its establishment by Cavalieri amounts precisely to this, that he thinks of both solids as built up of layers of thin leaves which, according to the hypothesis, are congruent in pairs, i.e., one of the bodies could be transformed into the other by translating its individual leaves (see Fig. 97); but this could not alter the volume, since this consists of the same summands before and after the translation.

Fig. 97.

Naïve sense perception leads in the same way to the differential quotient of a function, i. e., to the tangent to the curve. In this case, we can replace (and this is the way it was actually done) the curve by a polygonal line (see Fig. 98) which has on the curve a sufficient number of points, as vertices, taken close together. From the nature of our sense perception we can hardly distinguish the curve from this aggregate of points and still less from the polygonal line. The tangent is now defined outright as the line joining two successive points, that is, as the prolongation of one of the sides of the polygon.

Fig. 98.

From the abstract logical standpoint, this line remains only a secant, no matter how close together the points are taken; and the tangent is only the limiting position approached by the secant when the distance between the points approaches zero. Again, from this naïve standpoint, the circle of curvature is thought of as the circle which passes through three successive polygon vertices, whereas exact procedure defines the circle of curvature as the limiting position of this circle when the three points approach each other.

The force of conviction inherent in such naïve guiding reflections is, of course, different for different individuals. Many—and I include myself here—find them very satisfying. Others, again, who are gifted only on the purely logical side, find them thoroughly meaningless and are unable to see how anyone can consider them as a basis for mathe-

---

[1] Bononiae, 1635. First edition, 1653.

matical thought. Yet considerations of this sort have often formed the beginnings of new and fruitful speculations.

Moreover, these naïve methods always rise to unconscious importance whenever in mathematical physic, mechanics, or differential geometry a preliminary theorem is to be set up. You all know that they are very serviceable then. To be sure, the pure mathematician is not sparing of his scorn on these occasions. When I was a student it was said that the differential, for a physicist, was a piece of brass which he treated as he did the rest of his apparatus.

In this connection, I should like to commend the Leibniz notation, the leading one today, because it combines with a suitable suggestion of naïve intuition, a certain reference to the abstract *limit process* which is implicit in the concept. Thus, the Leibniz symbol dy/dx, for the differential quotient, reminds one, first that it comes from a *quotient*; but the d, as opposed to the $\varDelta$ which is the usual symbol for finite difference, indicates that something new has been added, namely, the *passage to the limit*. In the same way, the integral symbol $\int y\, dx$ suggests the origin of the integral from a *sum* of small quantities. However, one does not use the usual sign $\varSigma$ for a sum, but rather a conventionalized S^*, which indicates here that something new has entered the process of summation.

We shall now discuss with some detail the logical foundation of differential and integral claculus, and at the same time consider it in its historical development.

1. The principal idea, as the subject is taught, in general, at the university (I need only briefly to refresh your memory here) is that infinitesimal calculus is *only an application of the general notion of limit*. The differential quotient is defined as the limit of the quotient of corresponding finite increments of variable and function

$$\frac{dy}{dx} = \lim_{\varDelta x = 0} \frac{\varDelta y}{\varDelta x}$$

provided that this limit exists; and not at all as a quotient in which dy and dx have an independent meaning. In the same way, the integral is defined as the limit of a sum:

$$\int_a^b y\, dx = \lim_{\varDelta x_i = 0} \sum_{(i)} y_i \cdot \varDelta x_i,$$

where the $\varDelta x_i$ are finite parts of the interval $a \leq x \leq b$, the y_i corresponding arbitrary values of the function in that interval, and all the $\varDelta x_i$ are to converge toward zero; but $y\, dx$ does not have any actual significance as, say, a summand of a sum. These designations are retained for the reasons of expediency which we mentioned above.

---

* It is remarkable that many are unaware that $\int$ has this meaning.

2. The conception as we have thus characterized it is set forth in precise form by Newton himself. I refer you to a place in his principal work, the *Philosophiae Naturalis Principia Mathematica*[1] of 1687: "Ultimae rationes illae, quibuscum quantitates evanescunt, revera non sunt rationes quantitatum ultimarum, sed limites, ad quos quantitatum sine limite descrescentium rationes semper appropinquant, et quos propius assequi possunt, quam pro data quavis differentia, nunquam vero transgredi neque prius attingere quam quantitates diminuuntur in infinitum." Moreover, Newton avoids the infinitesimal calculus, as such, in the discussions in this work, although he certainly had used it in deriving his results. For, the fundamental work in which he developed his method of infinitesimal calculus was written in 1671, although it did not appear until 1736. It bears the title *Methodus Fluxionum et Serierum Infinitarum*[2].

In this, Newton develops the new calculus in numerous examples, without going into fundamental explanations. He makes connection here with a phenomenon of daily life which suggests a passage to a limit. If one considers, namely, a motion $x = f(t)$ on the x axis in the time t, then every one has a notion as to what is meant by the velocity of this motion. If we analyze this motion it turns out that we mean the limiting value of the difference quotient $\Delta x/\Delta t$. Newton made this velocity of x with respect to the time the basis of his developments. He called it the "fluxion" of x and wrote it $\dot{x}$. He considered all the variables x, y as dependent on this fundamental variable t, the time. Accordingly the differential quotient dy/dx appears as the quotient of two fluxions $\dot{y}/\dot{x}$ which we now should write more fully $(dy/dt : dx/dt)$.

3. These ideas of Newton were accepted and developed by a long series of mathematicians of the eighteenth century, who built up the infinitesimal calculus, with more or less precision, upon the notion of limit. I shall select only a few names: C. Maclaurin, in his *Treatise of Fluxions*[3], which as a textbook certainly had a wide influence; then d'Alembert, in the great French *Encyclopédie Méthodique*; and finally Kästner[4], in Göttingen, in his lectures and books. Euler belongs primarily in this group although, with him, other tendencies also came to the front.

4. It was necessary to fill out an essential gap in all these developments, before one could speak of a consistent system of infinitesimal calculus. To be sure, the differential quotient was defined as a limit, but there was lacking a method for estimating, from it, the increment

---

[1] New edition by W. Thomson and H. Blackburn, Glasgow, 1871, p. 38.

[2] Newtoni, J., *Opuscula Mathematica, philosophica, et philologica*, vol. I, p. 29. Lausanne, 1744.

[3] Edinburgh, 1742.

[4] Kästner, A. G., *Anfangsgründe der Analysis des Unendlichen*, Göttingen, 1760.

of the function in a finite interval. This was supplied by the *mean value theorem*; and it was Cauchy's great service to have recognized its fundamental importance and to have made it the starting point accordingly of differential calculus. And it is not saying too much if, because of this, we adjudge Cauchy as the founder of exact infinitesimal calculus in the modern sense. The fundamental work in this connection, based on his Paris lectures, is his *Résumé des Leçons sur le Calcul Infinitésimal*[1], together with its second edition, of which only the first part, *Leçons sur le Calcul Différentiel*[2], was published.

The *mean-value theorem*, as you know, may be stated as follows. *If a continuous function f (x) possesses a differential quotient f'(x) everywhere in a given interval, then there must be a point $x + \vartheta h$ between x and $x + h$ such that*

$$f(x+h) = f(x) + h \cdot f'(x + \vartheta h), \qquad (0 < \vartheta < 1).$$

Note here the appearance of that ϑ, peculiar to the mean value theorems, and which to beginners often seems so strange at first. Geometrically.

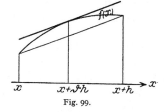
Fig. 99.

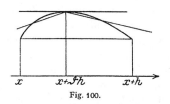
Fig. 100.

the theorem is fairly obvious. It says, merely, that between the points x and $x + h$ on the curve there is a point $x + \vartheta h$ on the curve at which the tangent is parallel to the secant joining the points x and $x + h$ (see Fig. 99).

5. How can one give an exact arithmetic proof of the mean value theorem, without appealing to geometric intuition? Such a proof could only mean, of course, throwing the theorem back upon arithmetic definitions of variable, function, continuity etc., which would have to be set up in advance in abstract, precise form. For this reason such a rigorous proof had to wait for Weierstrass and his followers, to whom, also, we owe the spread of the modern arithmetic concept of the number continuum. I shall try to give you the characteristic points of the argument.

In the first place, it is easy to make this theorem depend on the case where the secant is horizontal, i.e. $f(x) = f(x + h)$ (see Fig. 100). One must then prove the existence of a place where the tangent is

---

[1] Paris, 1823. Œuvres complètes, 2nd series, vol. 4, Paris, 1899.
[2] Paris, 1829. Œuvres complètes, 2nd series, vol. 4, Paris, 1899.

horizontal. To do this we can use the theorem of Weierstrass that every function which is continuous throughout a closed interval actually reaches a maximum, and also a minimum value, at least once in that interval. Because of our assumption, one of these extreme values of our function must lie *within* the interval $(x, x + h)$, provided we exclude the trivial case in which $f(x)$ is a constant. Let us suppose that there is a maximum (the case of a minimum is treated in the same way) and that it occurs at the place $x + \vartheta h$. It follows that $f(x)$ cannot have larger values, either to the right or to the left, i.e., the difference quotient to the right is negative, or zero, and to the left, positive or zero. Since the differential quotient exists, by hypothesis, at every place in the interval, its value at $x + \vartheta h$ can be looked upon as the limit of values which are either not positive or not negative, according as one thinks of it as a progressive or a regressive derivative. Therefore it must have the value zero, the tangent at $x = \vartheta h$ is horizontal, and the theorem is proved.

The scientific mathematics of today is built upon the series of developments which we have been outlining. But an essentially different conception of infinitesimal calculus has been running parallel with this through the centuries.

1. What precedes harks back to old metaphysical speculations concerning the structure of the continuum according to which this was made up of ultimate indivisible infinitely small parts. There were already, in ancient times, suggestions of these indivisibles and they were widely cultivated by the scholastics and still further by the jesuit philosophers. As a characteristic example I recall the title of Cavalieri's book, mentioned on p. 210 *Geometria Indivisibilibus Continuorum Promota*, which indicates its true nature. As a matter of fact, he considers intuitive mathematical approximation in a secondary way only. He actually considers space as consisting of ultimate indivisible parts, the "indivisibilia". In this connection it would be interesting and important to know the various analyses to which the notion of the continuum has been subjected in the course of centuries (and milleniums).

2. Leibniz, who shares with Newton the distinction of having invented the infinitesimal calculus, also made use of such ideas. The primary thing for him was not the differential quotient thought of as a limit. The differential dx of the variable x had for him actual existence as an ultimate indivisible part of the axis of abscissas, as a quantity smaller than any finite quantity and still not zero ("actually" infinitely small). In the same way the differentials of higher order $d^2x, d^3x, \ldots$ are defined as infinitely small quantities of second, third, ... order, each of which is "infinitely small in comparison with the preceding". Thus one had a series of systems of qualitatively different magnitudes. According to the theory of indivisibles, the area bounded by the curve

$y = y(x)$ and the axis of abscissas is the direct sum of all the individual ordinates. It is because of this view that Leibniz, in his first manuscript on integral calculus (1675), writes $\int y$ and not $\int y\,dx$.

This point of view, however, is by no means the only one which interested Leibniz. Sometimes he uses the notion of mathematical approximation, where, for example, the differential dx is a finite segment but so small that, for that interval, the curve is not appreciably different from the tangent. The above metaphysical speculations are surely only idealizations of these simple psychological facts.

But there is a third direction for the mathematical ideas of Leibniz, one which is especially characteristic of him. It is his formal point of view. I have frequently reminded you that we can look upon Leibniz as the founder of formal mathematics. His thought here is as follows. It makes no difference what meaning we attach to the differentials, or whether we attach any meaning whatever to them. If we define appropriate rules of operation for them, and if we employ these rules properly, it is certain that something reasonable and correct will result. Leibniz refers repeatedly to the analogy with complex numbers, concerning which he had corresponding notions. As to these rules of operation for differentials he was concerned chiefly with the formula

$$f(x + dx) - f(x) = f'(x) \cdot dx.$$

The mean value theorem shows that this is correct only if one writes $f'(x + \vartheta \cdot dx)$ instead of $f'(x)$; but the error which one commits by writing $f'(x)$ outright is infinitely small, of higher (second) order, and such quantities are to be neglected (this is the most important formal rule) in operations with differentials.

The chief publications of Leibniz are contained in that famous first scientific journal, the *Acta Eruditorum*[1]; in the years 1684, 1685, and 1712. In the first volume, you find, under the title *Nova methodus pro maximis et minimis* (p. 467 et seq.), the very first publication concerning differential calculus. In this Leibniz merely develops the rules for differentiation. The later works give also expositions of principles, where preference is given to the formal standpoint. In this connection, the short article of the year 1712[2], one of the last years of his life, was especially characteristic. In this he speaks outright of theorems and definitions which are only *"toleranter vera"* or French *"passables"*: "Rigorem quidem non sustinent, habent tamen usum magnum in calculando et ad artem inveniendi universalesque conceptus valent." He has reference here to complex numbers as well as to the infinite. If

---

[1] Translated, in part, in Ostwalds Klassikern No. 162. Edited by G. Kowalewski, Leipzig, 1908. Also in Leibniz, Mathematische Schriften. Edited by K. J. Gerhardt, from 1849 on.

[2] Observatio...; et de vero sensu methodi infinitesimalis, p. 167—169.

we speak, perhaps, of the infinitely small, then "commoditati expressionis seu breviloquio mentalis inservimus, sed non nisi toleranter vera loquimur, quae explicatione rigidantur."

3. From Leibniz as center the new calculus spread rapidly over the continent and we find each of his three points of view represented. I must mention here the first textbook of differential calculus that ever appeared, the *Analyse des Infiniment Petits pour l'Intelligence des Courbes*[1] by Marquis de l'Hospital, a pupil of Johann Bernoulli, who for his part, had absorbed the new ideas from Leibniz with surprising speed and had himself published the first textbook on the integral calculus[2]. Both books represent the point of view of mathematics of approximation. For example, a curve is thought of as a polygon with short sides, a tangent as the prolongation of one of these sides. In Germany, the differential calculus according to Leibniz was spread widely by Christian Wolff, of Halle, who published the contents of his lectures in *Elementa matheseos universal*[3]. He introduces the differentials of Leibniz immediately, at the beginning of the differential calculus, although he emphasizes particularly that they have no *actual equivalent of any kind*. And, indeed, as an aid to our intuition he develops his views concerning the infinitely small in a manner which savors thoroughly of mathematics of approximation. Thus he says, by way of example, that for purposes of practical measurement, the height of a mountain is not noticeably changed by adding or removing a particle of dust.

4. You will also frequently find the metaphysical view which ascribes an actual existence to the differentials. It has always had supporters, especially on the philosophic side, but also among mathematical physicists. One of the most prominent here is Poisson, who, in the preface to his celebrated *Traité de Mécanique*[4], expressed himself strongly to the effect that the infinitely small magnitudes are not merely an aid in investigation but that they have a thoroughly real existence.

5. Due probably to the philosophic tradition, this concept went over into textbook literature and plays a marked rôle there even today. As an example, I mention the textbook by Lübsen *Einleitung in die Infinitesimalrechnung*[5] which appeared first in 1855 and which had for a long time an extraordinary influence among a large part of the public. Everyone, in my day, certainly had Lübsen's book in his hand, either when he was a pupil, or later, and many received from it the first

---

[1] Paris, 1696; second edition, 1715.

[[2] Translated in Ostwalds Klassikern No. 194. Edited by G. Kowalewski. Joh. Bernoulli's Differentialrechnung was discovered and discussed a short time ago by P. Schafheitlin. Verhandlungen der Naturforscher-Gesellschaft in Basel, vol. 32 (1921).]

[3] Appeared first in 1710. — Editio nova Hallae, Magdeburgiae, 1742, p. 545.

[4] Part I, second edition, p. 14. Paris, 1833.

[5] Eighth edition, Leipzig, 1899.

stimulation to further mathematical study. Lübsen defined the differential quotient first by means of the limit notion; but along side of this he placed (after the second edition) what he considered to be the *true infinitesimal calculus* — a mystical scheme of operating with infinitely small quantities. These chapters are marked with an asterisk to indicate that they bring nothing new in the way of result. The differentials are introduced as ultimate parts which arise, for example, by continued halving of a finite quantity an infinite, non assignable number of times; and each of these parts "although different from absolute zero is nevertheless not assignable, but an infinitesimal magnitude, a breath, an instant". And then follows an English quotation: "An infinitesimal is the spirit of a departed quantity" (p. 59, 60). Then in another place (p. 76): "The infinitesimal method is, as you see, very subtle, but correct. If this is not manifest from what has preceded, together with what follows, it is the fault only of inadequate exposition." It is certainly very interesting to read these passages.

As companion piece to this I put before you the sixth edition of the widely used *Lehrbuch der Experimentalphysik* by Wüllner[1]. The first volume contains a brief preliminary exposition of infinitesimal calculus for the benefit of those students of natural science or medicine who have not acquired, at the gymnasium, that knowledge of calculus which is indispensable for physics. Wüllner begins (p. 31) with the explanation of the meaning of the infinitely small quantity dx, then follows with the explanation for the second differential d^2x, which, of course, is more difficult. I urge you to read this introduction with the eye of the mathematician and to reflect upon the absurdity of suppressing infinitesimal calculus in the schools because it is too difficult, and then of expecting a student in his first semester to gain an understanding of it from this ten page presentation, which is not only far from satisfying, but very hard to read!

The reason why such reflections could so long hold their place abreast of the mathematically rigorous method of limits, must be sought probably in the widely felt need of penetrating beyond the abstract logical formulation of the method of limits to the intrinsic nature of continuous magnitudes, and of forming more definite images of them than were supplied by emphasis solely upon the psychological moment which determined the concept of limit. There is one formulation which is characteristic, which is due, I believe, to the philosopher Hegel, and which formerly was frequently used in textbooks and lectures. It declares that *the function $y = f(x)$ represents the being, the differential quotient dy/dx, however, the becoming, of things.* There is assuredly something impressive in this, but one must recognize clearly that such

---

[1] Leipzig, 1907.

words do not promote further mathematical development because this must be based upon precise concepts.

In the most recent mathematics, "actually" infinitely small quantities have come to the front again, but in entirely different connection, namely in the geometric investigations of Veronese and also in Hilbert's *Grundlagen der Geometrie*[1]. The guiding thought of these investigations can be stated briefly as follows: A geometry is considered in which $x = a$ (a an ordinary real number) determines not only *one* point on the x axis, but infinitely many points, whose abscissas differ by finite multiples of infinitely small quantities of different orders $\eta, \zeta, \ldots$ A point is thus determined only when one assigns

$$x = a + b\eta + c\zeta + \cdots,$$

where $a, b, c, \ldots$ are ordinary real numbers, and the $\eta, \zeta, \ldots$ actually infinitely small quantities of decreasing orders. Hilbert uses this guiding idea by subjecting these new quantities $\eta, \zeta, \ldots$ to such axiomatic assumptions as will make it evident that one can operate with them consistently. To this end it is of chief importance to determine appropriately the relation as to size between x and a second quantity $x_1 = a_1 + b_1\eta + c_1\zeta + \cdots$. The first assumption is that $x >$ or $< x_1$ if $a >$ or $< a_1$; but if $a = a_1$, the determination as to size rests with the second coefficient, so that $x \gtreqless x_1$ according as $b \gtreqless b_1$; and if, in addition, $b = b_1$, the decision lies with the c, etc. These assumptions will be clearer to you if you refrain from attempting to associate with the letters any sort of concrete representation.

Now it turns out that, after imposing upon these new quantities this rule, together with certain others, it is possible to operate with them as with finite numbers. One essential theorem, however, which holds in the system of ordinary real numbers, now loses its validity, namely the theorem: *Given two positive numbers e, a, it is always possible to find a finite integer n such that $n \cdot e > a$, no matter how small e is nor how large a may be.* In fact, it follows immediately from the above definition that an arbitrary finite multiple $n \cdot \eta$ of η is smaller than any positive finite number a, and it is precisely this property that characterizes the η as an infinitely small quantity. In the same way $n \cdot \zeta < \eta$, that is, ζ is an infinitely small quantity of higher order than η.

This number system is called non-Archimedean. The above theorem concerning finite numbers is called, namely, the *axiom of Archimedes*, because he emphasized it as an unprovable assumption, or as a fundamental one which did not need proof, in connection with the numbers which he used. The denial of this axiom characterizes the possibility of actually infinitely small quantities. The name *Archimedean axiom*, however, like most personal designations, is historically inexact. Euclid

---

[1] Fifth edition, Leipzig, 1922.

gave prominence to this axiom more than half a century before Archimedes; and it is said not to have been invented by Euclid, either, but, like so many of his theorems, to have been taken over from Eudoxus of Knidos. The study of non-Archimedean quantities[1], which have been used especially as coordinates in setting up a non-Archimedean geometry, aims at deeper knowledge of the nature of continuity and belongs to the large group of investigations concerning the logical dependence of different axioms of ordinary geometry and arithmetic. For this purpose, the method is always to set up artificial number systems for which only a part of the axioms hold, and to infer the logical independence of the remaining axioms from these.

The question naturally arises whether, starting from such number systems, it would be possible to modify the traditional foundations of infinitesimal calculus, so as to include actually infinitely small quantities in a way that would satisfy modern demands as to rigor; in other words, to construct a non-Archimedean analysis. The first and chief problem of this analysis would be to prove the mean-value theorem

$$f(x+h) - f(x) = h \cdot f'(x + \vartheta h)$$

from the assumed axioms. I will not say that progress in this direction is impossible, but it is true that none of the investigators who have busied themselves with actually infinitely small quantities have achieved anything positive.

I remark for your orientation that, sincy Cauchy's time, the words *infinitely small* are used in modern textbooks in a somewhat changed sense. One never says, namely, that a quantity *is* infinitely small, but rather that it *becomes* infinitely small; which is only a convenient expression implying that the quantity decreases without bound toward zero.

We must bear in mind the reaction which was evoked by the use of infinitely small quantities in infinitesimal calculus. People soon sensed the mystical, the unproven, in these ideas, and there arose often a prejudice, as though the differential calculus were a *particular philosophical system* which could not be proved, which could only be believed or, to put it bluntly, a *fraud*. One of the keenest critics, in this sense, was the philosopher Bishop Berkeley, who in the little book *The Analyst*[2] assailed in an amusing manner the lack of clearness which prevailed in the mathematics of his time. Claiming the privilege of exercising the same freedom in criticizing the principles and methods of mathematics "which the mathematicians employed with respect to the mysteries of religion", he launched a violent attack upon all the methods of the new

---

[1] The so-called horn-shaped angles, known already to Euclid, are examples of non-Archimedean quantities. Compare also the excursus, in the second volume of this work, in connection with the critique of Euclid's *Elements*.]

[2] London, 1734.

analysis, the calculus with fluxions as well as the operation with differentials. He came to the conclusion that the entire structure of analysis was obscure and thoroughly unintelligible.

Similar views have often maintained themselves even up to the present time, especially on the philosophical side. This is due, perhaps, to the fact that acquaintance here is confined to the operation with differentials; the rigorous method of limits, a rather recent development, has not been comprehended. As an example, let me quote from Baumann's *Raum, Zeit und Mathematik*[1] which appeared in the sixties: "Thus we discard the logical and metaphysical justification, which Leibniz gave to calculus, but we decline to touch this calculus itself. We look upon it as an ingenious invention which has turned out well in practice; as an art rather than a science. It cannot be constructed logically. It does not follow from the elements of ordinary mathematics . . ."

This reaction against differentials accounts also for the attempt by Lagrange, already mentioned, in his *Théorie des Fonctions Analytiques*, published in 1797, to eliminate from the theory not only infinitely small quantities, but also every passage to the limit. He confined himself, namely, to those functions which are defined by power series

$$f(x) = a_0 + a_1 x + a_2 x^2 + a_3 x^3 + \cdots,$$

and he defines formally the "derived function $f'(x)$" (he avoids characteristically the expression differential quotient and the sign dy/dx) by means of a new power series

$$f'(x) = a_1 + 2a_2 x + 3a_3 x^2 + \cdots.$$

Consequently he talks of *derivative calculus* instead of *differential calculus*.

This presentation, of course, could not be permanently satisfactory. In the first place, the concept of function used here is, as we have shown, much too limited. More than that, however, such thoroughly formal definitions make a deeper comprehension of the nature of the differential coefficient impossible, and take no account of what we called the *psychological moment*—they leave entirely unexplained just why one should be interested in a series obtained in such a peculiar way. Finally, one can get along without giving any thought to a limit process only by disregarding entirely the convergence of these series and the question within what limits of error they can be replaced by finite sums. As soon as one begins a consideration of these problems, which is essential, of course, for any actual use of the series, it is necessary to have recourse precisely to that notion of limit, the avoidance of which was the purpose of inventing the system.

---

[1] Vol. 2, p 55. Berlin, 1869.

It would be fitting, perhaps, to say a few words about the differences of opinion concerning the foundations of calculus, as these come up, even today, beyond the narrow circle of professional mathematicians. I believe that we can often find here the preliminary conditions for understanding, in considerations very similar to those which we set forth respecting the foundations of arithmetic (p. 13). In every branch of mathematical knowledge one must separate sharply the question as to the inner logical consistency of its structure from that as to the justification for applying its axiomatically and (so to speak) arbitrarily formulated notions and theorems to objects of our external or internal perception. George Cantor[1] makes the distinction, with reference to whole numbers, between *immanent reality*, which belongs to them by virtue of their logical definability, and *transient reality*, which they possess by virtue of their applicability to concrete things. In the case of infinitesimal calculus, the first problem is completely solved by means of those theories which the science of mathematics has developed in logically complete manner (through the use of the concept of limit). The second question belongs entirely to the theory of knowledge, and the mathematician contributes only to its precise formulation when he separates from it and solves the first part. No pure mathematical work can, from its very nature, supply any immediate contribution to its solution. (See the analogous remarks on arithmetic, p. 13 et seq.) All disputes concerning the foundations of infinitesimal calculus labor under the disadvantage that these two entirely different phases of the problem have not been sharply enough separated. In fact, the first, the purely mathematical part, is established here precisely as in all other branches of mathematics, and the difficulties lie in the second, the philosophical part. The value of investigations which press forward in this second direction takes on especial importance in view of these considerations; but it becomes imperative to make them depend upon exact knowledge of the results of the purely mathematical work upon the first problem.

I must conclude with this excursus our short historical sketch of the development of infinitesimal calculus. In it I was obliged of course to confine myself to an emphasis of the most important guiding notions. It should be extended, naturally, by a thorough-going study of the entire literature of that period. You will find many interesting references in the lecture given by Max Simon at the Frankfurt meeting of the natural scientists of 1896: *Zur Geschichte und Philosophie der Differentialrechnung*.

If we now examine, finally, the attitude towards infinitesimal calculus in school instruction, we shall see that the course of its historical development is mirrored there to a certain extent. In earlier years,

---

[1] Mathematische Annalen, vol. 21 (1883), p. 562.

where infinitesimal calculus was taught in the schools, there was by no means a clear notion of its exact scientific structure as based on the method of limits. At least this was manifest in the textbooks, and it was doubtless the same in the schools. This method cropped up in a vague way at most, whereas operations with infinitely small quantities and sometimes also derivative calculus, in the sense of Lagrange, came to the front. Such instruction, of course, lacked not only rigor but intelligibility as well, and it is easy to see why a marked aversion arose to the treatment of infinitesimal calculus at all in the schools. This culminated in the seventies and eighties in an official order forbidding this instruction even in the "real" institutions.

To be sure this did not entirely prevent (as I indicated earlier) the using of the method of limits in the schools, where it was necessary—one merely avoided that name, or one even thought sometimes that something else was being taught. I shall mention here only three examples which most of you will recall from your school days.

a) The well known calculation of the perimeter and the area of the circle by an approximation which uses the inscribed and circumscribed regular polygons is obviously nothing but an *integration*. It was employed, even in ancient times, and was used particularly by Archimedes; in fact, it is owing to its classical antiquity that is has been retained in the schools.

b) Instruction in physics, and particularly in mechanics, necessarily involves the notions of *velocity* and *acceleration*, and their use in various deductions, including the *laws of falling bodies*. But the derivation of these laws is essentially identical with the *integration of the differential equation* $z'' = g$ by means of the function $z = \frac{1}{2} g t^2 + a t + b$, where a, b are constants of integration. The schools must solve this problem, under pressure of the demands of physics, and the means which they employ are more or less exact methods of integration, of course disguised.

c) In many North German schools the *theory of maxima and minima* was taught according to a method which bore the name of Schellbach, the prominent mathematical pedagogue of whom you all must have heard. According to this method one puts

$$\lim_{x = x_1} \left(\frac{f(x) - f(x_1)}{x - x_1} \right) = 0$$

in order to obtain the extremes of the function $y = f(x)$. But that is precisely the method of differential calculus, only that the word differential quotient is not used. It is certain that Schellbach used the above expression only because differential calculus was prohibited in the schools and he nevertheless did not want to miss these important notions. His pupils, however, took it over unchanged, called it by his name, and so it came about that methods which Fermat, Leibniz, and

Newton had possessed were put before the pupils under the name of Schellbach!

Let me now indicate, finally, the attitude toward these things of our *reform tendency*, which is now gaining ground more and more in Germany, as well as elsewhere, especially in France, and which we hope will control the mathematical instruction of the next decades. *We desire that the concepts which are expressed by the symbols $y = f(x)$, dy/dx, $\int y\, dx$ be made familiar to pupils, under these designations; not, indeed, as a new abstract discipline, but as an organic part of the total instruction; and that one advance slowly, beginning with the simplest examples.* Thus one might begin, with pupils of the age of fourteen and fifteen, by treating fully the functions $y = ax + b$ (a, b definite numbers) and $y = x^2$, drawing them on cross-section paper, and letting the concepts *slope* and *area* develop slowly. But one should hold to concrete examples. During the next three years this knowledge could be gathered together and treated as a whole, the result being that the pupils would come into complete possession of the beginnings of infinitesimal calculus. It is essential here to make it clear to the pupil that he is dealing, not with something mystical, but with simple things that anyone can understand.

The urgent necessity of such reforms lies in the fact that they are concerned with those mathematical notions which govern completely the applications of mathematics which are being made today in every possible field, and without which all studies at the university, even the simplest studies in experimental physics, are suspended in mid air. I can be content with these few hints, chiefly because this subject is fully discussed in Klein-Schimmack (referred to on p. 3).

In order to supplement these general considerations with something which again is concrete I shall now discuss in some detail an especially important subject in infinitesimal calculus.

2. Taylor's Theorem

I shall proceed here in a manner analogous to the plan I followed with trigonometric series. I shall depart, namely, from the usual treatment in the textbooks by bringing to the foreground the *finite series*, so important in practice, and by aiding the intuitive grasp of the situation by means of graphs. In this way it will all seem elementary and easily comprehensible.

We begin with the question whether we can make a suitable approximation to an arbitrary curve $y = f(x)$, for a short distance, by means of curves of the simplest kind. The most obvious thing is to replace the curve in the neighborhood of a point $x = a$ by its rectilinear tangent

$$y = A + Bx,$$

just as, in physics and in other applications, we often discard the higher powers of the independent variable in a series development (see Fig. 101). In a similar manner we can obtain better approximations by making use of parabolas of second, third, ... order

$$y = A + Bx + Cx^2, \qquad y = A + Bx + Cx^2 + Dx^3, \ldots$$

or, in analytic terms, by using polynomials of higher degree. Polynomials are especially suitable because they are so easy to calculate. We shall give all these curves a special position, so that at the point $x = a$ they lie as close as possible to the curve, i.e., so that they shall be parabolas of osculation. Thus the quadratic parabola will coincide with $y = f(x)$ not only in its ordinate but also in its first and second derivatives (i.e., it will "osculate"). A simple calculation shows that the analytic expression for the parabola having osculation of order n will be

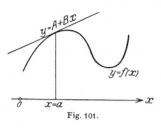

Fig. 101.

$$y = f(a) + \frac{f'(a)}{1}(x-a) + \frac{f''(a)}{1 \cdot 2}(x-a)^2 + \cdots + \frac{f^{(n)}(a)}{1 \cdot 2 \ldots n}(x-a)^n,$$
$$(n = 1, 2, 3, \ldots)$$

and these are precisely the first $n + 1$ terms of Taylor's series.

The investigation as to whether and how far these polynomials represent usable curves of approximation will be started by a somewhat experimental method, such as we used in the case (p. 194) of the trigonometric series. I shall show you a few drawings of the first osculating parabolas of simple curves, which were made[1] by Schimmack. The first are the four following functions, all having a singularity at $x = -1$, drawn with their parabolas of osculation at $x = 0$ (see Figs. 102, 103, 104, 105).

1. $\log(1 + x) \approx x - \dfrac{x^2}{2} + \dfrac{x^3}{3} - + \cdots$,

2. $(1 + x)^{\frac{1}{2}} \approx 1 + \dfrac{x}{2} - \dfrac{x^2}{8} + \dfrac{x^3}{16} - + \cdots$,

3. $(1 + x)^{-1} \approx 1 - x + x^2 - x^3 + - \cdots$,

4. $(1 + x)^{-2} \approx 1 - 2x + 3x^2 - 4x^3 + - \cdots$.

In the interval $(-1, +1)$ the parabolas approach the original curve more and more as the order increases; but to the right of $x = +1$ they deviate from it increasingly, now above, now below, in a striking way.

---

[1] Four of these drawings accompanied Schimmack's report on the Göttingen Vacation Course, Easter, 1908: *Über die Gestaltung des mathematischen Unterrichts im Sinne der neueren Reformideen*, Zeitschrift für den Mathematischen und naturwissenschaftlichen Unterricht, vol. 39 (1908), p. 513; also separate reprints. Leipzig, 1908.

At the singular point $x = -1$, in Cases 1, 3, 4, where the original function becomes infinite, the ordinates of the successive parabolas are increasingly large. In Case 2, where the branch of the original curve which appears in the drawing, ends in $x = -1$ at a vertical tangent,

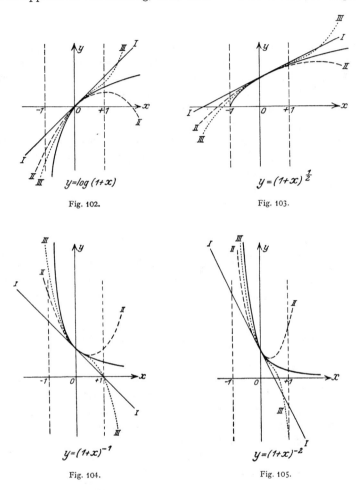

Fig. 102. $y = \log(1+x)$

Fig. 103. $y = (1+x)^{\frac{1}{2}}$

Fig. 104. $y = (1+x)^{-1}$

Fig. 105. $y = (1+x)^{-2}$

all the parabolas extend beyond this point but approach the original curve more and more at $x = -1$, by becoming ever steeper. At the point $x = +1$, symmetrical to $x = -1$, the parabolas in the first two cases approach the original curve more and more closely. In Case 3, their ordinates are alternately equal to 1 and 0, while that of the original curve has the value $\frac{1}{2}$. In Case 4, finally, the ordinates increase indefinitely with the order, and alternate in sign.

We shall examine, now, sketches of the osculating parabolas of two integral transcendental functions (see Fig. 106, 107)

5. $\quad e^x \approx 1 + \dfrac{x}{1!} + \dfrac{x^2}{2!} + \dfrac{x^3}{3!} + \cdots,$

6. $\quad \sin x \approx x - \dfrac{x^3}{3!} + \dfrac{x^5}{5!} - \dfrac{x^7}{7!} + \cdots.$

You notice that as their order increases, the parabolas give usable aproximations to the original curve for a greater and greater interval. It is especially striking in the case of $\sin x$ how the parabolas make the effort to share more and more oscillations with the sine curve.

I call your attention to the fact that the drawing of such curves in simple cases is perhaps suitable material even for the schools. After we have thus assembled our experimental material we must consider it mathematically.

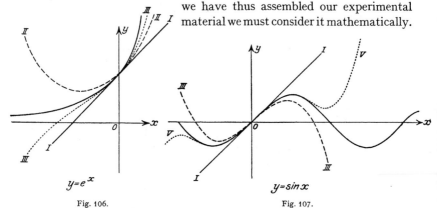

Fig. 106. $y = e^x$

Fig. 107. $y = \sin x$

The first question here is the extremely important one in practice as to the closeness with which the n-th parabola of osculation represents the original curve. This implies an estimate of the remainder for the values of the ordinate, and is connected naturally with the passage of n to infinity. *Can the curve be represented exactly, at least for a part of its course, by an infinite power series?*

It will be sufficient to state the commonest of the theorems concerning the remainder:

$$R_n(x) = f(x) - \left\{ f(a) + \frac{x-a}{1!} f'(a) + \cdots + \frac{(x-a)^{n-1}}{(n-1)!} f^{(n-1)}(a) \right\}.$$

The proof of the theorem is given in all the books and I shall revert to it later, anyway, from a more general standpoint. The theorem is: *There is a value ξ between a and x such that R_n can be represented in the form*

$$R_n(x) = \frac{(x-a)^n}{n!} f^{(n)}(\xi), \qquad (a < \xi < x).$$

The question as to the justification of the transition to an infinite series is now reduced to that as to whether this $R_n(x)$ has the limit 0 or not when n becomes infinite.

Returning to our examples, it appears, as you can verify by reading anywhere, that in Cases 5 and 6 the infinite series converges for all values of x. In Cases 1 to 4, it turns out that the series converges, *between* -1 and $+1$, to the original function, but that it diverges *outside* this interval. For $x = -1$ we have, in Case 2, convergence to the function value; in Cases 1, 3, 4, the limiting value of the series as well as that of the function is infinite, so that one could speak of convergence here also, but it is not customary to use this word with a series that has a definitely infinite limit. For $x = +1$, finally, we have convergence in the first two examples, divergence in the last two. All this is in fullest agreement with our graphs.

We may now raise the question, as we did with the trigonometric series, as to the limiting positions toward which the approximating parabolas converge, thought of as complete curves. They cannot, of course, break off suddenly at $x = \pm 1$. For the case of log $(1 + x)$ I have sketched for you the limit curve (Fig. 108). The even and odd parabolas

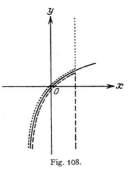

Fig. 108.

have different limiting positions, (indicated in the figure by dashes and dots) which consist of the logarithm curve between -1 and $+1$ together with the lower and upper portions, respectively, of the vertical line $x = +1$. The other three cases are similar.

The theoretical consideration of Taylor's series cannot be made complete without going over to the complex variable. It is only then that one can understand the sudden ceasing of the power series to converge at places where the function is entirely regular. To be sure, one might be satisfied, in the case of our examples, by saying that the series cannot converge any farther to the right than to the left, and that the convergence must cease at the left because of the singularity at $x = -1$. But such reasoning would not fit a case like the following. The Taylor's series development for the branch of $\tan^{-1} x$ which is regular for all real x

$$\tan^{-1} x \approx x - \frac{x^3}{3} + \frac{x^5}{5} - + \cdots$$

converges only in the interval $(-1, +1)$, and the parabolas of osculation converge alternately to two different limiting positions (see Fig. 109). The first consists, in the figure, of the long dotted parts of the vertical lines $x = +1$, $x = -1$ together with the portion of the inverse tangent curve lying between these verticals. The second limiting position is obtained from the first by taking the short dotted parts of the vertical lines instead of the long dotted parts. The convergence is toward the first of these limit curves when we take an odd number of terms in the series, toward the second when we take an even number. In the figure, the long dotted curve represents $y = x - x^3/3 + x^5/5$, the short dotted curve is $y = x - x^3/3$. The sudden cessation of convergence at the thoroughly regular points $x = \pm 1$ is incomprehensible if we limit ourselves to real values of x and notice the behavior of the function. The explanation is to be found in the important theorem on the circle of convergence, the most beautiful of Cauchy's function-theoretic achievements, which can be stated as follows. *If one marks on the complex x plane all the singular points of the analytic function $f(x)$, when $f(x)$ is single-valued, and on the Riemann surface belonging to $f(x)$ when $f(x)$ is many-valued, then the Taylor's series corresponding to a regular point $x = a$ converges inside the largest circle about a which has no singular point in its interior (i.e., so that at least one singular point lies on its* circumference). The series converges for no point outside this circle (see Fig. 110).

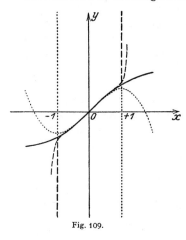

Fig. 109.

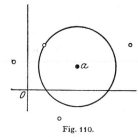

Fig. 110.

Now our example $\tan^{-1} x$ has, as you know, singularities at $x = \pm i$, and the circle of convergence of the development in powers of x is consequently the unit circle about $x = 0$. The convergence must cease therefore at $x = \pm 1$, since the real axis leaves the circle of convergence at these points (see Fig. 111).

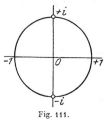

Fig. 111.

Finally, as to the convergence of the series *on the unit circle itself*, I shall give you the reference which came up when we were talking about the connection between power series and trigonometric series. The convergence depends upon whether or not the real and the imaginary part of the function, in view of the singularities that must exist on the circle of convergence, can be developed there into a convergent trigonometric series.

I should like now to enliven the discussion of Taylor's theorem by showing its relations to the problems of interpolation and of finite differences. There, also, we are concerned with the approximation to a given curve by means of a parabola. But instead of trying to make the parabola fit as closely as possible at *one* point, we require it to cut the given curve in a number of preassigned points; and the question is, again, as to how far this interpolation parabola gives a tolerable approximation. In the simplest case, this amounts to replacing the curve by a secant instead of the tangent (see Fig. 112). Similarly one passes a quadratic parabola through three points of the given curve, then a cubic parabola through four points, and so on.

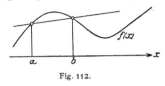

Fig. 112.

This is a natural way of approaching interpolation, one that is very often employed, e. g., in the use of logarithmic tables There we assume that the logarithmic curve runs rectilinearly between two given tabular values and we interpolate "linearly" in the well known way, which is facilitated by the difference tables. If this approximation is not close enough, we apply quadratic interpolation.

From this broad statement of the general problem, we get a determination of the osculating parabolas in Taylor's theorem as a special case, that is, when we simply allow the intersections with the interpolation parabolas to coincide. To be sure, the replacing of the curve by these osculating parabolas is not properly expressed by the word "interpolation", except that one includes "extrapolation" in the problem of interpolation. For example, the curve is compared not only with the part of the secant lying between its points of intersection, but also with the part beyond. For the entire process the comprehensive word approximation seems more suitable.

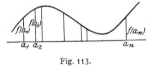

Fig. 113.

I shall now give the most important formulas of interpolation. Let us first determine the parabolas of order $n-1$ which cut the given function in the points $x = a_1, a_2, \ldots, a_n$, that is, whose ordinates in these points are $f(a_1), f(a_2), \ldots, f(a_n)$ (see Fig. 113). This problem, as you know, is solved by Lagrange's interpolation formula

$$(1) \quad \begin{cases} y = \dfrac{(x-a_2)(x-a_3)\cdots(x-a_n)}{(a_1-a_2)(a_1-a_3)\cdots(a_1-a_n)} \cdot f(a_1) \\ \quad + \dfrac{(x-a_1)(x-a_3)\cdots(x-a_n)}{(a_2-a_1)(a_2-a_3)\cdots(a_2-a_n)} \cdot f(a_2) \\ \quad + \cdots\cdots\cdots\cdots\cdots\cdots\cdots\cdots\cdots \end{cases}$$

It contains n terms with the factors $f(a_1), f(a_2), \ldots, f(a_n)$. The numerators lack in succession the factors $(x-a_1), (x-a_2), \ldots, (x-a_n)$. It is easy to verify the correctness of the formula. For, each summand of y, and hence y itself, is a polynomial in x of degree $n-1$. If we put $x = a_1$ all the fractions vanish except the first, which reduces to 1, so that we get $y = f(a_1)$. Similarly we get $y = f(a_2)$ for $x = a_2$, etc.

From this formula it is easy to derive, by specialization, one that is often called Newton's formula. This has to do with the case where the abscissas $a_1, \ldots, a_n$ are equidistant (see Fig. 114). As the notation of the calculus of finite differences is advantageous here we shall first introduce it.

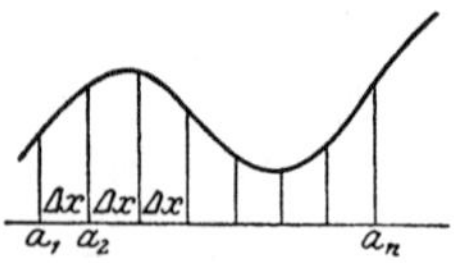

Let Δx be any increment of x and let $\Delta f(x)$ be the corresponding increment of $f(x)$ so that

$$f(x + \Delta x) = f(x) + \Delta f(x).$$

Fig. 114.

Now $\Delta f(x)$ is also a function of x which, if we change x by Δx, will have a definite difference called the second difference, $\Delta^2 f(x)$, so that

$$\Delta f(x + \Delta x) = \Delta f(x) + \Delta^2 f(x).$$

In the same way we have

$$\Delta^2 f(x + \Delta x) = \Delta^2 f(x) + \Delta^3 f(x), \quad \text{etc.}$$

This notation is precisely analogous to that of differential calculus, except that one is concerned here with finite quantities and there is no passing to the limit.

From the above definitions of differences there follows at once for the values of f at the successive equidistant places

(2) $\begin{cases} f(x + \Delta x) = f(x) + \Delta f(x), \\ f(x + 2\Delta x) = f(x + \Delta x) + \Delta f(x + \Delta x) \\ \qquad\qquad = f(x) + 2\Delta f(x) + \Delta^2 f(x), \\ f(x + 3\Delta x) = f(x + 2\Delta x) + \Delta f(\Delta x + 2\Delta x) \\ \qquad\qquad = f(x) + 3\Delta f(x) + 3\Delta^2 f(x) + \Delta^3 f(x), \\ f(x + 4\Delta x) = f(x) + 4\Delta f(x) + 6\Delta^2 f(x) + 4\Delta^3 f(x) + \Delta^4 f(x). \end{cases}$

This table could be continued, the values at equidistant points being expressed by means of successive differences taken at the initial point and involving the binomial coefficients as factors.

Newton's formula for the interpolation parabola of order $(n-1)$ belonging to the n equidistant points of the x axis,

$$a_1 = a, \quad a_2 = a + \Delta x, \ldots, a_n = a + (n-1)\Delta x,$$

that is, which has at these points the same ordinates as $f(x)$, will be

(3) $\begin{cases} y = f(a) + \dfrac{(x-a)}{1!} \dfrac{\Delta f(a)}{\Delta x} + \dfrac{(x-a)(x-a-\Delta x)}{2!} \dfrac{\Delta^2 f(a)}{(\Delta x)^2} + \cdots \\ \qquad + \dfrac{(x-a)(x-a-\Delta x)\cdots(x-a-(n-2)\Delta x)}{(n-1)!} \dfrac{\Delta^{n-1} f(a)}{(\Delta x)^{n-1}}. \end{cases}$

This is, in fact, a polynomial in x of order $n-1$. For $x = a$ it reduces to $f(a)$; for $x = a + \Delta x$ all the terms, except the first two, become zero and there remains $y = f(a) + \Delta f(a)$, which by (2) is equal to $f(a + \Delta x)$; and so on. Thus the table (2) yields a polynomial which assumes the correct values at all the n places.

If we wish to use this interpolation formula to real advantage, however, we must know something as to the correctness with which it represents $f(x)$, that is, we must be able to *estimate the remainder*. Cauchy gave[1] the formula for this in 1840, and I should like to derive it. I shall start from the more general Lagrange formula. Let x be any value between the values $a_1, a_2, \ldots, a_n$, or beyond them (interpolation or extrapolation). We denote by $P(x)$ the ordinate of the interpolation parabola given by the formula and by $R(x)$ the remainder

(4) $$f(x) = P(x) + R(x).$$

According to the definition of $P(x)$ the remainder R vanishes for $x = a_1, a_2, \ldots, a_n$ and we therefore set

$$R(x) = \frac{(x - a_1)(x - a_2) \cdots (x - a_n)}{n!} \psi(x).$$

It is convenient to take out the factor $n!$. Then it turns out, in complete analogy with the remainder term of Taylor's series, *that $\psi(x)$ is equal to the n-th derivative of $f(x)$ taken for a value $x = \xi$ lying between the $n - 1$ points $a_1, a_2, \ldots, a_n, x$*. This assertion that the deviation of $f(x)$ from the polynomial of order $n - 1$ depends upon the entire course of the function $f^{(n)}(x)$ seems entirely plausible, if we reflect that $f(x)$ is equal to that polynomial when $f^{(n)}(x)$ vanishes.

As to the proof of the *remainder formula*, we derive it by the following device. Let us set up, as a function of a new variable z, the expression

$$F(z) = f(z) - P(z) - \frac{(z - a_1)(z - a_2) \cdots (z - a_n)}{n!} \psi(x),$$

where x remains as a parameter in $\psi(x)$. Now $F(a_1) = F(a_2) = \cdots = F(a_n) = 0$, since $P(a_1) = f(a_1)$, $P(a_2) = f(a_2), \ldots, P(a_n) = f(a_n)$ by definition. Furthermore $F(x) = 0$ because the last summand goes over into $R(x)$, for $z = x$, so that the right side vanishes by (4). We know, therefore, $n + 1$ zeros $z = a_1, a_2, \ldots, a_n, x$, of $F(z)$. Now apply the *extended mean-value theorem*, which one gets by repeated application of the ordinary theorem (p. 213), namely: *If a continuous function, together with its first n derivatives, vanishes at $n + 1$ points, then the n-th derivative vanishes at one point, at least, which lies in the interval containing all the zeros*. Hence if $f(z)$, and therefore also $F(z)$, has n continuous derivatives, there must be a value ξ between the extremes of the values $a_1, a_2, \ldots, a_n, x$ for which

$$F^{(n)}(\xi) = 0.$$

---

[1] Comptes Rendus, vol. 11, pp. 775—789.—Œuvres, 1st series, vol. 5, pp. 409 to 424, Paris, 1885.

But we have
$$F^{(n)}(z) = f^{(n)}(z) - \psi(x),$$
since the polynomial $P(z)$ of degree $n-1$ has 0 for its n-th derivative and since only $z^n \cdot \psi(x)/n!$, the highest term of the last summand, has an n-th derivative which does not vanish. Therefore we have, finally
$$F^{(n)}(\xi) = f^{(n)}(\xi) - \psi(x) = 0, \quad \text{or} \quad \psi(x) = f^{(n)}(\xi),$$
which we wished to prove.

I shall write down Newton's interpolation formula with its remainder term

(5) $\begin{cases} f(x) = f(a) + \dfrac{x-a}{1!} \dfrac{\Delta f(a)}{\Delta x} + \dfrac{(x-a)(x-a-\Delta x)}{2!} \dfrac{\Delta^2 f(a)}{(\Delta x)^2} + \cdots \\ \quad + \dfrac{(x-a) \cdots [x-a-(n-2)\Delta x]}{(n-1)!} \dfrac{\Delta^{(n-1)} f(a)}{(\Delta x)^{n-1}} \\ \quad + \dfrac{(x-a) \cdots [x-a-(n-1)\Delta x]}{n!} f^{(n)}(\xi), \end{cases}$

where ξ is a mean value in the interval containing the $n-1$ points a, $a+\Delta x, a+2\Delta x, \ldots, a+(n-1)\Delta x, x$. The formula (5) is, in fact, indispensable in the applications. I have already alluded to linear interpolation when logarithmic tables are used. If $f(x) = \log x$ and $n = 2$, we find, from (5)
$$\log x = \log a + \frac{x-a}{1!} \frac{\Delta \log a}{\Delta x} - \frac{(x-a)(x-a-\Delta x)}{2!} \frac{M}{\xi^2}.$$
Since $d^2 \log x/dx^2 = -M/x^2$ where M is the modulus of the logarithmic system. Hence we have an expression for the error which we commit when we interpolate linearly between the tabular logarithms for a and $a + \Delta x$. This error has different signs according as x lies between a and $a + \Delta x$ or outside this interval. Every one who has to do with logarithmic tables should really know this formula.

I shall not devote any more attention to applications, but shall now draw your attention to the marked analogy between the interpolation formula of Newton and the formula of Taylor. There is a substantial reason for this analogy. It is easy to give an exact deduction of Taylor's theorem from the Newtonian formula, corresponding to the passage to the limit from interpolation parabolas to osculating parabolas. Thus, if we keep x, a, and n fixed and let Δx converge to zero, then, since $f(x)$ has n derivatives, the $n-1$ difference quotients in (5) go over into the derivatives
$$\lim_{\Delta x = 0} \frac{\Delta f(a)}{\Delta x} = f'(a), \quad \lim_{\Delta x = 0} \frac{\Delta^2 f(a)}{\Delta x^2} = f''(a), \ldots$$
In the last term of (5), the value of ξ can change with decreasing Δx. Since all the other terms on the right have definite limits, however, and the left side has the fixed value $f(x)$ during the entire limit process, it follows that the values of $f^{(n)}(\xi)$ must converge to a definite value

and that this value, furthermore, must, because of the continuity of $f^{(n)}$, be a value of this function for some place between a and x. If we denote this again by ξ we have

$$f(a) = f(a) + \frac{x-a}{1!}f'(a) + \cdots + \frac{(x-a)^{n-1}}{(n-1)!}f^{(n-1)}(a) + \frac{(x-a)^n}{n!}f^{(n)}(\xi),$$
$$(a < \xi < x).$$

Thus we have obtained a complete proof of Taylor's theorem with the remainder term and at the same time have given it an ordered place in the theory of interpolation.

It seems to me that this proof of Taylor's theorem, which brings it into wider relation with very simple questions and which provides such a smooth passage to the limit, is the very best possible one. But all the mathematicians to whom these things are familiar (it is remarkable that they are unknown to many, including perhaps even some writers of textbooks) do not think so. They are accustomed to confront a passage to a limit with a very grave face and would therefore prefer a direct proof of Taylor's theorem to one linking it with the calculus of finite differences.

I must emphasize however that, as a matter of history, the source of Taylor's theorem is actually the calculus of finite differences. I have already mentioned that Brook Taylor first published it in his *Methodus incrementorum*[1]. He first deduces Newton's formula, without the remainder, of course, and then puts in it $\Delta x = 0$ and $n = \infty$. He thus gets correctly from first terms of Newton's formula the first terms of his new series:

$$f(x) = f(a) + \frac{x-a}{1!}\frac{df(a)}{da} + \frac{(x-a)^2}{2!}\frac{d^2f(a)}{da^2} + \cdots$$

The continuation of this series, according to the same law, seems to him self evident, and he gives no thought either to a remainder term or to convergence. We have here, in fact, a *passage to the limit of unexampled audacity*. The earlier terms, in which $x - a - \Delta x$, $x - a - 2\Delta x$, ... appear, offer no difficulty, because these finite multiples of Δx approach zero with Δx; but with increasing n there appear terms in ever increasing number, presenting factors $x - a - k\Delta x$ with larger and larger k, and one is not justified in treating these forthwith in the same way and in assuming that they go over into a convergent series.

Taylor really operates here with infinitely small quantities (differentials) in the same unquestioning way as the Leibnizians. It is interesting to reflect that although, as a young man of twenty-nine, he was under the eye of Newton, he departed from the latter's method of limits.

You will find an excellent critical presentation of the entire development of Taylor's theorem in Alfred Pringsheim's memoir: *Zur Geschichte*

---

[1] Londini, 1715, p. 21—23.

des Taylorschen Lehrsatzes[1]. I should like to speak here about the customary distinction between Taylor's series and that of Maclaurin. As is well known, many textbooks make a point of putting $a = 0$ and of calling the obvious special case of Taylor's series which thus arises:

$$f(x) = f(0) + \frac{x}{1!}f'(0) + \frac{x^2}{2!}f''(0) + \cdots$$

the series of Maclaurin; and many persons may think that this distinction is important. Anybody who understands the situation however sees that it is comparitively unimportant mathematically. But it is not so well known that, considered historically, it is pure nonsense. For Taylor had undoubted priority with his general theorem, deduced in the way indicated above. More than this, he emphasizes at a later place in his book (p. 27) the special form of the series for $a = 0$ and remarks that it could be derived directly by the method which is called today that of undetermined coefficients. Furthermore, Maclaurin took over[2] this deduction in 1742 in his *Treatise of Fluxions* (which we mentioned on p. 212) where he quoted Taylor expressly and made no claim whatever of offering anything new. But the quotation seems to have been disregarded and the author of the book seems to have been looked upon as the discoverer of the theorem. Errors of this sort are common. It was only later that people went back to Taylor and named the general theorem, at least, after him. It is difficult, if not impossible, to overcome such deeprooted absurdities. At best, one can only spread the truth in the small circle of those who have historical interests.

I shall now supplement our discussion of infinitesimal calculus with some remarks of a general nature.

3. Historical and Pedagogical Considerations

I should like to remind you, first of all, that the bond which Taylor established between difference calculus and differential calculus held for a long time. These two branches always went hand in hand, still in the analytic developments of Euler, and the formulas of differential calculus appeared as limiting cases of elementary relations that occur in the difference calculus. This natural connection was first broken by the oft mentioned formal definitions of Lagrange's derivative calculus. I should like to show you a compilation from the end of the eighteenth century which, closely following Lagrange, brings together all the facts then known about infinitesimal calculus, namely the *Traité du Calcul Différentiel et du Calcul Intégral* of Lacroix[3]. As a characteristic sample from this work, consider the definition of the derivative (vol. I, p. 145):

---

[1] Bibliotheca Mathematica, 3rd series, vol. I (1900), p. 433—479.
[2] Edinburgh, 1742, vol. II, p. 610.
[3] Three volumes, Paris, 1797—1800, with many later editions.

A function $f(x)$ is defined by means of a power series. By using the binomial theorem (and rearranging the terms) one has

$$f(x+h) = f(x) + hf'(x) + \tfrac{1}{2} h^2 f''(x) + \cdots.$$

Lacroix now denotes the term of this series which is linear in h by $df(x)$, and, writing dx for h itself, he has for the derivative, which he calls *differential coefficient*

$$\frac{df(x)}{dx} = f'(x).$$

Thus this formula is deduced in a manner thoroughly superficial even if unassailable. Within the range of these thoughts, Lacroix could not, of course, use the calculus of differences as a starting point. However, since this branch seemed to him too important in practice to be omitted, he adopted the expedient of developing it independently, which he did very thoroughly in a third volume, but without any connecting bridge between it and differential calculus.

This "large Lacroix" is historically significant as the proper source of the many textbooks of infinitesimal calculus which appeared in the nineteenth century. In the first rank of these I should mention his own textbook, the "small Lacroix"[1].

Since the twenties of the last century the textbooks have been strongly influenced also by the method of limits which Cauchy raised to such an honorable place. Here we should first think of the many French textbooks, most of which, as *Cours d'Analyse de l'Ecole Polytechnique*, were prepared expressly for university instruction. Directly or indirectly, German textbooks also have depended on them, with the single exception, perhaps, of the one by Schlömilch. From the long list of books, I shall single out only Serret's *Cours de Calcul Différentiel et Intégral*, which appeared first in 1869 in Paris. It was translated into German in 1884 by Axel Harnack and has been since then one of our most widely used textbooks. It suffered as to symmetry at the hands of a long series of revisers. The editions[2] which have appeared since 1906, however, have been subjected to a thoroughgoing revision by G. Scheffers of Charlottenburg, the result being a homogeneous work. I am glad to mention also an entirely new French book, the *Cours d'Analyse Mathématique* by Goursat[3] in three volumes, which is fuller in many ways than Serret and contains, in particular, a long series of entirely modern developments. Furthermore it is a very readable book.

---

[1] *Traité Elémentaire du Calcul Différentiel et Intégral*. Two volumes, Paris, 1797.

[2] Since 1906: Serret, J. A., u. G. Scheffers, *Lehrbuch der Differential- und Integralrechnung*, vol. I, sixth edition. Leipzig 1915; vol. II, 6—7 edition; vol. III, fifth edition, 1914.

[3] Paris 1902—1907, vol. I, third edition. 1917; vol. II, third edition. 1918; vol. III, second edition. 1915. (Translated into English: vol. I by E. R. Hedrick, 1904, Ginn and Co.; vol. II by E. R. Hedrick and O. Dunkel, 1916, Ginn and Co.)

In all these recent books, the derivative and the integral are based entirely upon the concept of limit. There is never any question as to difference calculus or interpolation. One sees the thing in a clearer light, perhaps, in this way, but, on the other hand, the field of view is considerably narrowed,—as it is when we use a microscope. Difference calculus is now left entirely to the practical calculators, who are obliged to use it, especially the astronomers; and the mathematician hears nothing of it. We may hope that the future will bring a change[1] here.

As a conclusion of my discussion of infinitesimal calculus I should like to bring up again for emphasis four points, in which my exposition differs especially from the customary presentation in the textbooks:

1. *Illustration of abstract considerations by means of figures* (curves of approximation, in the case of Fourier's and Taylor's series).

2. *Emphasis upon its relation to neighboring fields*, such as calculus of differences and of interpolation, and finally to philosophical investigations.

3. *Emphasis upon historical growth.*

4. *Exhibition of samples of popular literature to mark the difference between the notions of the public, as influenced by this literature and those of the trained mathematician.*

It seems to me extremely important that precisely the prospective teacher should take account of all of these. As soon as you begin teaching you will be confronted with the popular views. If you lack orientation, if you are not well informed concerning the intuitive elements of mathematics as well as the vital relations with neighboring fields, if, above all, you do not know the historical development, your footing will be very insecure. You will then either withdraw to the ground of the most modern pure mathematics, and fail to be understood in the school, or you will succumb to the assault, give up what you learned at the university, and even in your teaching allow yourself to be buried in the traditional routine. The discontinuity between school and university, of which I have often spoken, is greatest precisely in the field of infinitesimal calculus. I hope that my words may contribute to its removal and that they may provide you with useful armor in your teaching.

This brings me to the end of the conventional analysis. By way of supplement I shall discuss a few theories of modern mathematics to which I have referred occasionally and with which I think the teacher should have some acquaintance.

---

[1] In order to make a beginning here, I induced Friesendorff and Prümm to translate Markoffs *Differenzenrechnung* into German (Leipzig, 1896). There is a series of articles in the Enzyklopädie. A work on *Differenzenrechnung* by E. Nörlund has just appeared (Berlin, Julius Springer, 1924) which exhibits the subject in new light.

Supplement

I. Transcendence of the Numbers e and π

The first topic which I shall discuss will be the numbers e and π. In particular, I wish to prove that they are transcendental numbers.

Interest in the number π, in geometric form, dates from ancient times. Even then it was usual to distinguish between the problem of its approximate calculation and that of its exact theoretical construction; and one had certain fundamentals for the solution of both problems. Archimedes made an essential advance, in the first, with his process of approximating to the circle by means of inscribed and circumscribed polygons. The second problem soon centered in the question as to whether or not it was possible to construct π with ruler and compasses. This was attempted in all possible ways with never a suspicion that the reason for continued failure was the impossibility of the construction. An account of some of the early attempts has been published by Rudio[1]. The quadrature of the circle still remains one of the most popular problems, and many persons, as I have already remarked, seek salvation in its solution, without knowing, or believing, that modern science has long since settled the question.

In fact, these ancient problems are completely solved today. One is sometimes inclined to doubt whether human knowledge really can advance, and in some fields the doubt may be justified. In mathematics, however, there are indeed advances of which we have here an example.

The foundations upon which the modern solution of these problems rests date from the period between Newton and Euler. A valuable tool for the approximate calculation of π was supplied by infinite series, a tool which made possible an accuracy adequate for all needs. The most elaborate result obtained was that of the Englishman Shanks, who calculated π to 707 places[2]. One can ascribe this feat to a sportsmanlike interest in making a record, since no applications could ever require such accuracy.

On the theoretical side, we find the number e, the base of the system of natural logarithms, coming into the investigations during the same

---

[1] *Der Bericht des Simplicius über die Quadraturen des Antiphon und Hippokrates.* Leipzig, 1908.

[2] See Weber-Wellstein, vol. 1, p. 523.

period. The remarkable relation $e^{i\pi} = -1$ was discovered and a means was developed in the integral calculus which, as we shall see, was of importance for the final solution of the question as to the quadrature of the circle. The decisive step in the solution of the problem was taken by Hermite[1] in 1873, when he proved the transcendence of e. He did not succeed in proving the transcendence of π. That was done by Lindemann[2] in 1882.

These results represent an essential generalization of the classical problem. That was concerned only with the construction of π by means of ruler and compasses, which amounts, analytically, as we saw (p. 51) to representing π by a finite succession of square roots and rational numbers. But the modern results prove not merely the impossibility of this representation; they show far more, namely, that π (and likewise e) is transcendental, that is, that it satisfies no algebraic relation whatever whose coefficients are integers. In other words, neither e nor π can be the root of an algebraic equation with integral coefficients:

$$a_0 + a_1 x + a_2 x^2 + \cdots + a_n x^n = 0$$

no matter how large the integers $a_0, \ldots, a_n$ or the degree n. It is essential that the coefficients be integers. It would suffice however to say fractions, since we could make them integral by multiplying through by a common denominator.

I pass now to the *proof of the transcendence of e*, in which I shall follow the simplified method given by Hilbert in Volume 43 of the Mathematische Annalen (1893). We shall show that the assumption of an equation

(1) $\quad a_0 + a_1 e + a_2 e^2 + \cdots + a_n e^n = 0$, where $a_0 \neq 0$,

in which $a_0, \ldots, a_n$ are integers, leads to a contradiction. This will involve the use of only the simplest properties of whole numbers. We shall need, namely, from the theory of numbers, only the most elementary theorems on divisibility, in particular, that an integer can be separated into prime factors in only one way, and, second, that the number of primes is infinite.

The plan of the proof is as follows. We shall set up a method which enables one to approximate especially well to e and powers of e, by means of rational numbers, so that we have

(2) $\quad e = \dfrac{M_1 + \varepsilon_1}{M}, \quad e^2 = \dfrac{M_2 + \varepsilon_2}{M}, \quad \ldots, \quad e^n = \dfrac{M_n + \varepsilon_n}{M}$

where $M, M_1, M_2, \ldots, M_n$ are integers, and $\varepsilon_1/M, \varepsilon_2/M, \ldots, \varepsilon_n/M$ are

---

[1] Comptes Rendus, vol. 77 (1873), p. 18—24, 74—79, 226—233, 285—293; = Werke III (1912), p. 150.

[2] Sitzungsberichte der Berliner Akademie, 1882, p. 679, and Mathematische Annalen, vol. 20 (1882), p. 213.

very small positive fractions. Then the assumed equation (1), after multiplication by M, takes the form

(3) $\quad [a_0 M + a_1 M_1 + a_2 M_2 + \cdots + a_n M_n] + [a_1 \varepsilon_1 + a_2 \varepsilon_2 + \cdots + a_n \varepsilon_n] = 0.$

The first parenthesis is an integer, and we shall prove that it is not zero. As for the second parenthesis, we shall show that $\varepsilon_1, \ldots, \varepsilon_n$ can be made so small that it will be a positive proper fraction. Then we shall have the obvious contradiction that an integer $a_0 M + a_1 M_1 + \cdots + a_n M_n$ which is not zero, increased by a proper fraction $a_1 \varepsilon_1 + \cdots + a_n \varepsilon_n$ is zero. This will show the impossibility of (1).

In the course of the discussion which I have just outlined we shall make use of the theorem that if an integer is not divisible by a definite number, the integer cannot be zero (for zero is divisible by every number). We shall show, namely, that $M_1, \ldots, M_n$ are divisible by a certain prime number p, but that $a_0 M$ does not contain p, and that, therefore, $a_0 M + a_1 M_1 + \cdots + a_n M_n$ is not divisible by p, and hence is different from zero.

The principal aid in carrying out the indicated proof comes from a certain definite integral which was devised by Hermite for this purpose and which we shall call Hermite's integral. The key to this proof lies in its structure. This integral, whose value, as we shall see, is a positive whole number and which we shall use to define M, is

(4) $\quad M = \displaystyle\int_0^\infty \frac{z^{p-1}[(z-1)(z-2)\cdots(z-n)]^p e^{-z}}{(p-1)!} dz,$

where n is the degree of the assumed equation (1), and p is an odd prime which we shall determine later. From this integral we shall get the desired approximation (2) to the powers e^ν ($\nu = 1, 2, \ldots, n$) by breaking the interval of integration of the integral $M \cdot e^\nu$ at the point ν and setting

(4a) $\quad M_\nu = e^\nu \displaystyle\int_\nu^\infty \frac{z^{p-1}[(z-1)\cdots(z-n)]^p e^{-z}}{(p-1)!} dz,$

(4b) $\quad \varepsilon_\nu = e^\nu \displaystyle\int_0^\nu \frac{z^{p-1}[(z-1)\cdots(z-n)]^p e^{-z}}{(p-1)!} dz.$

Let us now take up the details of the proof.

1. We start with the well known formula from the beginnings of the theory of the gamma function:

$$\int_0^\infty z^{\varrho-1} e^{-z} dz = \Gamma(\varrho),$$

We shall need this formula only for integral values of ϱ, in which case $\Gamma(\varrho) = (\varrho - 1)!$, and I shall deduce it under this restriction. If we integrate by parts we have, for $\varrho > 1$:

$$\int_0^\infty z^{\varrho-1} e^{-z} dz = \left[-z^{\varrho-1} e^{-z}\right]_0^\infty + \int_0^\infty (\varrho - 1) z^{\varrho-2} e^{-z} dz$$

$$= (\varrho - 1) \int_0^\infty z^{\varrho-2} e^{-z} dz.$$

The integral on the right is of the same form as the one on the left, except that the exponent of z is reduced. If we apply this process repeatedly we must eventually come to z^0, since ϱ is an integer; and since $\int_0^\infty e^{-z} dz = 1$, we obtain finally

(5) $$\int_0^\infty z^{\varrho-1} e^{-z} dz = (\varrho - 1)(\varrho - 2) \ldots 3 \cdot 2 \cdot 1 = (\varrho - 1)!$$

Thus for integral ϱ the integral is a whole number which increases very rapidly when ϱ increases.

To make this result clear geometrically, let us draw the curve $y = z^{\varrho-1} e^{-z}$ for different values of ϱ. The value of the integral will then be represented by the area under the curve extending to infinity (see Fig. 115). The larger ϱ is the more closely the curve hugs the z axis at the origin, but the more rapidly it rises beyond $z = 1$. The curve has a maximum at $z = \varrho - 1$, for all values of ϱ; in other words the maximum occurs farther and farther to the right as ϱ increases; and its value also increases with ϱ. To the right of the maximum, the factor e^{-z} prevails so that the curve falls, approaching the z axis asymptotically. It is thus comprehensible that the area (our integral) always remains finite but increases rapidly with ϱ.

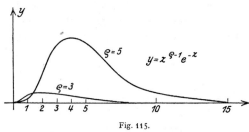

Fig. 115.

2. With this formula we can now easily evaluate our Hermite integral. Developing the integrand by the polynomial theorem

$$[z-1)(z-2) \ldots (z-n)]^p = [z^n + - \cdots + (-1)^n n!]^p$$
$$= z^{np} + - \cdots + (-1)^n (n!)^p,$$

where only the terms involving the highest and the lowest powers of z have been written down, the integral becomes

$$M = \frac{(-1)^n (n!)^p}{(p-1)!} \int_0^\infty z^{p-1} e^{-z} dz + \sum_{\varrho=p+1}^{np+p} \frac{C_\varrho}{(p-1)!} \int_0^\infty z^{\varrho-1} e^{-z} dz.$$

The C_ϱ are integral constants, by the polynomial theorem. Now we can apply formula (5) to each of these integrals and obtain

$$M = (-1)^n (n!)^p + \sum_{\varrho=p+1}^{np+p} C_\varrho \frac{(\varrho-1)!}{(p-1)!}.$$

The summation index ϱ is always larger than p and consequently $(\varrho-1)!/(p-1)!$ is an integer and one which contains p as a factor, so that we can take p as a factor out of the entire sum:

$$M = (-1)^n (n!)^p + p[C_{p+1} + C_{p+2}(p+1) + C_{p+3}(p+1)(p+2) + \cdots].$$

Now, so far as divisibility by p is concerned, M must behave like the first summand $(-1)^n(n!)^p$. And since p is a prime number it will not be a divisor of this summand if it is not a divisor of any of its factors $1, 2, \ldots, n$, which will certainly be the case if $p > n$. But this condition can be satisfied in an unlimited number of ways, since the number of primes is infinite. Consequently we can bring it about that $(-1)^n(n!)^p$, and hence M, is not divisible by p.

Since furthermore $a_0 \neq 0$, we can see to it, at the same time, that a_0 is not divisible by p by selecting p larger also than $|a_0|$, which is, of course, possible, by what was said above. But then the product $a_0 \cdot M$ will not be divisible by p, and that is what we wished to show.

3. Now we must examine the numbers M_ν ($\nu = 1, 2, \ldots, n$), defined in (4a) (p. 239). Putting the factor e^ν under the sign of integration and introducing the new variable of integration $\zeta = z - \nu$, which varies from 0 to ∞ when z runs from ν to ∞, we have

$$M_\nu = \int_0^\infty \frac{(\zeta+\nu)^{p-1}[(\zeta+\nu-1)(\zeta+\nu-2)\cdots\zeta\cdots(\zeta+\nu-n)]^p e^{-\zeta}}{(p-1)!} d\zeta.$$

This expression has a form entirely analogous to the one considered before for M and we can treat it in the same way. If we multiply out the factors of the integrand there will result an aggregate of powers with integral coefficients of which the lowest will be ζ^p. The integral of the numerator will thus be a combination of the integrals

$$\int_0^\infty \zeta^p e^{-\zeta} d\zeta, \quad \int_0^\infty \zeta^{p+1} e^{-\zeta} d\zeta, \quad \ldots, \quad \int_0^\infty \zeta^{(n+1)p-1} e^{-\zeta} d\zeta,$$

and since these are, by (5), equal to $p!, (p+1)!, \ldots$ the numerator will be $p!$ multiplied by a whole number A, so that we have

$$M_\nu = \frac{p! A_\nu}{(p-1)!} = p \cdot A_\nu, \qquad (\nu = 1, 2, \ldots, n).$$

In other words, every M^ν is a whole number which is divisible by p.

This, combined with the result of No. 2, proves the statement made on p. 239 that $a_0 M + a_1 M_1 + \cdots + a_n M_n$ is not divisible by p and is therefore different from zero.

4. The second part of the proof has to do with the sum $a_1 \varepsilon_1 + \cdots + a_n \varepsilon_n$, where, by (4b),

$$\varepsilon_\nu = \int_0^\nu \frac{z^{p-1}[(z-1)(z-2)\cdots(z-n)]^p e^{-z+\nu}}{(p-1)!} dz.$$

We must show that these ε_ν can be made arbitrarily small by an appropriate choice of p. To this end we use the fact that we can make p as large as we chose; for the only conditions thus far imposed upon p are that it should be a prime number larger than n and also larger than $|a_0|$, and these can be satisfied by arbitrarily large prime numbers.

Let us examine the graph of the integrand. At $z = 0$ it will be tangent to the z axis, but at $z = 1, 2, \ldots, n$ (in Fig. 116, $n = 3$) it will be tangent to the z axis and also cut it, since p is odd. As we shall see soon, the presence in the denominator of $(p-1)!$ brings it about that for large p the curve departs but little from the z axis in the interval $(0, n)$, so that it seems plausible that the integrals ε_ν should be very small. For $z > n$ the curve rises and runs asymptotically like the former curve $z^{\varrho-1}e^{-z}$ [for $\varrho = (n+1)p$] and finally approaches the z axis. It was for this reason that the value M of the integral (when the interval of integration was from 0 to ∞) increased so rapidly with p.

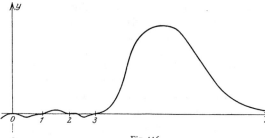

Fig. 116.

In actually estimating the integrals we can be satisfied with a rough approximation. Let G and g_ν be the maxima of the absolute values of the functions $z(z-1)\cdots(z-n)$ and $(z-1)(z-2)\cdots(z-n)e^{-z+\nu}$ respectively in the interval $(0, n)$:

$$\left.\begin{array}{l} |z(z-1)\cdots(z-n)| \leq G \\ |(z-1)(z-2)\cdots(z-n)e^{-z+\nu}| \leq g_\nu \end{array}\right\} \text{ for } 0 \leq z \leq n.$$

Since the integral of a function is never larger than the integral of its absolute value, we have, for each ε_ν

(6) $$|\varepsilon_\nu| \leq \left\{ \int_0^\nu \frac{G^{p-1} g_\nu}{(p-1)!} dz = \frac{G^{p-1} g_\nu \cdot \nu}{(p-1)!} \right\}.$$

Now G, g_ν, and ν are fixed numbers independent of p, but the number $(p-1)!$ in the denominator increases ultimately more rapidly than the power G^{p-1}, or, more exactly, the fraction $G^{p-1}/(p-1)!$ becomes, for sufficiently large p, smaller than any preassigned number, however small. Thus, because of (6), we can actually make each of the n numbers ε_ν arbitrarily small by choosing p sufficiently large.

It follows immediately from this that we can also make the sum of n terms $a_1 \varepsilon_1 + \cdots + a_n \varepsilon_n$ arbitrarily small. We have, in fact

$$|a_1\varepsilon_1 + a_2\varepsilon_2 + \cdots + a_n\varepsilon_n| \leq |a_1||\varepsilon_1| + |a_2||\varepsilon_2| + \cdots + |a_n||\varepsilon_n|$$

and by (6)

$$\leq (|a_1| \cdot 1 \cdot g_1 + |a_2| \cdot 2 g_2 + \cdots + |a_n| \cdot n \cdot g_n) \cdot \frac{G^{p-1}}{p-1!}.$$

Since the parenthesis has a value which is independent of p, we can, by virtue of the factor $G^{p-1}/(p-1)!$, make the entire right hand side, and hence also $a_1 \varepsilon_1 + a_2 \varepsilon_2 + \cdots + a_n \varepsilon_n$, as small as we choose, and, in particular, smaller than unity.

With this we have shown, as we agreed to do (p. 239), that the assumption of the equation (3)

$$[a_0 M + a_1 M_1 + \cdots + a_n M_n] + [a_1 \varepsilon_1 + \cdots + a_n \varepsilon_n] = 0$$

leads to a contradiction, namely that a non vanishing integer increased by a proper fraction gives zero. And since this equation cannot exist the transcendence of e is proved.

Proof of the Transcendence of π

We turn now to the proof of the transcendence of the number π. This proof is somewhat more difficult than the foregoing, but it is still fairly easy. It is only necessary to begin at the right end, which is indeed the art of all mathematical discovery.

The problem, as Lindemann considered it, was the following: It has been shown thus far that an equation $\sum_{\nu=0}^{n} a_\nu e^\nu = 0$ cannot exist if the coefficients a_ν and the exponents ν of e are ordinary whole numbers. Would it not be possible to prove a similar thing where a_ν and ν are arbitrary algebraic numbers? He succeeded in doing this; in fact, his most general theorem concerning the exponential function is as follows: *An equation $\sum_{\nu=1}^{n} a_\nu e^{b_\nu} = 0$ cannot exist if the a_ν, b_ν are algebraic numbers, whereby the a_ν are arbitrary, the b_ν different from one another.* The transcendence of π is then a corollary to this theorem. For, as is well known, $1 + e^{i\pi} = 0$; and if π were an algebraic number, $i\pi$ would be also, and the existence of this equation would contradict the above theorem of Lindemann.

I shall now prove in detail only a certain special case of Lindemann's theorem, one which carries with it, however, the transcendence of π. I shall follow again, in the main, Hilbert's proof in Volume 43 of the Mathematische Annalen, which is essentially simpler than Lindemann's, and which is an exact generalization of the discussion which we have given for e.

The starting point is the relation

(1) $$1 + e^{i\pi} = 0.$$

If, now, π satisfies any algebraic equation with integral coefficients then $i\pi$ also satisfies such an equation. Let $\alpha_1, \alpha_2, \ldots, \alpha_n$ be all the roots, including $i\pi$ itself, of this last equation. Then we must also have, because of (1):

$$(1 + e^{\alpha_1})(1 + e^{\alpha_2}) \cdots (1 + e^{\alpha_n}) = 0.$$

Multiplying out we obtain

(2) $$\begin{cases} 1 + (e^{\alpha_1} + e^{\alpha_2} + \cdots + e^{\alpha_n}) + (e^{\alpha_1+\alpha_2} + e^{\alpha_1+\alpha_3} + \cdots + e^{\alpha_{n-1}+\alpha_n}) \\ \quad + \cdots + (e^{\alpha_1+\alpha_2+\cdots+\alpha_n}) = 0. \end{cases}$$

Now some of the exponents which appear here might, by chance, be zero. Everytime that this occurs the left hand sum has a positive summand 1, and we combine these, together with the 1 that appears formally, into a positive integer a_0, which is certainly different from zero. The remaining exponents, all different from zero, we denote by $\beta_1, \beta_2, \ldots, \beta_N$ and we write, accordingly, instead of (2),

(3) $$a_0 + e^{\beta_1} + e^{\beta_2} + \cdots + e^{\beta_N} = 0, \quad \text{where} \quad a_0 > 0.$$

Now $\beta_1, \ldots, \beta_N$ are the roots of an algebraic equation with integral coefficients. For, from the equation whose roots are $\alpha_1, \ldots, \alpha_n$ we can construct one of the same character whose roots are the two term sums $\alpha_1 + \alpha_2, \alpha_1 + \alpha_3, \ldots$, then another for the three term sums $\alpha_1 + \alpha_2 + \alpha_3, \alpha_1 + \alpha_2 + \alpha_4, \ldots$ and so on; finally, $\alpha_1 + \alpha_2 + \cdots + \alpha_n$ is itself rational and satisfies therefore a linear integral equation. By multiplying together all these equations, we obtain again an equation with integral coefficients, which might have some zero roots, and whose remaining roots are $\beta_1, \ldots, \beta_N$. Omitting the power of the unknown which corresponds to the zero roots there will remain for the N quantities β an algebraic equation of degree N with integral coefficients and absolute term different from zero

(4) $$b_0 + b_1 z + b_2 z^2 + \cdots + b_N z^N = 0, \quad \text{where} \quad b_0, b_N \neq 0.$$

We now have to prove the following special case of Lindemann's theorem. *An equation of the form* (3), *with integral non-vanishing a_0, cannot exist if $\beta_1, \ldots, \beta_N$ are the roots of an algebraic equation of degree N, with integral coefficients.* This theorem includes the transcendence of π.

The proof involves the same steps as the one already given for the transcendence of e. Just as we could there approximate closely to the powers $e^1, e^2, \ldots, e^n$ by means of rational numbers, so we shall be concerned here with the best possible approximation to the powers of e which appear in (3), and we shall write, in the old notation,

(5) $$e^{\beta_1} = \frac{M_1 + \varepsilon_1}{M}, \quad e^{\beta_2} = \frac{M_2 + \varepsilon_2}{M}, \quad \ldots, \quad e^{\beta_N} = \frac{M_N + \varepsilon_N}{M};$$

where the denominator M is again an ordinary integer but $M_1, \ldots, M_N$ are not integers as formerly, but are integral algebraic numbers, and the $\beta_1, \ldots, \beta_N$, which in general can now be complex, are in absolute value very small. It is here that the difficulty in this proof lies, as compared to the earlier one. The sum of all the $M_1, \ldots, M_N$ will again, however, be an integer, and we shall be able to arrange it so that the first summand in the equation:

(6) $\quad [a_0 M + M_1 + M_2 + \cdots + M_N] + [\varepsilon_1 + \varepsilon_2 + \cdots + \varepsilon_N] = 0,$

[into which (3) goes over when we multiply by M and use (5)] will be a non-vanishing integer, while the second summand will be, in absolute value, smaller than unity. Essentially, this will be the same contradiction which we used before. It will show the impossibility of (6) and (3) and so complete our proof. As to detail, we shall again show that $M_1 + M_2 + \cdots + M_N$ is divisible by a certain prime number p, but that $a_0 M$ is not, which will show that the first summand in (6) cannot vanish; then we shall choose p so large that the second summand will be arbitrarily small.

1. Our first concern is to define M by a suitable generalization of Hermite's integral. A hint here lies in the fact that the zeros of the factor $(z-1)(z-2)\ldots(z-n)$ in Hermite's integral were the exponents of e in the hypothetical algebraic equation. Hence we now replace that factor by the product made by using the exponents in (3), i. e., the solutions in (4):

(7) $\quad (z-\beta_1)(z-\beta_2)\ldots(z-\beta_N) = \dfrac{1}{b_N}[b_0 + b_1 z + \cdots + b_N z^N].$

It turns out to be essential here to put in a suitable power of b_N as factor, which was unnecessary before because $(z-1)\ldots(z-n)$ was integral. We set then finally

(8) $\quad M = \displaystyle\int_0^\infty \dfrac{e^{-z} z^{p-1} dz}{(p-1)!} [b_0 + b_1 z + \cdots + b_N z^N]^p b_N^{(N-1)p-1}.$

2. Just as before, we now develop the integrand of M according to powers of z. The lowest power, that belonging to z^{p-1}, gives then:

$$\int_0^\infty \dfrac{e^{-z} z^{p-1} dz}{(p-1)!} b_0^p b_N^{(N-1)p-1} = b_0^p b_N^{(N-1)p-1},$$

where the integral has been evaluated by means of the gamma formula (p. 239). The remaining summands in the integrand contain either z^p or still higher powers, so that the integrals contain the factor $p!/(p-1)!$, multiplied by integers, and are thus all divisible by p. Consequently M is an integer which is certainly not divisible by p, i.e., provided the prime number p is not a divisor of either b_0 or b_N. But since these two numbers are both different from zero, we can bring this about by choosing p so that $p > |b_0|$ and also $p > |b_N|$.

Since $a_0 > 0$ it follows that $a_0 M$ is not divisible by p if we impose the additional condition $p > a_0$. Inasmuch as the number of primes is infinite we can satisfy all these conditions in an unlimited number of ways.

3. We must now set up M_ν and ε_ν. Here we must modify our earlier plan because the β_ν, which now take the place of the old ν, can be complex; in fact one them is $i\pi$. If we are to split the integral M as we did before we must first determine the path of integration in the complex plane. Fortunately the integrand of our integral is a finite single-valued function of the variable of integration z, regular everywhere except at $z = \infty$, where it has an essential singularity. Instead of integrating from 0 to ∞ along the real axis we can choose any other path from 0 to ∞, provided it ultimately runs asymptotically parallel to the positive real half axis. This is necessary if the integral is to have a meaning at all, in view of the behavior of e^{-z} in the complex plane.

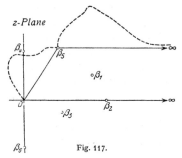

Fig. 117.

Let us now mark the N points β_1, β_2, ..., β_N in the plane and recall that we shall obtain the same value for M if we first integrate rectilinearly from 0 to one of the points β_N and then to ∞ along a parallel to the real axis (see Fig. 117). Along this path we can separate M into the two characteristic parts: The rectilinear path from 0 to β_N supplies the ε_ν which will become arbitrarily small with increasing p; the parallel from β_N to ∞ will supply the integral algebraic number M_N:

(8a) $\quad \varepsilon_\nu = e^{\beta_\nu} \int_0^{\beta_\nu} \frac{e^{-z} z^{p-1} dz}{(p-1)!} [b_0 + b_1 z + \cdots + b_N z^N]^p b_N^{(N-1)p-1}$,

$\hfill (\nu = 1, 2, \ldots, N),$

(8b) $\quad M_\nu = e^{\beta_\nu} \int_{\beta_\nu}^\infty \frac{e^{-z} z^{p-1} dz}{(p-1)!} [b_0 + b_1 z + \cdots + b_N z^N]^p b_N^{(N-1)p-1}$.

These assumptions satisfy (5). Our choice of a rectilinear path of integration was made solely for convenience; a curvilinear path from 0 to β_ν would, of course yield the same value for ε_ν, but it is easier to estimate the integral when the path is straight. Similarly, we might choose, instead of the horizontal path from β_ν to ∞, an arbitrary curve provided only that it approached the horizontal asymptotically; but that would be unnecessarily inconvenient.

4. I will discuss first the estimation of the ε_ν, because this involves nothing new if we only recall that the absolute value of a complex integral cannot be larger than the maximum of the absolute value of the integrand, multiplied by the length of the path of integration,

which, in our case, is $|\beta_\nu|$. The upper limit for ε_ν would be, then, the product of $G^{p-1}/(p-1)!$ by factors which are independent of p, where G denotes the maximum of $|z(b_0 + b_1 z + \cdots + b_N z^N) b_N^{N-1}|$ in a region which contains all the segments joining 0 with the β_ν. From this one may infer, as we did before, (p. 243), that, by sufficiently increasing p, the value of each ε_ν and, therefore, the value of $\varepsilon_1 + \cdots + \varepsilon_N$ can be made as small as we please and, in particular, smaller than unity.

5. It is only in the discussion of the M_ν that essentially new considerations enter, and these are, to be sure, only generalizations of our former reasoning, due to the fact that integral algebraic numbers take the place now of what were then integral rational numbers. We shall consider, as a whole, the sum:

$$\sum_{\nu=1}^{N} M_\nu = \sum_{\nu=1}^{N} e^{\beta_\nu} \int_{\beta_\nu}^{\infty} \frac{e^{-z} z^{p-1} dz}{(p-1)!} [b_0 + b_1 z + \cdots + b_N z^N]^p b_N^{(N-1)p-1}.$$

If we make use of (7) (p. 245) and replace, in each summand of the above summation, the polynomial in z by the product of the factors $(z - \beta_1) \cdots (z - \beta_N)$, and introduce the new variable of integration $\zeta = z - \beta_\nu$, which will run through real values from 0 to ∞, we obtain

(9) $\left\{ \begin{aligned} \sum_{\nu=1}^{N} M_\nu &= \sum_{\nu=1}^{N} \int_0^\infty \frac{e^{-\zeta} d\zeta}{(p-1)!} (\zeta+\beta_\nu)^{p-1}(\zeta+\beta_\nu-\beta_1)^p \cdots \zeta^p \cdots (\zeta+\beta_\nu-\beta_N)^p b_N^{Np-1} \\ \text{which may be written} \quad &= \int_0^\infty \frac{e^{-\zeta} d\zeta}{(p-1)!} \zeta^p \cdot \Phi(\zeta), \end{aligned} \right.$

where we use the abbreviation

(9') $\left\{ \begin{aligned} \Phi(\zeta) = \sum_{\nu=1}^{N} b_N^{Np-1}(\zeta + \beta_\nu)^{p-1}(\zeta + \beta_\nu - \beta_1)^p \cdots \\ (\zeta + \beta_\nu - \beta_{\nu-1})^p (\zeta + \beta_\nu - \beta_{\nu+1})^p \cdots (\zeta + \beta_\nu - \beta_N)^p. \end{aligned} \right.$

This sum $\Phi(\zeta)$, like each of its N summands, is a polynomial in ζ. In each of the summands, one of the N quantities $\beta_1, \ldots, \beta_N$ plays a marked role; but if we consider the polynomial in ζ obtained by multiplying out in $\Phi(\zeta)$, we see that these N quantities appear, without preference, in the coefficients of the different powers of ζ. In other words, each of these coefficients is a symmetric function of $\beta_1, \ldots, \beta_N$. The multiplying out of the individual factors by the multinomial theorem permits the further inference that these functions $\beta_1, \ldots, \beta_N$ are rational integral functions with rational integral coefficients. But according to a well known theorem in algebra, any rational symmetric function, with rational coefficients, of all the roots of a rational equation is itself always a rational number; and since the $\beta_1, \ldots, \beta_N$ are all the roots of equation (4), the coefficients of $\Phi(\zeta)$ are actually rational numbers.

But, more than this, we need rational integral numbers. These are supplied by the power of b_N which occurs as a factor of $\Phi(\zeta)$. We can, in fact, distribute this power among all the linear factors which occur there and write

$$(9'') \quad \begin{cases} \Phi(\zeta) = \sum_{\nu=1}^{N} (b_N\zeta + b_N\beta_\nu)^{p-1}(b_N\zeta + b_N\beta_\nu - b_N\beta_1)^p \cdots (b_N\zeta + b_N\beta_\nu - b_N\beta_{\nu-1})^p \\ \qquad (b_N\zeta + b_N\beta_\nu - b_N\beta_{\nu+1})^p \cdots (b_N\zeta + b_N\beta_\nu - b_N\beta_N)^p. \end{cases}$$

In analogy with what we had earlier, the coefficients of ζ, when this polynomial is calculated, are rational integral symmetric functions of the products $b_N\beta_1, b_N\beta_2, \ldots, b_N\beta_N$, with rational integral coefficients. But these N products are roots of the equation into which (4) goes if we replace z by z/b_N:

$$b_0 + b_1 \frac{z}{b_N} + \cdots + b_{N-1} \left(\frac{z}{b_N}\right)^{N-1} + b_N \left(\frac{z}{b_N}\right)^N = 0.$$

If we multiply through by b_N^{N-1} this equation goes over into:

$$(10) \quad b_0 b_N^{N-1} + b_1 b_N^{N-2} \cdot z + \cdots + b_{N-2} b_N z^{N-2} + b_{N-1} z^{N-1} + z^N = 0,$$

that is, an equation with integral coefficients when the coefficient of the highest power is unity. Numbers which satisfy such an equation are called *integral algebraic numbers*, and we have the following refinement of the theorem mentioned above: *Rational integral symmetric functions, with rational integral coefficients, of all the roots of an integral equation whose highest coefficient is unity (i.e., of integral algebraic numbers) are themselves rational integral numbers.* You will find this theorem in textbooks on algebra; and if it is not always enunciated in this precise form you can, nevertheless, by following the proof, convince yourselves of its correctness.

Now the coefficients of the polynomial $\Phi(\zeta)$ actually satisfied the assumptions of this theorem so that they are rational integral numbers which we shall denote by $A_0, A_1, \ldots, A_{Np-1}$. We have, then,

$$\sum_{\nu=1}^{N} M_\nu = \int_0^\infty \frac{e^{-\zeta}\zeta^p \, d\zeta}{(p-1)!} (A_0 + A_1\zeta + \cdots + A_{Np-1}\zeta^{Np-1}).$$

With this we have essentially reached our goal. For, if we carry out the integrations in the numerator, using our gamma formula (p. 239), we obtain factors $p!, (p+1)!, (p+2)!\ldots$, since each term contains as factor a power of p of degree p or higher; and after division by $(p-1)!$ there remains everywhere as factor a multiple of p, while the other factors are rational integral numbers (the $A_0, A_1, A_2, \ldots$). Thus $\sum_{\nu=1}^{N} M_\nu$ is certainly a rational integral number divisible by p.

We saw (p. 246) that $a_0 \cdot M$ was not divisible by p, so that

$$a_0 M + \sum_{\nu=1}^{N} M_\nu$$

is necessarily a rational integral number which is not divisible by p and hence, in particular, different from zero. Therefore the equation (6):

$$\left\{a_0 M + \sum_{\nu=1}^{N} M_\nu\right\} + \left\{\sum_{\nu=1}^{N} \varepsilon_\nu\right\} = 0$$

cannot exist, for a non vanishing integer added to $\sum_{\nu=1}^{N} \varepsilon_\nu$, which was shown in No. 4 (p. 247) to be smaller than unity in absolute value, cannot yield zero. But this proves the special case of Lindemann's theorem which we enunciated above (p. 244) and which carries with it the transcendence of π.

I should like to mention here another interesting special case of the general Lindemann theorem, namely, that in the equation $e^\beta = b$ the numbers b, β cannot both be algebraic, with the trivial exception $\beta = 0$, $b = 1$. In other words, the exponential function of an algebraic argument β as well as the natural logarithm of an algebraic number b is, with this one exception, transcendental. This statement includes the transcendence of both e and π, the former for $\beta = 1$, the latter for $b = -1$ (because $e^{i\pi} = -1$). The proof of this theorem can be effected by an exact generalization of the last discussion. One would start from $b - e^\beta$ instead of from $1 + e^\alpha$ as before. It would be necessary to take into account not only all the roots of the algebraic equation for β, but also all the roots of the equation for b, in order to arrive at an equation analogous to (3), so that one would need more notation and the proof would be apparently less perspicuous; but it would require no essentially new ideas.

I shall not go farther into these proofs, but I should like to point out graphically the significance of the last theorem concerning the exponential function. Let us think of all points with an algebraic abscissa as marked off on the x axis ⊢⊣⊢⊣⊢⊣⊢⊣⊢⊣⊢⊣⊢⊣⊢⊣⊢⊣→ x. We know that even the rational numbers, and hence, with greater reason, the algebraic numbers lie everywhere dense on the x axis. One might think at first that the algebraic numbers would exhaust the real numbers. But our theorem declares that this is not the case; that between the algebraic numbers there are infinitely many other numbers, viz. the transcendental numbers; and that we have examples of them in unlimited quantity in $e^{\text{algbr. no.}}$, in log (algebr. no.), and in every algebraic function of these transcendental numbers. It will be more obvious, perhaps, if we write the equation in the form $y = e^x$ and draw the curve in an xy plane (see Fig. 118). If we now mark all the algebraic numbers on the x axis and on the y axis and consider all the points in the plane that have both an algebraic x and an algebraic y, the xy plane will be "densely" covered with these algebraic points. In spite of this dense distribution, the exponential curve $y = e^x$ does not contain a single algebraic point except the one $x = 0$, $y = 1$. Of all the other number pairs x, y which satisfy $y = e^x$, one, at least, is transcendental. This course of the exponential curve is certainly a most remarkable fact.

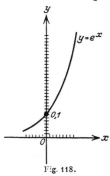

Fig. 118.

The full significance of these theorems which reveal the existence in great quantity of numbers which are not only not rational but which cannot be represented by algebraic operations upon whole numbers—their significance for our concept of the number continuum is tremendous. What would Pythagorus have sacrificed after such a discovery if the irrational seemed to him to merit a hecatomb!

It is remarkable how little in general these questions of transcendence are grasped and assimilated, although they are so simple when one has thought them through. I continually have the experience, in an examination, that the candidate cannot even explain the notion "transcendence". I often get the answer that a transcendental number satisfies no algebraic equation, which, of course, is entirely false, as the example $x - e = 0$ shows. The essential thing, that the coefficients in the equation must be rational, is overlooked.

If you will think our transcendence proofs through again you will be able to grasp these simple elementary steps as a whole, and to make them permanently your own. You need to impress upon your memory only the Hermite integral; then everything develops itself naturally. I should like to emphasize the fact that in these proofs we have used the integral concept (or, speaking geometrically, the idea of area) as something in its essence thoroughly elementary, and I believe that this has contributed materially to the clearness of the proofs. Compare in this respect, the presentation in Volume I of Weber-Wellstein, or in my own little book, *Vorträge über augewählte Fragen der Elementargeometrie*[1], where, as in the older school books, the integral sign is avoided and its use replaced by approximate calculation of series developments. I think that you will admit that the proofs there are far less clear and easy to grasp.

These discussions concerning the distribution of the algebraic numbers within the realm of real numbers lead us naturally to that second modern field to which I have often referred during these lectures, and which I shall now consider in some detail.

II. The Theory of Assemblages

The investigations of George Cantor, the founder of this theory, had their beginning precisely in considerations concerning the existence of

---

[1] Referred to p. 55.

transcendental numbers[1]. They permit one to view this matter in an entirely new light.

If the brief survey of the theory of assemblages which I shall give you has any special character, it is this, that I shall bring the treatment of concrete examples more into the foreground than is usually done in those very general abstract presentations which too often give this subject a form that is hard to grasp and even discouraging.

1. The Power of an Assemblage

With this end in view, let me remind you that in our earlier discussions we have often had to do with different characteristic totalities of numbers which we can now call assemblages of numbers. If I confine myself to real numbers, these assemblages are

1. The positive integers.
2. The rational numbers.
3. The algebraic numbers.
4. All real numbers.

Each of these assemblages contains infinitely many numbers. Our first question is whether, in spite of this, we cannot compare the magnitude or the range of these assemblages in a definite sense, i.e., whether we cannot call the *"infinity"* of one *greater than, equal to,* or *less than* that of another. It is the great achievement of Cantor to have cleared up and answered this really quite indefinite question, by setting up precise concepts. Above all we have to consider his concept of power or cardinal number: *Two assemblages have equal power (are equivalent) when their elements can be put into one-to-one correspondence, i.e., when the two assemblages can be so related to each other that to each element of the one there correponds one element of the other, and conversely.* If such a mutual correspondence is not possible the two assemblages are of *different power*; if it turns out that, no matter how one tries to set up a correspondence, there are always elements of one of the assemblages left over, this one has the greater power.

Let us now apply this principle to the four examples given above. It might seem, at first, that the power of the positive integers would be smaller than that of the rational numbers, the power of these smaller than that of the algebraic numbers, and this finally smaller than that of all real numbers; for each of these assemblages arises from the preceding by the addition of new elements. But such a conclusion would be too hasty. For although *the power of a finite assemblage is always greater than the power of a part of it*, this theorem is by no means valid for infinite assemblages. This discrepancy, after all, need not cause

---

[1] See Journal für Mathematik, vol. 77 (1873), p. 258.

surprise, since we are concerned in the two cases with entirely different fields. Let us examine a simple example which will show clearly that an infinite assemblage and a part of it can actually have the same power, the aggregate, namely, of all positive integers and that of all positive even integers

$$1, \quad 2, \quad 3, \quad 4, \quad 5, \quad 6, \quad \ldots,$$
$$\updownarrow \quad \updownarrow \quad \updownarrow \quad \updownarrow \quad \updownarrow \quad \updownarrow$$
$$2, \quad 4, \quad 6, \quad 8, \quad 10, \quad 12, \quad \ldots$$

The correspondence indicated by the double arrows is obviously of the sort prescribed above, in that each element of one assemblage corresponds to one and only one of the other. Therefore, by Cantor's definition, the assemblage of the positive integers and the partial assemblage of the even integers have the same power.

You see that the question as to the powers of our four assemblages is not so easily disposed of. The simple answer, which then appears the more remarkable, consists in Cantor's great discovery of 1873: *The three assemblages, the positive integers, the rational, and the algebraic numbers, have the same power; but the assemblage of all real numbers has another, namely, a larger power.* An assemblage whose elements can be put into one-to-one correspondence with the series of positive integers (which has therefore the same power) is called *denumerable*. The above theorem can therefore be stated as follows: *The assemblage of the rational as well as of the algebraic numbers is denumerable; that of all real numbers is not denumerable.*

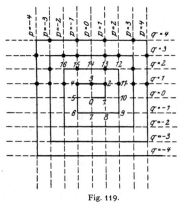

Fig. 119.

Let us first give the proof for rational numbers, which is no doubt familiar to some of you. Every rational number (we shall include the negative ones) can be expressed uniquely in the form p/q, where p and q are integers without a common divisor, where, say, q is positive, while p may also be zero or negative. In order to bring all these fractions p/q into a single series, let us mark in a $p\,q$ plane all points with integral coordinates (p, q), so that they appear as points on a spiral path as shown in Fig. 119. Then we can number all these pairs (p, q) so that only one number will be assigned to each and all integers will be used (see Fig. 119). Now delete from this succession all the pairs (p, q) which do not satisfy the above prescription (p prime to q and $q > 0$) and number anew only those which remain (indicated in the figure by heavy points). We get thus a series which begins as follows:

$$\underbrace{1}\ \underbrace{2}\ \underbrace{3}\ \underbrace{4}\ \underbrace{5}\ \underbrace{6}\ \underbrace{7}\ \underbrace{8}\ \underbrace{9}\ \underbrace{10}\ \underbrace{11}$$
$$1\ \ 0\ -1\ \ 2\ \ \tfrac{1}{2}\ -\tfrac{1}{2}\ -2\ \ 3\ \ \tfrac{3}{2}\ \ \tfrac{2}{3}\ \ \tfrac{1}{3}\ \ldots,$$

one in which a positive integer is assigned to each rational number and a rational number to each positive integer. This shows that the rational numbers are denumerable. This arrangement of the rational numbers

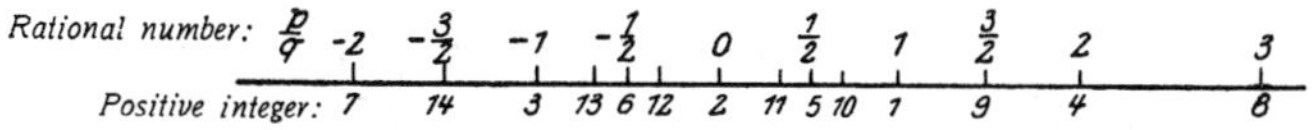

Rational number: $\tfrac{p}{q}$ -2 $-\tfrac{3}{2}$ -1 $-\tfrac{1}{2}$ 0 $\tfrac{1}{2}$ 1 $\tfrac{3}{2}$ 2 3
Positive integer: 7 14 3 13 6 12 2 11 5 10 1 9 4 8

Fig. 120.

into a denumerable series requires, of course, a complete dislocation of their rank as to size, as is indicated in Fig. 120, where the rational points, laid off on the axis of abscissas, are marked with the order of their appearance in the artificial series.

We come, secondly, to the algebraic numbers. I shall confine myself here to real numbers, although the inclusion of complex numbers would not make the discussion essentially more difficult. Every real algebraic number satisfies a real integral equation

$$a_0 \omega^n + a_1 \omega^{n-1} + \cdots + a_{n-1} \omega + a_n = 0,$$

which we shall assume to be *irreducible*, i.e., we shall omit any rational factors of the left-hand member, and also any common divisors of $a_1, a_1, \ldots, a_n$. We assume also that a_0 is always positive. Then, as is well known, every algebraic ω satisfies but one irreducible equation with integral coefficients, in this normal form; and conversely, every such equation has as roots at most n real algebraic numbers, but perhaps fewer, or none at all. If, now, we could bring all these algebraic equations into a denumerable series we could obviously infer that their roots and hence all real algebraic numbers are denumerable.

Cantor succeeded in doint this by assigning to each equation a definite number, its index,

$$N = n - 1 + a_0 + |a_1| + \cdots + |a_{n-1}| + |a_n|,$$

and by separating all such equations into a denumerable succession of classes, according as the index $N = 1, 2, 3, \ldots$ In no one of these equations can either the degree n or the absolute value of any coefficient exceed the finite limit N, so that, in every class, there can be only a finite number of equations, and hence, in particular, only a finite number of irreducible equations. One can easily determine the coefficients by trying out all possible solutions for a given N and can, in fact, write down at once the beginning of the series of equations for small values of N.

Now let us consider that, for each value of the index N, the real roots of the finite number of corresponding irreducible equations have been determined, and arranged according to size. Take first the roots, thus ordered, belonging to index one, then those belonging to index two, and so on, and number them in that order. *In this way we shall have shown, in fact, that the assemblage of real algebraic numbers is denumerable*, for we come in this way to every real algebraic number and, on the other hand, we use all the positive integers. In fact one could, with sufficient patience, determine say the 7563-rd algebraic number of the scheme, or the position of a given algebraic number, however complicated.

Here, again, our "denumeration" disturbs completely the natural order of the algebraic numbers, although that order is preserved among the numbers of like index. For example, two algebraic numbers so nearly equal as 2/5 and 2001/5000 have the widely separated indices 7 and 7001 respectively; whereas $\sqrt{5}$, as root of $x^2 - 5 = 0$, has the same index, 7, as 2/5.

Before we go over to the last example, I should like to give you an auxiliary theorem which will supply us with another denumerable assemblage, as well as with a method of proof that will be useful to us later on. If we have two denumerable assemblages

$$a_1, a_2, a_3, \ldots \quad \text{and} \quad b_1, b_2, b_3, \ldots,$$

then the assemblage of all a and all b which arises by combining these two is also denumerable. For one can write this assemblage as follows:

$$a_1, b_1, a_2, b_2, a_3, b_3, \ldots,$$

and we can at once set this into a one-to-one relation with the series of positive integers. Similarly, if we combine $3, 4, \ldots$, or any finite number of denumerable assemblages, we obtain likewise a denumerable assemblage.

But it does not seem quite so obvious, and this is to be our auxiliary theorem, that *the combination of a denumerable infinity of denumerable assemblages yields also a denumerable assemblage*. To prove this, let us denote the elements of the first assemblage by $a_1, a_2, a_3, \ldots$, those of the second by $b_1, b_2, b_3, \ldots$, those of the third by $c_1, c_2, c_3, \ldots$, and so on, and let us imagine these assemblages written under one another. Then we need only choose the elements of this totality according to successive diagonals, as indicated in the following scheme:

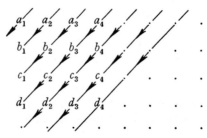

The resulting arrangement

$$\underset{a_1}{1} \quad \underset{a_2}{2} \quad \underset{b_1}{3} \quad \underset{a_3}{4} \quad \underset{b_2}{5} \quad \underset{c_1}{6} \quad \underset{a_4}{7} \quad \underset{b_3}{8} \quad \underset{c_2}{9} \quad \underset{d_1}{10} \quad \underset{a_5}{11} \quad \ldots$$

reaches ultimately every one of the numbers $a, b, c, \ldots$ and brings it into correspondence with a definite positive integer, which proves the theorem. In view of this scheme one could call the process a "counting by diagonals".

The large variety of denumerable assemblages which has thus been brought to our knowledge might incline us to the belief that all infinite assemblages are denumerable. To show that this is not true we shall prove the second part of Cantor's theorem, namely, that the continuum of all real numbers is certainly not denumerable. We shall denote it by $\mathfrak{C}_1$ because we shall have occasion later to speak of multi-dimensional continua.

$\mathfrak{C}_1$ is defined as the totality of all finite real values x, where we may think of x as an abscissa on a horizontal axis. We shall first show that the assemblage of all inner points on the unit segment $0 < x < 1$ has the same power as $\mathfrak{C}_1$. If we represent the first assemblage on the x axis and the second on the y axis, at right angles to it, then a one-to-one correspondence between them will be established by a rising monotone curve of the sort sketched in Fig. 121 (e.g., a branch of the curve $y = -(1/\pi) \cdot \tan^{-1} x$). It is permissible, therefore, to think of the assemblage of all real numbers between 0 and 1 as standing for $\mathfrak{C}_1$ and we shall do this from now on.

The proof by which I shall show you that $\mathfrak{C}_1$ is not denumerable is the one which Cantor gave in 1891 at the meeting of the natural scientists in Halle. It is clearer and more susceptible of generalization than the one which he published in 1873. The essential thing in it is the so-called "diagonal process", by which a real number is disclosed that cannot possibly be contained in any assumed denumerable arrangement of all real numbers. This is a contradiction, and $\mathfrak{C}_1$ cannot, therefore, be denumerable.

We write all our numbers $0 < x < 1$ as decimals and think of them as forming a denumerable sequence

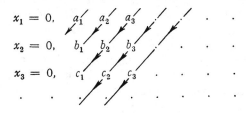

where a, b, c are the digits $0, 1, \ldots, 9$ in every possible choice and arrangement. Now we must not forget that our decimal notation is not uniquely definite. In fact according to our definition of equality we have $0.999\ldots = 1.000\ldots$, and we could write every terminating decimal as a non-terminating one in which all the digits, after a certain one, would be nines. This is one of the first assumptions in calculating with decimals (see p. 34). In order, then, to have a unique notation, let us assume that we are employing only infinite, non-terminating decimals; that all terminating ones have been converted into such as have an endless succession of nines; and that only infinite decimals appear in our scheme above.

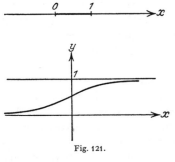

Fig. 121.

In order now to write down a decimal x' which shall be different from every real number in the table, we fix our attention on the digits $a_1, b_2, c_3, \ldots$ of the diagonal of the table (hence the name of the process). For the first decimal place of x' we select a digit a_1' different from a_1; for the second place a digit b_2' different from b_2; for the third place a digit c_3' different from c_3; and so on:

$$x' = 0.\ a_1'\ b_2'\ c_3' \ldots$$

These conditions for $a_1', b_2', c_3', \ldots$ allow sufficient freedom to insure that x' is actually a proper decimal fraction, not, e.g., $0.999\ldots = 1$, and that it shall not terminate after a finite number of digits; in fact, we can select $a_1', b_2', c_3', \ldots$ always different from 9 and 0. The x' is certainly different from x_1 since they are unlike in the first decimal place, and two infinite decimals can be equal only when they coincide in every decimal place. Similarly $x' \neq x_2$, on account of the second place; $x' \neq x_3$, because of the third place; etc. That is, x', a proper decimal fraction, is different from all the numbers $x_1, x_2, x_3, \ldots$ of the denumerable tabulation. Thus the promised contradiction has appeared and we have proved that the continuum $\mathfrak{C}_1$ is not denumerable.

This theorem assures us, a priori, the existence of transcendental numbers; for the totality of algebraic numbers was denumerable and could not therefore exhaust the non-denumerable continuum of all real numbers. But, whereas all the earlier discussions exhibited only a denumerable infinity of transcendental numbers, it follows here that the power of this assemblage is actually greater, so that it is only now that we get the correct general view. To be sure, those special examples were of service in giving life to an otherwise somewhat abstract picture[1].

[1 The existence of transcendental numbers was first proved by Liouville; in an article which appeared in 1851 in vol. 16, series 1, of the Journal des mathématiques, he gave an elementary method for constructing such numbers.]

Now that we have disposed of the one dimensional continuum it is very natural to inquire about the two-dimensional continuum. Everybody had supposed that there were more points in the plane than in the straight line, and it attracted much attention when Cantor showed[1] that the power of the two dimensional continuum $\mathfrak{C}_2$ was exactly the same as that of the one dimensional $\mathfrak{C}_1$. Let us take for $\mathfrak{C}_2$ the square with side of unit length, and for $\mathfrak{C}_1$ the unit segment (see Fig. 122). We shall show that the points of these two aggregates can be put into a one-to-one relation. The fact that this statement seems so paradoxical depends probably on our difficulty in freeing our mental picture of a certain continuity in the correspondence. But the relation which we shall establish will be as discontinuous or, if you please, as inorganic as it is possible to be. It will disturb everything which one thinks of as characteristic for the plane and the linear manifold as such, with the exception of the "power". It will be as though one put all the points of the square into a sack and shook them up thoroughly.

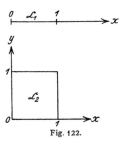

Fig. 122.

The assemblage of the points of the square coincides with that of all pairs of decimal fractions

$$x = 0.\, a_1 a_2 a_3 \ldots, \quad y = 0.\, b_1 b_2 b_3 \ldots,$$

all of which we shall suppose to be non-terminating. We exclude points on the boundary for which one of the coordinates (x, y) vanishes, i.e., we exclude the two sides which meet at the origin, but we include the other two sides. It is easy to show that this has no effect on the power. The fundamental idea of the Cantor proof is to combine these two decimal fractions into a new decimal fraction z from which one can obtain (x, y) again uniquely and which will take just once all the values $0 < z \leqq 1$ when the point (x, y) traverses the square once. If we then think of z as an abscissa, we have the desired one-to-one correpondence between the square $\mathfrak{C}_2$ and the segment $\mathfrak{C}_1$, whereby the agreement concerning the square carries with it the inclusion of the end $z = 1$ of the segment.

One might try to effect this combination by setting

$$z = 0.\, a_1 b_1 a_2 b_2 a_3 b_3 \ldots,$$

from which one could in fact determine (x, y) uniquely by selecting the odd and even decimal places respectively. But there is an objection to this, due to the ambiguous notation for decimal fractions. This z, namely, would not traverse the whole of $\mathfrak{C}_1$ when we chose for (x, y)

---

[1] Journal für Mathematik, vol. 84 (1878), p. 242 et seq.

all possible pairs of non-terminating decimals, that is, when we traversed all the points of $\mathfrak{C}_2$. For, although z is, to be sure, always non-terminating, there can be non-terminating values of z, such as

$$z = 0.\ c_1 c_2\ 0\ c_4\ 0\ c_6\ 0\ c_8 \ldots,$$

which correspond only to a terminating x or y, in the present case to the values

$$x = 0.\ c_1\ 000 \ldots, \quad y = 0.\ c_2 c_4 c_6 c_8 \ldots$$

This difficulty is best overcome by means of a device suggested by J. König of Budapest. He thinks of the a, b, c not as individual digits but as complexes of digits—one might call them "molecules" of the decimal fraction. A "molecule" consists of a single digit, different from zero, together with all the zeros which immediately precede it. Thus every non-terminating decimal must contain an infinity of molecules, since digits different from zero must always recur; and conversely. As an example, in

$$x = 0.320\ 8007\ 000\ 302\ 405 \ldots$$

we should take as molecules $a_1 = 3$, $a_2 = 2$, $a_3 = 08$, $a_4 = 007$, $a_5 = 0003$, $a_6 = 02$, $a_7 = 4, \ldots$

Now let us suppose, in the above rule for the relation between x, y and z, that the a, b, c stand for such molecules. Then there will correspond uniquely to every pair (x, y) a non-terminating z which would, in its turn, determine x and y. But now every z breaks up into an x and a y each with an infinity of molecules, and each z appears therefore just once when (x, y) run through all possible pairs of non terminating decimal fractions. This means, however, that the unit segment and the square have been put into one-to-one correspondence, i.e., they have the same power.

In an analogous way, of course, it can be shown that the continuum of $3, 4, \ldots$ dimensions has the same power as the one dimensional segment. It is more remarkable, however, that the continuum $\mathfrak{C}_\infty$, of infinitely many dimensions, or more exactly, of denumerably infinitely many dimensions, has also the same power. This infinite dimensional space is defined as the totality of the systems of values which can be assumed by the denumerable infinity of variables

$$x_1, x_2, \ldots, x_n, \ldots$$

when each, independently of the others, takes on all real values. This is really only a new form of expression for a concept that has long been in use in mathematics. When we talk of the totality of all power series or of all trigonometric series, we have, in the denumerable infinity of coefficients, really nothing but so many independent variables which, to be sure, are for purposes of calculation restricted by certain requirements to ensure convergence.

Let us again confine ourselves to the "unit cube" of the $\mathfrak{C}_\infty$, i.e., to the totality of points which are subject to the condition $0 < x_n \leq 1$, and show that they can be put into one-to-one relation with the points of the unit segment $0 < z \leq 1$ of $\mathfrak{C}_1$. For convenience, we exclude all boundary points for which one of the coordinates x_m vanishes, as well as the end point $z = 0$, but admit the others. As before we start with the decimal fractional representation of the coordinates in $\mathfrak{C}_\infty$:

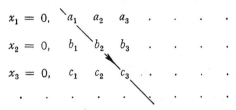

where we assume that the decimal fractions are all written in non-terminating form, and furthermore that the $a, b, c, \ldots$ are "decimal fraction molecules" in the sense indicated above, i.e., digit complexes which end with a digit which is different from zero, but which is preceded exclusively by zeros. Now we must combine all these infinitely many decimal fractions into a new one which will permit recognition of its components; or, if we keep to the chemical figure, we wish to form such a loose alloy of these molecular aggregates that we can easily separate out the components. This is possible by means of the "diagonal process" which we applied before (p. 254). From the above table we get, according to that plan

$$z = 0, a_1 a_2 b_1 a_3 b_2 c_1 a_4 b_3 c_2 d_1 a_5 \ldots,$$

which relates uniquely a point of $\mathfrak{C}_1$ to each point of $\mathfrak{C}_\infty$. Conversely we get in this way every point z of $\mathfrak{C}_1$, for from the non terminating decimal fraction for a given z we can derive, according to the above given scheme, an infinity of non-terminating decimals $x_1, x_2, x_3, \ldots$, out of which this z would arise by the method indicated. We have succeeded therefore in setting up a one-to-one correspondence between the unit cube in $\mathfrak{C}_\infty$ and the unit segment in $\mathfrak{C}_1$.

Our results thus far show that there are at least two different powers:

1. That of the denumerable assemblages.
2. That of all continua $\mathfrak{C}_1, \mathfrak{C}_2, \mathfrak{C}_3, \ldots$, including $\mathfrak{C}_\infty$.

The question naturally arises whether there are still larger powers. The answer is that one can exhibit an assemblage having a still higher power, not merely as a result of abstract reasoning, but one lying quite within the range of concepts which have long been used in mathematics. This aggregate is, namely:

3. That of all possible real functions $f(x)$ of a real variable x.

It will be sufficient for our purpose to restrict the variable to the interval $0 < x < 1$. It is natural to think first of the aggregate of the continuous functions $f(x)$, but there is a remarkable theorem which states that the totality of all continuous functions has the same power as the continuum, and belongs therefore in group 2. We can reach a new, a higher power, only by admitting discontinuous functions of the most general kind imaginable, i.e., where the function value at any place x is entirely arbitrary and has no relation to neighboring values.

I shall first prove the theorem concerning the aggregate of continuous functions. This will involve a repetition and a refinement of the considerations which we adduced (p. 206) in order to make plausible the possibility of developing "arbitrary" functions into trigonometric series. At that time I remarked:

a) A continuous function $f(x)$ is determined if one knows the values $f(r)$ at all rational values of r.

b) We know now that all rational values r can be brought into a denumerable series $r_1, r_2, r_3, \ldots$

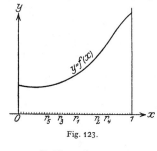

Fig. 123.

c) Consequently $f(x)$ is determined when one knows the denumerable infinity of quantities $f(r_1), f(r_2), f(r_3), \ldots$ Moreover, these values cannot, of course, be assumed arbitrarily if we are to have a single-valued continuous function. The assemblage then of all possible systems of values $f(r_1), f(r_2), \ldots$ must contain a sub-assemblage whose power is the same as that of the assemblage of all continuous functions (see Fig. 123).

d) Now the magnitudes $f_1 = f(r_1)$, $f_2 = f(r_2), \ldots$ can be considered as the coordinates of a $\mathfrak{C}_\infty$, since they make up a denumerable infinity of continuously varying magnitudes. Hence, in view of the theorem already proved, the totality of all their possible systems of values has the power of the continuum.

e) Since the assemblage of continuous functions is contained in an assemblage which is equivalent to the continuum, it must itself be equivalent to a sub-assemblage of the continuum.

f) But it is not hard to see that, conversely, the entire continuum can be put into one-to-one correspondence with a part of the assemblage of all continuous functions. For this purpose, we need to consider only the functions defined by $f(x) = k = $ const., where k is a real parameter. If k traverses the continuum $\mathfrak{C}_1$ then $f(x)$ will describe an assemblage which is in one-to-one correspondence with $\mathfrak{C}_1$ but which is only a part of the totality of all continuous functions.

g) Now we must make use of an important general theorem of the theory of assemblages, the so-called theorem of equivalence, due to

F. Bernstein[1]: *If each of two assemblages is equivalent to a part of the other then the two assemblages are equivalent.* This theorem is very plausible. The proof of it would take us too far afield.

h) According to e) and f) the continuum $\mathfrak{C}_1$ and the aggregate of all continuous functions satisfy the conditions of the theorem of equivalence. They are therefore of like power, and our theorem is proved.

Let us now go over to the proof of our first theorem, that the assemblage of all possible functions that are really entirely arbitrary has a power higher than that of the continuum. The proof is an immediate application of Cantor's diagonal process.

a) Assume the theorem to be false, i.e., that the assemblage of all functions can be put into one-to-one correspondence with the continuum $\mathfrak{C}_1$. Suppose now, in this one-to-one relation, that the function $f(x, \nu)$ of x corresponds to the value $x = \nu$ in $\mathfrak{C}_1$, so that, while ν traverses the continuum $\mathfrak{C}_1$, $f(x, \nu)$ represents all possible functions of x. We shall reduce this supposition to an absurdity by actually setting up a function $F(x)$ which is different from all such functions $f(x, \nu)$.

b) For this purpose we construct the "diagonal function" of the tabulation of the $f(x, \nu)$, i.e., that function which, for every value $x = x_0$, has that value which the assumed correspondence imposes upon $f(x, \nu)$ when the parameter ν also has the value $\nu = x_0$, namely $f(x_0, x_0)$. Written as a function of x, this is simply the function $f(x, x)$.

c) Now we construct a function $F(x)$ which for every x is different from this $f(x, x)$:

$$F(x) \neq f(x, x) \text{ for every } x.$$

We can do this in the greatest variety of ways, since we admit the most completely discontinuous functions, whose value at any place can be arbitrarily determined. We might, for example, put

$$F(x) = f(x, x) + 1.$$

d) This $F(x)$ is actually different from every one of the functions $f(x, \nu)$. For, if $F(x) = f(x, \nu_0)$ for some $\nu = \nu_0$, the equality would hold also for $x = \nu_0$; that is, we should have $F(\nu_0) = f(\nu_0, \nu_0)$, which contradicts the assumption in c) concerning $F(x)$.

The assumption a) that the functions $f(x, \nu)$ could exhaust all functions is thus overthrown, and our theorem is proved.

It is interesting to compare this proof with the analogous one for the non-denumerability of the continuum. There we assumed the totality of decimal fractions arranged in a denumerable table; here we consider the function scheme $f(x, \nu)$. The singling out there of the diagonal elements corresponds to the construction here of the diagonal function $f(x, x)$; and in both cases the application was the same, namely

---

[1] First published in Borel's *Leçons sur la Théorie des Fonctions*, Paris, 1898, p. 103.

the setting up of something new, i.e., not contained in the table,—in the one case a decimal fraction, in the other a function.

You can readily imagine that similar considerations could lead us to assemblages of ever increasing power—beyond the three which we have already discussed. The most noteworthy thing in all these results is that there remain any abiding distinctions and gradations at all in the different infinite assemblages, notwithstanding our having subjected them to the most drastic treatment imaginable; treatment which disturbed special properties, such as order, and permitted only the ultimate elements, the atoms, to retain an independent existence as things which could be tossed about in the most arbitrary manner. And it is worth noting that the three gradations which we did establish were among things which have long been familiar in mathematics—integers, continua, and functions.

With this I shall close this first part of my discussion of the theory of assemblages, which has been devoted mainly to the concept of power. In a similar concrete manner, but with still greater brevity, I shall now tell you something about a farther chapter of this theory.

2. Arrangement of the Elements of an Assemblage

We shall now bring to the front just that thing which we have heretofore purposely neglected, the question, namely, how individual assemblages of the same power differ from one another by virtue of those relations as to the arrangement of the elements which are intrinsic in the assemblage. The most general one-to-one representations which we have admitted thus far disturbed all these relations,—think only of the representation of the square upon the segment. I desire to emphasize, especially, the significance of precisely this chapter of the theory of assemblages. It cannot possibly be the purpose of the theory of assemblages to banish the differences which have long been so familiar in mathematics, by introducing new concepts of a most general kind. On the contrary, this theory can and should aid us to understand those differences in their deepest essence, by exhibiting their properties in new light.

We shall try to make clear the different possible arrangements, by considering definite familiar examples. Beginning with denumerable assemblages, we note three examples of fundamentally different arrangement, so different that the equivalence of their powers was, as we saw, the result of a special and by no means obvious theorem. These examples are:

1. The assemblage of all positive integers.
2. The assemblage of all (negative and positive) integers.
3. The assemblage of all rational numbers and that of all algebraic numbers.

All these assemblages have a common property in the arrangement of their elements, which finds expression in the designation *simply ordered*, i. e., of two given elements, it is always known which precedes the other, or, put algebraically, which is the smaller and which the larger. Further, if three elements a, b, c are given, then, if a precedes b and b precedes c, a precedes c (if $a < b$ and $b < c$ then $a < c$).

But now as to the *characteristic differences*. In (1), there is a first element (one) which preceded all the others, but no last which follows all the others; in (2), there is neither a first nor a last element. Both (1) and (2) have this in common, that every element is followed by another definite one, and also that every element [except the first in (1)] is preceded by another definite one. In contrast with this, we find in (3) (as we saw p. 31) that between any two elements there are always infinitely many others—the elements are "everywhere dense", so that among the rational or the algebraic numbers lying between a and b there is neither a smallest nor a largest. The manner of arrangement in these three examples, the *type of arrangement* (Cantor's term *type of order* seems to me less expressive) is quite different, although the power is the same. One could raise the question here as to all the types of arrangement that are possible in denumerable assemblages, and that is what the students of the theory of assemblages actually do.

Let us now consider assemblages having the power of the continuum. In the continuum $\mathfrak{C}_1$ of all real numbers, we have a simply ordered assemblage; but in the multidimensional types $\mathfrak{C}_2, \mathfrak{C}_3, \ldots$ we have examples of an order no longer simple. In the case of $\mathfrak{C}_2$, for instance, two relations are necessary, instead of one, to determine the mutual position of two points.

The most important thing here is to analyze the concept of continuity for the one dimensional continuum. The recognition of the fact that continuity here depends on simple properties of the arrangement which is peculiar to C_1, is the first great achievement of the theory of assemblages toward the clarifying of traditional mathematical concepts. It was found,—namely, that all the continuity properties of the ordinary continuum flow from its being a simply—ordered assemblage with the following two properties:

1. *If we separate the assemblage into two parts A, B such that every element belongs to one of the two parts and all the elements of A precede all those of B, then either A has a last element or B a first element.* If we recall Dedekind's definition of irrational number (see p. 33 et seq.) we can express this by saying that every "cut" in our assemblage is produced by an actual element of the assemblage.

2. *Between any two elements of the assemblage there are always infinitely many others.*

This second property is common to the continuum and the denumerable assemblage of all rational numbers. It is the first property however that marks the distinction between the two. In the theory of assemblages it is customary to call all simply-ordered assemblages *continuous* if they possess the two preceding properties, for it is actually possible to prove for them all the thorems which hold for the continuum by virtue of its continuity.

Let me remind you that these properties of continuity can be formulated somewhat differently in terms of Cantor's *fundamental series*. A fundamental series is a simply-ordered denumerable series of elements $a_1, a_2, a_3, \ldots$ of an aggregate such that each element of the series precedes the following or each succeeds it:

$$a_1 < a_2 < a_3 < \ldots \text{ or } a_1 > a_2 > a_3 > \ldots$$

An element a of the aggregate is called a *limit element* of the fundamental series if (in the first sort) every element which precedes a but no element which follows a is ultimately passed by elements of the fundamental series; and similarly for the second sort. Now if every fundamental series in an aggregate has a limit element, the aggregate is called *closed*; if, conversely, every element of the aggregate is a limit element of a fundamental series, the aggregate is said to be *dense*. Now continuity, in the case of aggregates having the power of the continuum, consists essentially in the union of these two properties.

Let me remind you incidentally that when we were discussing the foundations of the calculus we spoke also of another continuum, the *continuum of Veronese*, which arose from the usual one by the addition of actually infinitely small quantities. This continuum constitutes a simply-ordered assemblage in as much as the succession of any two elements is determinate, but it has a type of arrangement entirely different from that of the customary $\mathfrak{C}_1$; even the theorem that every fundamental series has a limit element no longer holds in it.

We come now to the important question as to what representations preserve the distinctions among the continua $\mathfrak{C}_1, \mathfrak{C}_2, \ldots$ of different dimensions. We know, indeed, that the most general one-to-one representation obliterates every distinction. We have here the important theorem that *the dimension of the continuum is invariant with respect to every continuous one-to-one representation*, i.e., that it is impossible to effect a reversibly unique and continuous mapping of a $\mathfrak{C}_m$ upon a $\mathfrak{C}_n$ where $m \neq n$. One might be inclined to accept this theorem, without further ado, as self evident; but we must recall that our naïve intuition seemed to exclude the possibility of a reversibly unique mapping of $\mathfrak{C}_2$ upon $\mathfrak{C}_1$, and this should dispose us to caution in accepting its pronouncements.

I shall discuss in detail only the simplest case[1], which concerns the relation between the one-dimensional and the two-dimensional continual and I shall then indicate the difficulties in the way of an extension to the most general case. We shall prove, then, that *a reversibly unique, continuous relation between $\mathfrak{C}_1$ and $\mathfrak{C}_2$ is not possible*. Every word here is essential. We have seen, indeed, that we may not omit continuity; and that reversible uniqueness may not be omitted is shown by the example of the "Peano curve" which is doubtless familiar to some of you.

We shall need the following auxiliary theorem: *Given two one-dimensional continua $\mathfrak{C}_1, \mathfrak{C}_1'$ which are mapped continuously upon each other so that to every element of $\mathfrak{C}_1'$ there corresponds one and but one element of C_1, and to every element of C_1 there corresponds at most one element of $\mathfrak{C}_1'$*; if, then, a, b are two elements of $\mathfrak{C}_1$ to which two elements a', b' in $\mathfrak{C}_1'$ actually correspond, respectively, it follows that to every element c of $\mathfrak{C}_1$ lying between a and b there will correspond an element c' of $\mathfrak{C}_1'$ which lies between a' and b' (see Fig. 124). This is analogous to the familiar theorem that a continuous function $f(x)$ which takes two values a, b at the values $x = a'$, b' must take a value c, chosen arbitrarily between a and b, at some value c' between a' and b'; and it could be proved as an exact generalization of this theorem, by using the above definition of continuity. This would require one also to explain continuous mapping of a continuous assemblage in a manner analogous to the usual definition of continuous functions, and it can be done with the aid of the concept of arrangement. But this is not the place to amplify these ideas.

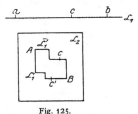

Fig. 125.

We shall give our proof as follows. We assume that a continuous reversibly unique mapping of the one dimensional segment $\mathfrak{C}_1$ upon the square $\mathfrak{C}_2$ has been effected (see Fig. 125). Let two elements a, b on $\mathfrak{C}_1$ correspond to the elements A, B, respectively, of $\mathfrak{C}_2$. Now we can join these elements A, B by two different paths within $\mathfrak{C}_2$, e.g., by the broken lines $\mathfrak{C}_1'$, $\overline{\mathfrak{C}_1'}$ drawn in the figure. To do this, it is not necessary to presuppose any special properties of $\mathfrak{C}_2$, such as the setting up of a coordinate system; we need merely use the concept of double order. Each of the paths $\mathfrak{C}_1'$ and $\overline{\mathfrak{C}_1'}$ will be a simply-ordered one-dimensional continuum like $\mathfrak{C}_1$, and because of the continuous reversibly unique relation between $\mathfrak{C}_1$ and $\mathfrak{C}_2$ there must correspond just one point on $\mathfrak{C}_1$ to each element of

---

[1] Brouwer, L. E. J. gave a proof for the general case in 1911, in volume 70, p. 161, of the Mathematische Annalen.

$\mathfrak{C}'_1$ and $\overline{\mathfrak{C}'_1}$; but to each element of $\mathfrak{C}_1$ there must correspond at most one on $\mathfrak{C}'_1$ or $\overline{\mathfrak{C}'_1}$. In other words, we have precisely the conditions of the above lemma, and it follows that to every point c in $\mathfrak{C}_1$ between a and b there corresponds not only a point c' of $\mathfrak{C}_1$ but also a point $\bar{}'$ of $\overline{\mathfrak{C}'_1}$. But this contradicts the assumed reversible uniqueness of the relation between $\mathfrak{C}_1$ and $\mathfrak{C}_2$. Consequently this mapping is not possible and the theorem is proved.

If one wished to extend these considerations to two arbitrary continua $\mathfrak{C}_m, \mathfrak{C}_n$, it would be necessary to know in advance something about the constitution of continua of general nature and of dimension $1, 2, 3, \ldots, m-1$, which can be embedded in $\mathfrak{C}_m$. As soon as m, $n \geqq 2$ one can not get along merely with the concept "between' as we could in the simplest case above. On the contrary, one is led to very difficult investigations which include, among the earliest cases, the abstruse fundamental geometric questions concerning the most general continuous one-dimensional assemblage of points in the plane, questions which only recently have been somewhat cleared up. One of these interesting questions is as to when such an assemblage of points should be called a curve.

I shall close with this my very special discussion of the theory of assemblages, in order to add a few remarks of a general nature. First, a word as to the general notions which Cantor had entertained concerning the position of the theory of assemblages with reference to geometry and analysis. These notions exhibit the theory of assemblages in a special light. The difference between the discrete magnitudes of arithmetic and the continuous magnitudes of geometry has always had a prominent place in history and in philosophical speculations. In recent times the discrete magnitude, as conceptually the simplest, has come into the foreground. According to this tendency we look upon natural numbers, integers, as the simplest given concepts; we derive from them in the familar way, rational and irrational numbers, and we construct the complete apparatus for the control of geometry by means of analysis, namely, analytic geometry. This tendency of modern development can be called that of arithmetizing geometry. The geometric idea of continuity is carried back to the idea of whole numbers. These lectures have, in the main, held to this direction.

Now, as opposed to this one-sided preference for integers, Cantor would (as he himself told me in 1903 at the meeting of the natural scientists in Cassel) achieve, in the theory of assemblages, "the genuine fusion of arithmetic and geometry". Thus the theory of integers, on one hand, as well as the theory of different point continua, on the other, and much more, would form a homogeneous group of equally important chapters in a general theory of assemblages.

I shall add a few general remarks concerning the relation of the theory of assemblages to geometry. In our discussion of assemblages we have considered:

1. The *power of an assemblage* as something that is unchanged by any reversibly unique mapping.

2. *Types of order of assemblages* which take account of the relations among the elements as to order. We were able here to characterize the notion of continuity, the different multiple arrangements or multi-dimensional continua, etc., so that the invariants of continuous mappings found their place here. When carried over to geometry, this gives the branch which, since Riemann, has been called analysis situs, that most abstract chapter of geometry, which treats those properties of geometric configurations which are invariant under the most general reversibly unique continuous mappings. Riemann had used the word manifold (Mannigfaltigkeit) in a very general sense. Cantor used it also, at first, but replaced it later by the more convenient word assemblage (Menge).

3. If we go over to *concrete geometry* we come to such differences as that between *metric* and *projective* geometry. It is not enough here to know, say, that the straight line is one-dimensional and the plane two-dimensional. We desire rather to construct or to compare figures, for which we need to use a fixed unit of measure or at least to choose a line in the plane, or a plane in space. In each of these concrete domains it is necessary, of course, to add a special set of axioms to the general properties of arrangement. This implies, of course, a further development of the theory of simply-ordered, doubly-ordered, ..., n-tuply-ordered, continuous assemblages.

This is not the place for me to go into these things in detail, especially since they must be taken up anyway in the succeeding volumes of the present work. I shall merely mention literature in which you can inform yourselves farther. Here, above all, I should speak of the reports in the Mathematische Enzyklopädie: Enriques, *Prinzipien der Geometrie* (III. A. B. 1) and v. Mangoldt, *Die Begriffe ,,Linie" und ,,Fläche"* (III. A. B. 2), which treat mainly the subject of axioms; also Dehn-Heegaard, *Analysis situs* (III. A. B. 3). The last article is written in rather abstract form. It begins with the most general formulation of the concepts and fundamental facts of analysis situs, as these were set up by Dehn himself, from which everything else is deduced then by pure logic. This is in direct opposition to the inductive method of presentation, which I always recommend. The article can be fully understood only by an advanced reader who has already thoroughly worked the subject through inductively.

As to literature concerning the theory of aggregates, I should mention, first of all, the report made by A. Schoenflies to the Deutsche Mathematikervereinigung, entitled: *Die Entwickelung der Lehre von den Punktmannigfaltigkeiten*[1]. The first part appeared in volume 8 of the Jahresbericht der deutschen Mathematikervereinigung; the second appeared recently as a second supplementary volume to the Jahresbericht. This work is really a report on the entire theory of aggregates, in which you will find information concerning numerous details. Alongside of this, I would mention the first systematic textbook on the theory of aggregates: *The Theory of Sets of Points*, by W. H. Young and his wife, Grace Chisholm Young (whom we mentioned p. 179).

In concluding this discussion of the theory of assemblages we must again put the question which accompanies all of our lectures: *How much of this can one use in the schools?* From the standpoint of mathematical pedagogy, we must of course protest against putting such abstract and difficult things before the pupils too early. In order to give precise expression to my own view on this point, I should like to bring forward the biogenetic fundamental law, according to which the individual in his development goes through, in an abridged series, all the stages in the development of the species. Such thoughts have become today part and parcel of the general culture of everybody. Now, I think that instruction in mathematics, as well as in everything else, should follow this law, at least in general. Taking into account the native ability of youth, instruction should guide it slowly to higher things, and finally to abstract formulations; and in doing this it should follow the same road along which the human race has striven from its naïve original state to higher forms of knowledge. It is necessary to formulate this principle frequently, for there are always people who, after the fashion of the mediaeval scholastics, begin their instruction with the most general ideas, defending this method as the "only scientific one". And yet this justification is based on anything but truth. To instruct scientifically can only mean to induce the person to think scientifically, but by no means to confront him, from the beginning, with cold, scientifically polished systematics.

An essential obstacle to the spreading of such a natural and truly scientific method of instruction is the lack of historical knowledge which so often makes itself felt. In order to combat this, I have made a point of introducing historical remarks into my presentation. By doing this I trust I have made it clear to you how slowly all mathematical ideas have come into being; how they have nearly always appeared first in rather prophetic form, and only after long development have crystallized into the rigid form so familiar in systematic presentation. It is my earnest hope that this knowledge may exert a lasting influence upon the character of your own teaching.

---

[1] 2 parts, Leipzig 1900 and 1908, A revision of the first half appeared in 1913 under the title: *Entwickelung der Mengenlehre und ihrer Anwendungen;* as a continuation of this, see H. Hahn: *Theorie der reellen Funktionen*, vol. I, Berlin, 1921.

Index of Names.

Abel 84, 138, 154.
d'Alembert 103, 212.
Archimedes 80, 209, 219, 222, 237.

Bachmann 39, 48.
Ball 74.
Baltzer 72.
Bauer 86.
Baumann 220.
Berkeley 219.
Bernoulli, Daniel 205.
—, Jacob 200.
—, Johann 200, 205, 216.
Bernstein 261.
Bessel 191.
Braunmühl 175.
Briggs 172, 173.
Brouwer 265.
Budan 94.
Bürgi 147.
Burkhardt 23, 29, 191, 205.

Cantor, Georg 12, 32, 35, 204, 221, 250, 266, 267.
Cardanus 55, 80, 134.
Cartesius, see Descartes.
Cauchy 84, 154, 202, 213, 219, 228, 231, 235.
Cavalieri 210, 214.
Cayley 68, 73, 74.
Chernac 40.
Chisholm 179.
Clebsch 84.
Coble 143.
Copernicus 81, 171.
Coradi 198.

Dedekind 13, 33.
Dehn 267.
Delambre 180, 181.
De Moivre 153, 168.
Descartes 81, 94.

Dirichlet 42, 199, 202, 203, 204, 206.
Dyck 94.

Enriques 55, 267.
Eratosthenes 40.
Eudoxus 219.
Euklid 32, 80, 219.
Euler 50, 56, 77, 82, 155, 166, 200, 202, 212, 234, 237.

Fejér 200.
Fermat 39, 48, 58.
Fourier 91, 201, 204, 206, 207, 222, 236.

Galle 17.
Gauss 39, 42, 50, 58, 76, 102, 154, 181.
Gibbs 199, 200.
Gordan 143.
Goursat 235.
Grassmann 12, 58, 64.
Gutzmer 2.

Hahn 268.
Hamilton 11, 58, 62, 73, 74.
Hammer 175.
Hankel 26, 56.
Harnack 235.
Hartenstein 99.
Heegard 267.
Hegel 217.
Heiberg 80, 209.
Hermite 238, 239, 245.
Hilbert 13, 14, 48, 218, 238, 247.
l'Hospita 216.

Jacobi 8

Kant 10
Kästner 76, 210, 212.
Kepler 208, 210.
Kimura 74.

König, J. 258.
Kowalewski 215, 216.
Kummer 48.

Lacroix 235.
Lagrange 66, 82, 83, 153, 200, 220, 222, 234.
Leibniz 13, 20, 56, 82, 200, 211, 214, 215, 220, 222.
Lie 84.
Lindemann 238, 243, 249.
Liouville 256.
Lübsen 216.
Lüroth 17.

Maclaurin 210, 212, 234.
Männchen 49.
Mangold 267.
Markoff 236.
Mehmke 95, 170.
Mercator, N. 81, 150, 168.
Michelson 198, 199.
Minkowski 11, 39.
Möbius 176, 177, 182.
Molk, J. 8.
Mollweide 181.
Monge 84.

Napier, see Neper.
Neper 81, 147, 150, 172, 173.
Netto 86.
Newton 81, 82, 151, 168, 210, 212, 222, 230, 233, 237.
Nörlund 236.

Odhner 17.
Ohm 76.
Ostrowski 103.

Peano 12, 265.
Peurbach 171.
Picard 84, 160.
Pitiscus 172, 174.

Index of Names.

Plato 80, 120.
Poincaré 11.
Poisson 216.
Pringsheim, A. 233.
Ptolemy 170.
Pythagorus 31, 250.

Regiomontanus 171.
Rhäticus 171.
Riemann 84, 159, 202, 267.
Runge 86, 92, 191, 198.

Schafheitlein 216.
Scheffers 155, 235.
Schellbach 222.

Schimmack 3, 194, 223, 224.
Schlömilch 235.
Schoenfliess 267.
Schubert 8.
Seeger 189.
Serret 86, 235.
Shanks 237.
Simon 5, 24, 85, 162, 221.
Stifel 146.
Stratton 198.
Study 175, 181.
Sturm, J. 94.

Tannery, J. 8.
Taylor 82, 153, 227, 232, 233, 234, 236.
Timerding 189.

Tropjke 28, 85, 170.

Vega 173.
Veronese 218, 264.
Vieta 25.
Vlacq 173.

Weber 4, 13, 23, 29, 86, 175, 182, 250.
Weierstrass 33, 84, 202, 203, 213.
Wilbraham 198.
Wolff, Chr. 216.
Wolfskehl 48.
Wüllner 217.

Young, G. Chisholm 180.
—, W. H. 268.

Index of Contents.

Abridged reckoning 10 et seq.
Actually infinitely small quantities 214, 218, 219.
Algorithmic method, see Processes of growth, plan C.
Analysis situs 267.
Applicability and logical consistency in infinitesimal calculus 221.
    in the theory of fractions 29.
        complex numbers 56—58.
        irrational numbers 33.
        natural numbers 14.
        negative numbers 23—25.
Applied mathematics 4, 15.
Approximation, mathematics of 36.
Archimedes, axiom of 218.
Arithmetization 266.
Arrangement within an assemblage 262.
Assemblage of continuous and real functions 206, 259—261.
— of algebraic and transcendental numbers 250, 254—256.

Branch points 107, 109.

Calculating machines 17—21.
— and formal rules of operation 21, 22.
Cardinal number 251.
Casus irreducibilis of the cubic equation 135.
Circular functions:
    analogy with hyperbolic functions 166.
    see also trigonometric functions.
Closed fundamental series 264.
Complex numbers, higher 58—75.
Consistency, proofs of 13, 25, 57.
— and applicability:
    of infinitesimal calculus 221.
    of the theory of fractions 30.
        complex numbers 55—58.
        irrational numbers 34.
        natural numbers 14.
        negative numbers 23.

Constructions with ruler and compasses 49.
Continued fractions 42—44.
Continuity, analysis of, based on theory of assemblages 263—266.
Curriculum proposals, the Meran 16.
Cut, after Dedekind 33.
Cyclometric functions:
    definition of, by means of quadrature of the circle 163—168.
Cyclotomic numbers 47.

Decimal system 6, 9, 20.
Dense 31, 249, 263, 264.
Denumerability of algebraic numbers 253 et seq.
— — rational numbers 252 et seq.
— — a denumerable infinity of denumerable assemblages 254.
Derivative calculus 220, 234.
Development of infinitesimal calculus 208—220.
Diagonal process 254, 259, 261.
Differences, calculus of 228, 230—232.
Differentials, calculation with:
    naive intuitional direction 208—210.
    direction of mathematics of approximation 215, 216.
    formal direction 215.
    speculative direction 214, 216, 217.
Dimension, invariance of the — of a continuum by reversibly unique mapping 264, 265.
Discriminant curve of the quadratic and cubic equation 92.
— surface of the biquadratic equation 98—101.

Equations:
    cyclotomic 50.
    pure 110—115, 131—134.
    reciprocal 51.
    of fifth degree 141—142.
    the dihedral — 115—120, 126.
    the tetrahedral — 120—130.

the octahedral — 120—130.
the icosahedral — 120—130.
Equivalence, of assemblages 251—262.
—, theorem of 260.
Exhaustion, method of 209.
Exponential function:
   definition by quadrature of hyperbola 149 et seq., 156—157.
   general — —, and e^{ix} 158—159, 160—161.
   series for e^x 152.
   function-theoretic discussion of 156 et seq.

Fermat, great theorem of 46—49.
Formal mathematics 24, 26, 29, 56.
Foundations of arithmetic:
   by means of intuition 11.
      formalism 13.
      logic 11.
      theory of point sets 12.
Fourier's series, see trigonometric series.
— integral 207.
Function, notion of:
   analytic function 200—201.
   arbitrary function 200.
   relation of the two in complex region 202—203.
   discontinuous real functions 204.
Functions, assemblage of continuous and real 206, 261—262.
Fundamental laws of addition and multiplication 8—10.
   logical foundation 10—16.
   consistency 13 et seq.
— regions on the sphere 111—114, 117—120.
— series, Cantor's 264.
— theorem of algebra 101—104.
Fractions, changing common into decimal 40.

Gamma function 239.
Graphical methods for equations in the complex field 102—133.
— — — determining the real solutions of equations 87—101.

Historical excursus on:
   relations between differential calculus and the calculus of finite differences 232—235.
   exponential function and logarithm 146—155.
   the notion of function 200—207.

infinitesimal calculus 207—223.
imaginary numbers 55, 75—76.
irrational numbers 31—34.
negative numbers 25—27.
Taylor's theorem 233—234.
transcendence of e and π 237—238.
trigonometric series 205—207.
trigonometric tables and logarithmic tables 170—174.
   the modern development and the general structure of mathematics 77—85.
Homogeneous variables in function theory 106—108.
Hyperbolic functions 164—166.
   analogy with circular functions 166.
   fundamental function for 166.

Impossibility, proofs of:
   general 51.
   construction of regular heptagon with ruler and compasses 51—55.
   trisection of an angle 114.
Induction, mathematical 11.
Infinitesimal calculus, invention and development of 207 et seq.
Instruction, reform in 5.
Interpolation:
   by means of polynomials after Lagrange 229.
   Newton 229—232.
   trigonometric 190—193.
Interpolation parabolas 229.
Investigation, mathematical 208.
Irreducibility:
   function-theoretic 113—114.
   number-theoretic 52.

Lagrange's interpolation formula 229.
Limit, method of 211—214.
Logarithm:
   base of the natural 150—151.
   calculation of 148 et seq., 172 et seq.
   definition of the natural — by means of quadrature of the hyperbola 149, 156.
   difference equation for the — 148.
   function-theoretic discussion of — 156—162.
   uniformization by means of — 133, 159.

Mean-value theorem of differential calculus 213—214; extension of same 231 et seq.

Newton's interpolation formula 229 to 232.
Nomographic scales for:
  order curves 89, 94.
  class curves 90, 95.
Non-denumerability of the continuum 256.
Non-Archimedean number system 218.
Normal class curve of biquadratic equation 96—98.
— curves as:
  class curves 90—93, 95, 97.
  order curves 89—90, 94.
— equations of the regular bodies: solution by separation and series 130—133.
    — — uniformization 133—138.
    — — radicals 138—141.
  reduction of general equations to normal equations 141—143.
Number, assemblage of continuous and real numbers 250, 251—253.
—, notion of 10.
—, transition from, to measure 28.
— pair 28, 56.
— scale 23, 26, 31.

Order, types of 263.
Osculating parabolas 224—226.
  limiting form of 227.

Peano curve 265.
Perception, inner 11.
— and logic 11.
Philologists, relation to 2.
Picard's theorem 160.
Point, the infinitely distant — of the complex plane 105.
Point lattice 43.
Power of the continuum of a denumerable infinity of dimensions 258.
  of a finite number of dimensions 257—258.
— of an assemblage 251—262.
— — the assemblage of all real functions 261.
  continuous functions 260.
Precision, mathematics of 36.
Prime numbers, existence of infinitely many 40.
— factor tables 40.
Principle of permanence 26.
Process of growth of mathematics:
  Plan A. Separating methods and disciplines; logical direction 75.
  Plan B. Fusing methods and disciplines; intuitive direction 77.
  Plan C. Algorithmic process; formal direction 79.
Psychologic moments in teaching 4, 10, 16, 28, 30, 34, 268.
Pythagorean numbers 44.

Quaternion 60—75.
— scalar part of — 60.
— vector part of — 60.
— tensor of — 63, 66, 72.
— versor of — 72.

Rational, in the sense of mathematics of approximation 36.
Reform, the Basel aims toward 2.
— movement:
  the beginnings of infinitesimal calculus in school instruction 223; see also curriculum proposals and reform in instruction.
— proposals:
  Dresden — — for training teachers 2.
Regular bodies, groups of 120—124.
Rieman surfaces 105—110.
— sphere 105—110.
Rotation of space 73.
— and expansion of space 67—73.

School instruction:
  treatment of fractions 27.
    rrational numbers 37.
    complex numbers 75.
    the pendulum 187—190.
  exposition of the formal rules of operation 10.
  introduction of negative numbers 22, 28.
    notion of function 205.
    infinitesimal calculus 221 et seq.
    exponent and logarithm 144—146, 155—156.
    operations with natural numbers 6—8.
    trigonometric solution of cubic equation 134—137.
    transition to operations with letters 8.
    uniformization of the pure equation by means of the logarithm 133—134.
    number-theoretic considerations 37—38.
— mathematics, contents of 4

Signs, rule of 24.
— quasi proof for 26.
Space perception 35.
Square root expressions:
   significance of for constructions with ruler and compasses 50.
   classification of 53.
Sturm's theorem, geometrical equivalent of 94.
Style of mathematical presentation 84.

Taylor's formula 223, 233.
   analogy with Newton's interpolation formula 232 et seq.
   remainder term 226, 231.
Teachers, academic education of 1.
—, academic and normal school training of 7.
Tensor 63, 66, 70, 72.
Terminology, different — in the schools:
   algebraic numbers 23.
   arithmetic 3.
   relative numbers 23.
—, misleading in:
   algebraically soluble 140.
   irreducible 136.
   root 140.
   Maclaurin's series 224.
Threshold of perception 35.
Transcendence of e 237—243.
— of π 243—249.
Triangle, notion of in spherical trigonometry:
   elementary 175.
   proper and improper 181—182.
   with Möbius 176, 177, 182—183.
   with Study 181.

triangular membranes 183—186.
Trigonometry, spherical 175—186.
   its place in geometry of hyperspace 178—182.
   supplementary relations of 183—186.
Trigonometric functions, see circular functions.
Trigonometric functions:
   calculation of 170—174
   definition by means of quadrature of circle 162 et seq.
   complex fundamental function for 165 et seq.
   real fundamental function for 166 et seq.
   function-theoretic discussion of 167—169.
   application of to spherical trigonometry 175—186.
   application of to oscillations of pendulum 186—190.
   application of to representation of periodic functions 190—200; see also trigonometric series.
— series 190—200.
   Gibb's phenomenon 199.
   approximating curves 194—196.
   convergence, proof of 196—198.
   trigonometric interpolation 190—193.
   behavior at discontinuities 197 et seq.

Uniformization 133, 138.
— by means of logarithm 134, 159.

Vector 60, 63—65.
Versor 72.

A CATALOG OF SELECTED
DOVER BOOKS
IN SCIENCE AND MATHEMATICS

Astronomy

BURNHAM'S CELESTIAL HANDBOOK, Robert Burnham, Jr. Thorough guide to the stars beyond our solar system. Exhaustive treatment. Alphabetical by constellation: Andromeda to Cetus in Vol. 1; Chamaeleon to Orion in Vol. 2; and Pavo to Vulpecula in Vol. 3. Hundreds of illustrations. Index in Vol. 3. 2,000pp. 6⅛ x 9¼.
Vol. I: 0-486-23567-X
Vol. II: 0-486-23568-8
Vol. III: 0-486-23673-0

EXPLORING THE MOON THROUGH BINOCULARS AND SMALL TELESCOPES, Ernest H. Cherrington, Jr. Informative, profusely illustrated guide to locating and identifying craters, rills, seas, mountains, other lunar features. Newly revised and updated with special section of new photos. Over 100 photos and diagrams. 240pp. 8¼ x 11. 0-486-24491-1

THE EXTRATERRESTRIAL LIFE DEBATE, 1750–1900, Michael J. Crowe. First detailed, scholarly study in English of the many ideas that developed from 1750 to 1900 regarding the existence of intelligent extraterrestrial life. Examines ideas of Kant, Herschel, Voltaire, Percival Lowell, many other scientists and thinkers. 16 illustrations. 704pp. 5⅜ x 8½. 0-486-40675-X

THEORIES OF THE WORLD FROM ANTIQUITY TO THE COPERNICAN REVOLUTION, Michael J. Crowe. Newly revised edition of an accessible, enlightening book recreates the change from an earth-centered to a sun-centered conception of the solar system. 242pp. 5⅜ x 8½. 0-486-41444-2

A HISTORY OF ASTRONOMY, A. Pannekoek. Well-balanced, carefully reasoned study covers such topics as Ptolemaic theory, work of Copernicus, Kepler, Newton, Eddington's work on stars, much more. Illustrated. References. 521pp. 5⅜ x 8½.
0-486-65994-1

A COMPLETE MANUAL OF AMATEUR ASTRONOMY: TOOLS AND TECHNIQUES FOR ASTRONOMICAL OBSERVATIONS, P. Clay Sherrod with Thomas L. Koed. Concise, highly readable book discusses: selecting, setting up and maintaining a telescope; amateur studies of the sun; lunar topography and occultations; observations of Mars, Jupiter, Saturn, the minor planets and the stars; an introduction to photoelectric photometry; more. 1981 ed. 124 figures. 25 halftones. 37 tables. 335pp. 6½ x 9¼. 0-486-40675-X

AMATEUR ASTRONOMER'S HANDBOOK, J. B. Sidgwick. Timeless, comprehensive coverage of telescopes, mirrors, lenses, mountings, telescope drives, micrometers, spectroscopes, more. 189 illustrations. 576pp. 5⅝ x 8¼. (Available in U.S. only.)
0-486-24034-7

STARS AND RELATIVITY, Ya. B. Zel'dovich and I. D. Novikov. Vol. 1 of *Relativistic Astrophysics* by famed Russian scientists. General relativity, properties of matter under astrophysical conditions, stars, and stellar systems. Deep physical insights, clear presentation. 1971 edition. References. 544pp. 5⅝ x 8¼. 0-486-69424-0

CATALOG OF DOVER BOOKS

Chemistry

THE SCEPTICAL CHYMIST: THE CLASSIC 1661 TEXT, Robert Boyle. Boyle defines the term "element," asserting that all natural phenomena can be explained by the motion and organization of primary particles. 1911 ed. viii+232pp. 5⅜ x 8½.
0-486-42825-7

RADIOACTIVE SUBSTANCES, Marie Curie. Here is the celebrated scientist's doctoral thesis, the prelude to her receipt of the 1903 Nobel Prize. Curie discusses establishing atomic character of radioactivity found in compounds of uranium and thorium; extraction from pitchblende of polonium and radium; isolation of pure radium chloride; determination of atomic weight of radium; plus electric, photographic, luminous, heat, color effects of radioactivity. ii+94pp. 5⅜ x 8½. 0-486-42550-9

CHEMICAL MAGIC, Leonard A. Ford. Second Edition, Revised by E. Winston Grundmeier. Over 100 unusual stunts demonstrating cold fire, dust explosions, much more. Text explains scientific principles and stresses safety precautions. 128pp. 5⅜ x 8½.
0-486-67628-5

THE DEVELOPMENT OF MODERN CHEMISTRY, Aaron J. Ihde. Authoritative history of chemistry from ancient Greek theory to 20th-century innovation. Covers major chemists and their discoveries. 209 illustrations. 14 tables. Bibliographies. Indices. Appendices. 851pp. 5⅜ x 8½.
0-486-64235-6

CATALYSIS IN CHEMISTRY AND ENZYMOLOGY, William P. Jencks. Exceptionally clear coverage of mechanisms for catalysis, forces in aqueous solution, carbonyl- and acyl-group reactions, practical kinetics, more. 864pp. 5⅜ x 8½.
0-486-65460-5

ELEMENTS OF CHEMISTRY, Antoine Lavoisier. Monumental classic by founder of modern chemistry in remarkable reprint of rare 1790 Kerr translation. A must for every student of chemistry or the history of science. 539pp. 5⅜ x 8½. 0-486-64624-6

THE HISTORICAL BACKGROUND OF CHEMISTRY, Henry M. Leicester. Evolution of ideas, not individual biography. Concentrates on formulation of a coherent set of chemical laws. 260pp. 5⅜ x 8½.
0-486-61053-5

A SHORT HISTORY OF CHEMISTRY, J. R. Partington. Classic exposition explores origins of chemistry, alchemy, early medical chemistry, nature of atmosphere, theory of valency, laws and structure of atomic theory, much more. 428pp. 5⅜ x 8½. (Available in U.S. only.)
0-486-65977-1

GENERAL CHEMISTRY, Linus Pauling. Revised 3rd edition of classic first-year text by Nobel laureate. Atomic and molecular structure, quantum mechanics, statistical mechanics, thermodynamics correlated with descriptive chemistry. Problems. 992pp. 5⅜ x 8½.
0-486-65622-5

FROM ALCHEMY TO CHEMISTRY, John Read. Broad, humanistic treatment focuses on great figures of chemistry and ideas that revolutionized the science. 50 illustrations. 240pp. 5⅜ x 8½.
0-486-28690-8

CATALOG OF DOVER BOOKS

Engineering

DE RE METALLICA, Georgius Agricola. The famous Hoover translation of greatest treatise on technological chemistry, engineering, geology, mining of early modern times (1556). All 289 original woodcuts. 638pp. 6¾ x 11. 0-486-60006-8

FUNDAMENTALS OF ASTRODYNAMICS, Roger Bate et al. Modern approach developed by U.S. Air Force Academy. Designed as a first course. Problems, exercises. Numerous illustrations. 455pp. 5⅜ x 8½. 0-486-60061-0

DYNAMICS OF FLUIDS IN POROUS MEDIA, Jacob Bear. For advanced students of ground water hydrology, soil mechanics and physics, drainage and irrigation engineering and more. 335 illustrations. Exercises, with answers. 784pp. 6⅛ x 9¼.
0-486-65675-6

THEORY OF VISCOELASTICITY (Second Edition), Richard M. Christensen. Complete consistent description of the linear theory of the viscoelastic behavior of materials. Problem-solving techniques discussed. 1982 edition. 29 figures. xiv+364pp. 6⅛ x 9¼. 0-486-42880-X

MECHANICS, J. P. Den Hartog. A classic introductory text or refresher. Hundreds of applications and design problems illuminate fundamentals of trusses, loaded beams and cables, etc. 334 answered problems. 462pp. 5⅜ x 8½. 0-486-60754-2

MECHANICAL VIBRATIONS, J. P. Den Hartog. Classic textbook offers lucid explanations and illustrative models, applying theories of vibrations to a variety of practical industrial engineering problems. Numerous figures. 233 problems, solutions. Appendix. Index. Preface. 436pp. 5⅜ x 8½. 0-486-64785-4

STRENGTH OF MATERIALS, J. P. Den Hartog. Full, clear treatment of basic material (tension, torsion, bending, etc.) plus advanced material on engineering methods, applications. 350 answered problems. 323pp. 5⅜ x 8½. 0-486-60755-0

A HISTORY OF MECHANICS, René Dugas. Monumental study of mechanical principles from antiquity to quantum mechanics. Contributions of ancient Greeks, Galileo, Leonardo, Kepler, Lagrange, many others. 671pp. 5⅜ x 8½. 0-486-65632-2

STABILITY THEORY AND ITS APPLICATIONS TO STRUCTURAL MECHANICS, Clive L. Dym. Self-contained text focuses on Koiter postbuckling analyses, with mathematical notions of stability of motion. Basing minimum energy principles for static stability upon dynamic concepts of stability of motion, it develops asymptotic buckling and postbuckling analyses from potential energy considerations, with applications to columns, plates, and arches. 1974 ed. 208pp. 5⅜ x 8½.
0-486-42541-X

METAL FATIGUE, N. E. Frost, K. J. Marsh, and L. P. Pook. Definitive, clearly written, and well-illustrated volume addresses all aspects of the subject, from the historical development of understanding metal fatigue to vital concepts of the cyclic stress that causes a crack to grow. Includes 7 appendixes. 544pp. 5⅜ x 8½. 0-486-40927-9

CATALOG OF DOVER BOOKS

ROCKETS, Robert Goddard. Two of the most significant publications in the history of rocketry and jet propulsion: "A Method of Reaching Extreme Altitudes" (1919) and "Liquid Propellant Rocket Development" (1936). 128pp. 5⅜ x 8½. 0-486-42537-1

STATISTICAL MECHANICS: PRINCIPLES AND APPLICATIONS, Terrell L. Hill. Standard text covers fundamentals of statistical mechanics, applications to fluctuation theory, imperfect gases, distribution functions, more. 448pp. 5⅜ x 8½.
0-486-65390-0

ENGINEERING AND TECHNOLOGY 1650–1750: ILLUSTRATIONS AND TEXTS FROM ORIGINAL SOURCES, Martin Jensen. Highly readable text with more than 200 contemporary drawings and detailed engravings of engineering projects dealing with surveying, leveling, materials, hand tools, lifting equipment, transport and erection, piling, bailing, water supply, hydraulic engineering, and more. Among the specific projects outlined-transporting a 50-ton stone to the Louvre, erecting an obelisk, building timber locks, and dredging canals. 207pp. 8⅜ x 11¼.
0-486-42232-1

THE VARIATIONAL PRINCIPLES OF MECHANICS, Cornelius Lanczos. Graduate level coverage of calculus of variations, equations of motion, relativistic mechanics, more. First inexpensive paperbound edition of classic treatise. Index. Bibliography. 418pp. 5⅜ x 8½. 0-486-65067-7

PROTECTION OF ELECTRONIC CIRCUITS FROM OVERVOLTAGES, Ronald B. Standler. Five-part treatment presents practical rules and strategies for circuits designed to protect electronic systems from damage by transient overvoltages. 1989 ed. xxiv+434pp. 6⅛ x 9¼. 0-486-42552-5

ROTARY WING AERODYNAMICS, W. Z. Stepniewski. Clear, concise text covers aerodynamic phenomena of the rotor and offers guidelines for helicopter performance evaluation. Originally prepared for NASA. 537 figures. 640pp. 6⅛ x 9¼.
0-486-64647-5

INTRODUCTION TO SPACE DYNAMICS, William Tyrrell Thomson. Comprehensive, classic introduction to space-flight engineering for advanced undergraduate and graduate students. Includes vector algebra, kinematics, transformation of coordinates. Bibliography. Index. 352pp. 5⅜ x 8½. 0-486-65113-4

HISTORY OF STRENGTH OF MATERIALS, Stephen P. Timoshenko. Excellent historical survey of the strength of materials with many references to the theories of elasticity and structure. 245 figures. 452pp. 5⅜ x 8½. 0-486-61187-6

ANALYTICAL FRACTURE MECHANICS, David J. Unger. Self-contained text supplements standard fracture mechanics texts by focusing on analytical methods for determining crack-tip stress and strain fields. 336pp. 6⅛ x 9¼. 0-486-41737-9

STATISTICAL MECHANICS OF ELASTICITY, J. H. Weiner. Advanced, self-contained treatment illustrates general principles and elastic behavior of solids. Part 1, based on classical mechanics, studies thermoelastic behavior of crystalline and polymeric solids. Part 2, based on quantum mechanics, focuses on interatomic force laws, behavior of solids, and thermally activated processes. For students of physics and chemistry and for polymer physicists. 1983 ed. 96 figures. 496pp. 5⅜ x 8½.
0-486-42260-7

Mathematics

FUNCTIONAL ANALYSIS (Second Corrected Edition), George Bachman and Lawrence Narici. Excellent treatment of subject geared toward students with background in linear algebra, advanced calculus, physics and engineering. Text covers introduction to inner-product spaces, normed, metric spaces, and topological spaces; complete orthonormal sets, the Hahn-Banach Theorem and its consequences, and many other related subjects. 1966 ed. 544pp. 6⅛ x 9¼. 0-486-40251-7

ASYMPTOTIC EXPANSIONS OF INTEGRALS, Norman Bleistein & Richard A. Handelsman. Best introduction to important field with applications in a variety of scientific disciplines. New preface. Problems. Diagrams. Tables. Bibliography. Index. 448pp. 5⅜ x 8½. 0-486-65082-0

VECTOR AND TENSOR ANALYSIS WITH APPLICATIONS, A. I. Borisenko and I. E. Tarapov. Concise introduction. Worked-out problems, solutions, exercises. 257pp. 5⅝ x 8¼. 0-486-63833-2

AN INTRODUCTION TO ORDINARY DIFFERENTIAL EQUATIONS, Earl A. Coddington. A thorough and systematic first course in elementary differential equations for undergraduates in mathematics and science, with many exercises and problems (with answers). Index. 304pp. 5⅜ x 8½. 0-486-65942-9

FOURIER SERIES AND ORTHOGONAL FUNCTIONS, Harry F. Davis. An incisive text combining theory and practical example to introduce Fourier series, orthogonal functions and applications of the Fourier method to boundary-value problems. 570 exercises. Answers and notes. 416pp. 5⅜ x 8½. 0-486-65973-9

COMPUTABILITY AND UNSOLVABILITY, Martin Davis. Classic graduate-level introduction to theory of computability, usually referred to as theory of recurrent functions. New preface and appendix. 288pp. 5⅜ x 8½. 0-486-61471-9

ASYMPTOTIC METHODS IN ANALYSIS, N. G. de Bruijn. An inexpensive, comprehensive guide to asymptotic methods–the pioneering work that teaches by explaining worked examples in detail. Index. 224pp. 5⅜ x 8½ 0-486-64221-6

APPLIED COMPLEX VARIABLES, John W. Dettman. Step-by-step coverage of fundamentals of analytic function theory–plus lucid exposition of five important applications: Potential Theory; Ordinary Differential Equations; Fourier Transforms; Laplace Transforms; Asymptotic Expansions. 66 figures. Exercises at chapter ends. 512pp. 5⅜ x 8½. 0-486-64670-X

INTRODUCTION TO LINEAR ALGEBRA AND DIFFERENTIAL EQUATIONS, John W. Dettman. Excellent text covers complex numbers, determinants, orthonormal bases, Laplace transforms, much more. Exercises with solutions. Undergraduate level. 416pp. 5⅜ x 8½. 0-486-65191-6

RIEMANN'S ZETA FUNCTION, H. M. Edwards. Superb, high-level study of landmark 1859 publication entitled "On the Number of Primes Less Than a Given Magnitude" traces developments in mathematical theory that it inspired. xiv+315pp. 5⅜ x 8½. 0-486-41740-9

CATALOG OF DOVER BOOKS

CALCULUS OF VARIATIONS WITH APPLICATIONS, George M. Ewing. Applications-oriented introduction to variational theory develops insight and promotes understanding of specialized books, research papers. Suitable for advanced undergraduate/graduate students as primary, supplementary text. 352pp. 5⅜ x 8½.
0-486-64856-7

COMPLEX VARIABLES, Francis J. Flanigan. Unusual approach, delaying complex algebra till harmonic functions have been analyzed from real variable viewpoint. Includes problems with answers. 364pp. 5⅜ x 8½. 0-486-61388-7

AN INTRODUCTION TO THE CALCULUS OF VARIATIONS, Charles Fox. Graduate-level text covers variations of an integral, isoperimetrical problems, least action, special relativity, approximations, more. References. 279pp. 5⅜ x 8½.
0-486-65499-0

COUNTEREXAMPLES IN ANALYSIS, Bernard R. Gelbaum and John M. H. Olmsted. These counterexamples deal mostly with the part of analysis known as "real variables." The first half covers the real number system, and the second half encompasses higher dimensions. 1962 edition. xxiv+198pp. 5⅜ x 8½. 0-486-42875-3

CATASTROPHE THEORY FOR SCIENTISTS AND ENGINEERS, Robert Gilmore. Advanced-level treatment describes mathematics of theory grounded in the work of Poincaré, R. Thom, other mathematicians. Also important applications to problems in mathematics, physics, chemistry and engineering. 1981 edition. References. 28 tables. 397 black-and-white illustrations. xvii + 666pp. 6⅛ x 9¼.
0-486-67539-4

INTRODUCTION TO DIFFERENCE EQUATIONS, Samuel Goldberg. Exceptionally clear exposition of important discipline with applications to sociology, psychology, economics. Many illustrative examples; over 250 problems. 260pp. 5⅜ x 8½.
0-486-65084-7

NUMERICAL METHODS FOR SCIENTISTS AND ENGINEERS, Richard Hamming. Classic text stresses frequency approach in coverage of algorithms, polynomial approximation, Fourier approximation, exponential approximation, other topics. Revised and enlarged 2nd edition. 721pp. 5⅜ x 8½. 0-486-65241-6

INTRODUCTION TO NUMERICAL ANALYSIS (2nd Edition), F. B. Hildebrand. Classic, fundamental treatment covers computation, approximation, interpolation, numerical differentiation and integration, other topics. 150 new problems. 669pp. 5⅜ x 8½. 0-486-65363-3

THREE PEARLS OF NUMBER THEORY, A. Y. Khinchin. Three compelling puzzles require proof of a basic law governing the world of numbers. Challenges concern van der Waerden's theorem, the Landau-Schnirelmann hypothesis and Mann's theorem, and a solution to Waring's problem. Solutions included. 64pp. 5⅜ x 8½.
0-486-40026-3

THE PHILOSOPHY OF MATHEMATICS: AN INTRODUCTORY ESSAY, Stephan Körner. Surveys the views of Plato, Aristotle, Leibniz & Kant concerning propositions and theories of applied and pure mathematics. Introduction. Two appendices. Index. 198pp. 5⅜ x 8½. 0-486-25048-2

CATALOG OF DOVER BOOKS

INTRODUCTORY REAL ANALYSIS, A.N. Kolmogorov, S. V. Fomin. Translated by Richard A. Silverman. Self-contained, evenly paced introduction to real and functional analysis. Some 350 problems. 403pp. 5⅜ x 8½. 0-486-61226-0

APPLIED ANALYSIS, Cornelius Lanczos. Classic work on analysis and design of finite processes for approximating solution of analytical problems. Algebraic equations, matrices, harmonic analysis, quadrature methods, much more. 559pp. 5⅜ x 8½. 0-486-65656-X

AN INTRODUCTION TO ALGEBRAIC STRUCTURES, Joseph Landin. Superb self-contained text covers "abstract algebra": sets and numbers, theory of groups, theory of rings, much more. Numerous well-chosen examples, exercises. 247pp. 5⅜ x 8½. 0-486-65940-2

QUALITATIVE THEORY OF DIFFERENTIAL EQUATIONS, V. V. Nemytskii and V.V. Stepanov. Classic graduate-level text by two prominent Soviet mathematicians covers classical differential equations as well as topological dynamics and ergodic theory. Bibliographies. 523pp. 5⅜ x 8½. 0-486-65954-2

THEORY OF MATRICES, Sam Perlis. Outstanding text covering rank, nonsingularity and inverses in connection with the development of canonical matrices under the relation of equivalence, and without the intervention of determinants. Includes exercises. 237pp. 5⅜ x 8½. 0-486-66810-X

INTRODUCTION TO ANALYSIS, Maxwell Rosenlicht. Unusually clear, accessible coverage of set theory, real number system, metric spaces, continuous functions, Riemann integration, multiple integrals, more. Wide range of problems. Undergraduate level. Bibliography. 254pp. 5⅜ x 8½. 0-486-65038-3

MODERN NONLINEAR EQUATIONS, Thomas L. Saaty. Emphasizes practical solution of problems; covers seven types of equations. ". . . a welcome contribution to the existing literature...."–*Math Reviews*. 490pp. 5⅜ x 8½. 0-486-64232-1

MATRICES AND LINEAR ALGEBRA, Hans Schneider and George Phillip Barker. Basic textbook covers theory of matrices and its applications to systems of linear equations and related topics such as determinants, eigenvalues and differential equations. Numerous exercises. 432pp. 5⅜ x 8½. 0-486-66014-1

LINEAR ALGEBRA, Georgi E. Shilov. Determinants, linear spaces, matrix algebras, similar topics. For advanced undergraduates, graduates. Silverman translation. 387pp. 5⅜ x 8½. 0-486-63518-X

ELEMENTS OF REAL ANALYSIS, David A. Sprecher. Classic text covers fundamental concepts, real number system, point sets, functions of a real variable, Fourier series, much more. Over 500 exercises. 352pp. 5⅜ x 8½. 0-486-65385-4

SET THEORY AND LOGIC, Robert R. Stoll. Lucid introduction to unified theory of mathematical concepts. Set theory and logic seen as tools for conceptual understanding of real number system. 496pp. 5⅜ x 8¼. 0-486-63829-4

CATALOG OF DOVER BOOKS

TENSOR CALCULUS, J.L. Synge and A. Schild. Widely used introductory text covers spaces and tensors, basic operations in Riemannian space, non-Riemannian spaces, etc. 324pp. 5⅜ x 8¼. 0-486-63612-7

ORDINARY DIFFERENTIAL EQUATIONS, Morris Tenenbaum and Harry Pollard. Exhaustive survey of ordinary differential equations for undergraduates in mathematics, engineering, science. Thorough analysis of theorems. Diagrams. Bibliography. Index. 818pp. 5⅜ x 8½. 0-486-64940-7

INTEGRAL EQUATIONS, F. G. Tricomi. Authoritative, well-written treatment of extremely useful mathematical tool with wide applications. Volterra Equations, Fredholm Equations, much more. Advanced undergraduate to graduate level. Exercises. Bibliography. 238pp. 5⅜ x 8½. 0-486-64828-1

FOURIER SERIES, Georgi P. Tolstov. Translated by Richard A. Silverman. A valuable addition to the literature on the subject, moving clearly from subject to subject and theorem to theorem. 107 problems, answers. 336pp. 5⅜ x 8½. 0-486-63317-9

INTRODUCTION TO MATHEMATICAL THINKING, Friedrich Waismann. Examinations of arithmetic, geometry, and theory of integers; rational and natural numbers; complete induction; limit and point of accumulation; remarkable curves; complex and hypercomplex numbers, more. 1959 ed. 27 figures. xii+260pp. 5⅜ x 8½. 0-486-63317-9

POPULAR LECTURES ON MATHEMATICAL LOGIC, Hao Wang. Noted logician's lucid treatment of historical developments, set theory, model theory, recursion theory and constructivism, proof theory, more. 3 appendixes. Bibliography. 1981 edition. ix + 283pp. 5⅜ x 8½. 0-486-67632-3

CALCULUS OF VARIATIONS, Robert Weinstock. Basic introduction covering isoperimetric problems, theory of elasticity, quantum mechanics, electrostatics, etc. Exercises throughout. 326pp. 5⅜ x 8½. 0-486-63069-2

THE CONTINUUM: A CRITICAL EXAMINATION OF THE FOUNDATION OF ANALYSIS, Hermann Weyl. Classic of 20th-century foundational research deals with the conceptual problem posed by the continuum. 156pp. 5⅜ x 8½. 0-486-67982-9

CHALLENGING MATHEMATICAL PROBLEMS WITH ELEMENTARY SOLUTIONS, A. M. Yaglom and I. M. Yaglom. Over 170 challenging problems on probability theory, combinatorial analysis, points and lines, topology, convex polygons, many other topics. Solutions. Total of 445pp. 5⅜ x 8½. Two-vol. set.
Vol. I: 0-486-65536-9 Vol. II: 0-486-65537-7

INTRODUCTION TO PARTIAL DIFFERENTIAL EQUATIONS WITH APPLICATIONS, E. C. Zachmanoglou and Dale W. Thoe. Essentials of partial differential equations applied to common problems in engineering and the physical sciences. Problems and answers. 416pp. 5⅜ x 8½. 0-486-65251-3

THE THEORY OF GROUPS, Hans J. Zassenhaus. Well-written graduate-level text acquaints reader with group-theoretic methods and demonstrates their usefulness in mathematics. Axioms, the calculus of complexes, homomorphic mapping, p-group theory, more. 276pp. 5⅜ x 8½. 0-486-40922-8

CATALOG OF DOVER BOOKS

Math–Decision Theory, Statistics, Probability

ELEMENTARY DECISION THEORY, Herman Chernoff and Lincoln E. Moses. Clear introduction to statistics and statistical theory covers data processing, probability and random variables, testing hypotheses, much more. Exercises. 364pp. 5⅜ x 8½. 0-486-65218-1

STATISTICS MANUAL, Edwin L. Crow et al. Comprehensive, practical collection of classical and modern methods prepared by U.S. Naval Ordnance Test Station. Stress on use. Basics of statistics assumed. 288pp. 5⅜ x 8½. 0-486-60599-X

SOME THEORY OF SAMPLING, William Edwards Deming. Analysis of the problems, theory and design of sampling techniques for social scientists, industrial managers and others who find statistics important at work. 61 tables. 90 figures. xvii +602pp. 5⅜ x 8½. 0-486-64684-X

LINEAR PROGRAMMING AND ECONOMIC ANALYSIS, Robert Dorfman, Paul A. Samuelson and Robert M. Solow. First comprehensive treatment of linear programming in standard economic analysis. Game theory, modern welfare economics, Leontief input-output, more. 525pp. 5⅜ x 8½. 0-486-65491-5

PROBABILITY: AN INTRODUCTION, Samuel Goldberg. Excellent basic text covers set theory, probability theory for finite sample spaces, binomial theorem, much more. 360 problems. Bibliographies. 322pp. 5⅜ x 8½. 0-486-65252-1

GAMES AND DECISIONS: INTRODUCTION AND CRITICAL SURVEY, R. Duncan Luce and Howard Raiffa. Superb nontechnical introduction to game theory, primarily applied to social sciences. Utility theory, zero-sum games, n-person games, decision-making, much more. Bibliography. 509pp. 5⅜ x 8½. 0-486-65943-7

INTRODUCTION TO THE THEORY OF GAMES, J. C. C. McKinsey. This comprehensive overview of the mathematical theory of games illustrates applications to situations involving conflicts of interest, including economic, social, political, and military contexts. Appropriate for advanced undergraduate and graduate courses; advanced calculus a prerequisite. 1952 ed. x+372pp. 5⅜ x 8½. 0-486-42811-7

FIFTY CHALLENGING PROBLEMS IN PROBABILITY WITH SOLUTIONS, Frederick Mosteller. Remarkable puzzlers, graded in difficulty, illustrate elementary and advanced aspects of probability. Detailed solutions. 88pp. 5⅜ x 8½. 65355-2

PROBABILITY THEORY: A CONCISE COURSE, Y. A. Rozanov. Highly readable, self-contained introduction covers combination of events, dependent events, Bernoulli trials, etc. 148pp. 5⅜ x 8¼. 0-486-63544-9

STATISTICAL METHOD FROM THE VIEWPOINT OF QUALITY CONTROL, Walter A. Shewhart. Important text explains regulation of variables, uses of statistical control to achieve quality control in industry, agriculture, other areas. 192pp. 5⅜ x 8½. 0-486-65232-7

CATALOG OF DOVER BOOKS

Math–Geometry and Topology

ELEMENTARY CONCEPTS OF TOPOLOGY, Paul Alexandroff. Elegant, intuitive approach to topology from set-theoretic topology to Betti groups; how concepts of topology are useful in math and physics. 25 figures. 57pp. 5⅜ x 8½. 0-486-60747-X

COMBINATORIAL TOPOLOGY, P. S. Alexandrov. Clearly written, well-organized, three-part text begins by dealing with certain classic problems without using the formal techniques of homology theory and advances to the central concept, the Betti groups. Numerous detailed examples. 654pp. 5¾ x 8½. 0-486-40179-0

EXPERIMENTS IN TOPOLOGY, Stephen Barr. Classic, lively explanation of one of the byways of mathematics. Klein bottles, Moebius strips, projective planes, map coloring, problem of the Koenigsberg bridges, much more, described with clarity and wit. 43 figures. 210pp. 5⅜ x 8½. 0-486-25933-1

THE GEOMETRY OF RENÉ DESCARTES, René Descartes. The great work founded analytical geometry. Original French text, Descartes's own diagrams, together with definitive Smith-Latham translation. 244pp. 5⅜ x 8½. 0-486-60068-8

EUCLIDEAN GEOMETRY AND TRANSFORMATIONS, Clayton W. Dodge. This introduction to Euclidean geometry emphasizes transformations, particularly isometries and similarities. Suitable for undergraduate courses, it includes numerous examples, many with detailed answers. 1972 ed. viii+296pp. 6⅛ x 9¼. 0-486-43476-1

PRACTICAL CONIC SECTIONS: THE GEOMETRIC PROPERTIES OF ELLIPSES, PARABOLAS AND HYPERBOLAS, J. W. Downs. This text shows how to create ellipses, parabolas, and hyperbolas. It also presents historical background on their ancient origins and describes the reflective properties and roles of curves in design applications. 1993 ed. 98 figures. xii+100pp. 6½ x 9¼. 0-486-42876-1

THE THIRTEEN BOOKS OF EUCLID'S ELEMENTS, translated with introduction and commentary by Sir Thomas L. Heath. Definitive edition. Textual and linguistic notes, mathematical analysis. 2,500 years of critical commentary. Unabridged. 1,414pp. 5⅜ x 8½. Three-vol. set.
Vol. I: 0-486-60088-2 Vol. II: 0-486-60089-0 Vol. III: 0-486-60090-4

SPACE AND GEOMETRY: IN THE LIGHT OF PHYSIOLOGICAL, PSYCHOLOGICAL AND PHYSICAL INQUIRY, Ernst Mach. Three essays by an eminent philosopher and scientist explore the nature, origin, and development of our concepts of space, with a distinctness and precision suitable for undergraduate students and other readers. 1906 ed. vi+148pp. 5⅜ x 8½. 0-486-43909-7

GEOMETRY OF COMPLEX NUMBERS, Hans Schwerdtfeger. Illuminating, widely praised book on analytic geometry of circles, the Moebius transformation, and two-dimensional non-Euclidean geometries. 200pp. 5⅜ x 8¼. 0-486-63830-8

DIFFERENTIAL GEOMETRY, Heinrich W. Guggenheimer. Local differential geometry as an application of advanced calculus and linear algebra. Curvature, transformation groups, surfaces, more. Exercises. 62 figures. 378pp. 5⅜ x 8½. 0-486-63433-7

CATALOG OF DOVER BOOKS

History of Math

THE WORKS OF ARCHIMEDES, Archimedes (T. L. Heath, ed.). Topics include the famous problems of the ratio of the areas of a cylinder and an inscribed sphere; the measurement of a circle; the properties of conoids, spheroids, and spirals; and the quadrature of the parabola. Informative introduction. clxxxvi+326pp. 5⅜ x 8½.
0-486-42084-1

A SHORT ACCOUNT OF THE HISTORY OF MATHEMATICS, W. W. Rouse Ball. One of clearest, most authoritative surveys from the Egyptians and Phoenicians through 19th-century figures such as Grassman, Galois, Riemann. Fourth edition. 522pp. 5⅜ x 8½. 0-486-20630-0

THE HISTORY OF THE CALCULUS AND ITS CONCEPTUAL DEVELOPMENT, Carl B. Boyer. Origins in antiquity, medieval contributions, work of Newton, Leibniz, rigorous formulation. Treatment is verbal. 346pp. 5⅜ x 8½. 0-486-60509-4

THE HISTORICAL ROOTS OF ELEMENTARY MATHEMATICS, Lucas N. H. Bunt, Phillip S. Jones, and Jack D. Bedient. Fundamental underpinnings of modern arithmetic, algebra, geometry and number systems derived from ancient civilizations. 320pp. 5⅜ x 8½. 0-486-25563-8

A HISTORY OF MATHEMATICAL NOTATIONS, Florian Cajori. This classic study notes the first appearance of a mathematical symbol and its origin, the competition it encountered, its spread among writers in different countries, its rise to popularity, its eventual decline or ultimate survival. Original 1929 two-volume edition presented here in one volume. xxviii+820pp. 5⅜ x 8½. 0-486-67766-4

GAMES, GODS & GAMBLING: A HISTORY OF PROBABILITY AND STATISTICAL IDEAS, F. N. David. Episodes from the lives of Galileo, Fermat, Pascal, and others illustrate this fascinating account of the roots of mathematics. Features thought-provoking references to classics, archaeology, biography, poetry. 1962 edition. 304pp. 5⅜ x 8½. (Available in U.S. only.) 0-486-40023-9

OF MEN AND NUMBERS: THE STORY OF THE GREAT MATHEMATICIANS, Jane Muir. Fascinating accounts of the lives and accomplishments of history's greatest mathematical minds–Pythagoras, Descartes, Euler, Pascal, Cantor, many more. Anecdotal, illuminating. 30 diagrams. Bibliography. 256pp. 5⅜ x 8½. 0-486-28973-7

HISTORY OF MATHEMATICS, David E. Smith. Nontechnical survey from ancient Greece and Orient to late 19th century; evolution of arithmetic, geometry, trigonometry, calculating devices, algebra, the calculus. 362 illustrations. 1,355pp. 5⅜ x 8½. Two-vol. set. Vol. I: 0-486-20429-4 Vol. II: 0-486-20430-8

A CONCISE HISTORY OF MATHEMATICS, Dirk J. Struik. The best brief history of mathematics. Stresses origins and covers every major figure from ancient Near East to 19th century. 41 illustrations. 195pp. 5⅜ x 8½. 0-486-60255-9

CATALOG OF DOVER BOOKS

Physics

OPTICAL RESONANCE AND TWO-LEVEL ATOMS, L. Allen and J. H. Eberly. Clear, comprehensive introduction to basic principles behind all quantum optical resonance phenomena. 53 illustrations. Preface. Index. 256pp. 5⅜ x 8½. 0-486-65533-4

QUANTUM THEORY, David Bohm. This advanced undergraduate-level text presents the quantum theory in terms of qualitative and imaginative concepts, followed by specific applications worked out in mathematical detail. Preface. Index. 655pp. 5⅜ x 8½. 0-486-65969-0

ATOMIC PHYSICS (8th EDITION), Max Born. Nobel laureate's lucid treatment of kinetic theory of gases, elementary particles, nuclear atom, wave-corpuscles, atomic structure and spectral lines, much more. Over 40 appendices, bibliography. 495pp. 5⅜ x 8½. 0-486-65984-4

A SOPHISTICATE'S PRIMER OF RELATIVITY, P. W. Bridgman. Geared toward readers already acquainted with special relativity, this book transcends the view of theory as a working tool to answer natural questions: What is a frame of reference? What is a "law of nature"? What is the role of the "observer"? Extensive treatment, written in terms accessible to those without a scientific background. 1983 ed. xlviii+172pp. 5⅜ x 8½. 0-486-42549-5

AN INTRODUCTION TO HAMILTONIAN OPTICS, H. A. Buchdahl. Detailed account of the Hamiltonian treatment of aberration theory in geometrical optics. Many classes of optical systems defined in terms of the symmetries they possess. Problems with detailed solutions. 1970 edition. xv + 360pp. 5⅜ x 8½. 0-486-67597-1

PRIMER OF QUANTUM MECHANICS, Marvin Chester. Introductory text examines the classical quantum bead on a track: its state and representations; operator eigenvalues; harmonic oscillator and bound bead in a symmetric force field; and bead in a spherical shell. Other topics include spin, matrices, and the structure of quantum mechanics; the simplest atom; indistinguishable particles; and stationary-state perturbation theory. 1992 ed. xiv+314pp. 6⅛ x 9¼. 0-486-42878-8

LECTURES ON QUANTUM MECHANICS, Paul A. M. Dirac. Four concise, brilliant lectures on mathematical methods in quantum mechanics from Nobel Prize-winning quantum pioneer build on idea of visualizing quantum theory through the use of classical mechanics. 96pp. 5⅜ x 8½. 0-486-41713-1

THIRTY YEARS THAT SHOOK PHYSICS: THE STORY OF QUANTUM THEORY, George Gamow. Lucid, accessible introduction to influential theory of energy and matter. Careful explanations of Dirac's anti-particles, Bohr's model of the atom, much more. 12 plates. Numerous drawings. 240pp. 5⅜ x 8½. 0-486-24895-X

ELECTRONIC STRUCTURE AND THE PROPERTIES OF SOLIDS: THE PHYSICS OF THE CHEMICAL BOND, Walter A. Harrison. Innovative text offers basic understanding of the electronic structure of covalent and ionic solids, simple metals, transition metals and their compounds. Problems. 1980 edition. 582pp. 6⅛ x 9¼. 0-486-66021-4

CATALOG OF DOVER BOOKS

HYDRODYNAMIC AND HYDROMAGNETIC STABILITY, S. Chandrasekhar. Lucid examination of the Rayleigh-Benard problem; clear coverage of the theory of instabilities causing convection. 704pp. 5⅜ x 8¼. 0-486-64071-X

INVESTIGATIONS ON THE THEORY OF THE BROWNIAN MOVEMENT, Albert Einstein. Five papers (1905-8) investigating dynamics of Brownian motion and evolving elementary theory. Notes by R. Fürth. 122pp. 5⅜ x 8½. 0-486-60304-0

THE PHYSICS OF WAVES, William C. Elmore and Mark A. Heald. Unique overview of classical wave theory. Acoustics, optics, electromagnetic radiation, more. Ideal as classroom text or for self-study. Problems. 477pp. 5⅜ x 8½. 0-486-64926-1

GRAVITY, George Gamow. Distinguished physicist and teacher takes reader-friendly look at three scientists whose work unlocked many of the mysteries behind the laws of physics: Galileo, Newton, and Einstein. Most of the book focuses on Newton's ideas, with a concluding chapter on post-Einsteinian speculations concerning the relationship between gravity and other physical phenomena. 160pp. 5⅜ x 8½. 0-486-42563-0

PHYSICAL PRINCIPLES OF THE QUANTUM THEORY, Werner Heisenberg. Nobel Laureate discusses quantum theory, uncertainty, wave mechanics, work of Dirac, Schroedinger, Compton, Wilson, Einstein, etc. 184pp. 5⅜ x 8½. 0-486-60113-7

ATOMIC SPECTRA AND ATOMIC STRUCTURE, Gerhard Herzberg. One of best introductions; especially for specialist in other fields. Treatment is physical rather than mathematical. 80 illustrations. 257pp. 5⅜ x 8½. 0-486-60115-3

AN INTRODUCTION TO STATISTICAL THERMODYNAMICS, Terrell L. Hill. Excellent basic text offers wide-ranging coverage of quantum statistical mechanics, systems of interacting molecules, quantum statistics, more. 523pp. 5⅜ x 8½. 0-486-65242-4

THEORETICAL PHYSICS, Georg Joos, with Ira M. Freeman. Classic overview covers essential math, mechanics, electromagnetic theory, thermodynamics, quantum mechanics, nuclear physics, other topics. First paperback edition. xxiii + 885pp. 5⅜ x 8½. 0-486-65227-0

PROBLEMS AND SOLUTIONS IN QUANTUM CHEMISTRY AND PHYSICS, Charles S. Johnson, Jr. and Lee G. Pedersen. Unusually varied problems, detailed solutions in coverage of quantum mechanics, wave mechanics, angular momentum, molecular spectroscopy, more. 280 problems plus 139 supplementary exercises. 430pp. 6½ x 9¼. 0-486-65236-X

THEORETICAL SOLID STATE PHYSICS, Vol. 1: Perfect Lattices in Equilibrium; Vol. II: Non-Equilibrium and Disorder, William Jones and Norman H. March. Monumental reference work covers fundamental theory of equilibrium properties of perfect crystalline solids, non-equilibrium properties, defects and disordered systems. Appendices. Problems. Preface. Diagrams. Index. Bibliography. Total of 1,301pp. 5⅜ x 8½. Two volumes. Vol. I: 0-486-65015-4 Vol. II: 0-486-65016-2

WHAT IS RELATIVITY? L. D. Landau and G. B. Rumer. Written by a Nobel Prize physicist and his distinguished colleague, this compelling book explains the special theory of relativity to readers with no scientific background, using such familiar objects as trains, rulers, and clocks. 1960 ed. vi+72pp. 5⅜ x 8½. 0-486-42806-0

CATALOG OF DOVER BOOKS

A TREATISE ON ELECTRICITY AND MAGNETISM, James Clerk Maxwell. Important foundation work of modern physics. Brings to final form Maxwell's theory of electromagnetism and rigorously derives his general equations of field theory. 1,084pp. 5⅜ x 8½. Two-vol. set. Vol. I: 0-486-60636-8 Vol. II: 0-486-60637-6

QUANTUM MECHANICS: PRINCIPLES AND FORMALISM, Roy McWeeny. Graduate student-oriented volume develops subject as fundamental discipline, opening with review of origins of Schrödinger's equations and vector spaces. Focusing on main principles of quantum mechanics and their immediate consequences, it concludes with final generalizations covering alternative "languages" or representations. 1972 ed. 15 figures. xi+155pp. 5⅜ x 8½. 0-486-42829-X

INTRODUCTION TO QUANTUM MECHANICS With Applications to Chemistry, Linus Pauling & E. Bright Wilson, Jr. Classic undergraduate text by Nobel Prize winner applies quantum mechanics to chemical and physical problems. Numerous tables and figures enhance the text. Chapter bibliographies. Appendices. Index. 468pp. 5⅜ x 8½. 0-486-64871-0

METHODS OF THERMODYNAMICS, Howard Reiss. Outstanding text focuses on physical technique of thermodynamics, typical problem areas of understanding, and significance and use of thermodynamic potential. 1965 edition. 238pp. 5⅜ x 8½. 0-486-69445-3

THE ELECTROMAGNETIC FIELD, Albert Shadowitz. Comprehensive undergraduate text covers basics of electric and magnetic fields, builds up to electromagnetic theory. Also related topics, including relativity. Over 900 problems. 768pp. 5⅝ x 8¼. 0-486-65660-8

GREAT EXPERIMENTS IN PHYSICS: FIRSTHAND ACCOUNTS FROM GALILEO TO EINSTEIN, Morris H. Shamos (ed.). 25 crucial discoveries: Newton's laws of motion, Chadwick's study of the neutron, Hertz on electromagnetic waves, more. Original accounts clearly annotated. 370pp. 5⅜ x 8½. 0-486-25346-5

EINSTEIN'S LEGACY, Julian Schwinger. A Nobel Laureate relates fascinating story of Einstein and development of relativity theory in well-illustrated, nontechnical volume. Subjects include meaning of time, paradoxes of space travel, gravity and its effect on light, non-Euclidean geometry and curving of space-time, impact of radio astronomy and space-age discoveries, and more. 189 b/w illustrations. xiv+250pp. 8⅜ x 9¼. 0-486-41974-6

STATISTICAL PHYSICS, Gregory H. Wannier. Classic text combines thermodynamics, statistical mechanics and kinetic theory in one unified presentation of thermal physics. Problems with solutions. Bibliography. 532pp. 5⅜ x 8½. 0-486-65401-X

Paperbound unless otherwise indicated. Available at your book dealer, online at **www.doverpublications.com**, or by writing to Dept. GI, Dover Publications, Inc., 31 East 2nd Street, Mineola, NY 11501. For current price information or for free catalogues (please indicate field of interest), write to Dover Publications or log on to **www.doverpublications.com** and see every Dover book in print. Dover publishes more than 500 books each year on science, elementary and advanced mathematics, biology, music, art, literary history, social sciences, and other areas.